Cognitive Ecology

Handbook of Perception and Cognition
2nd Edition

Series Editors
Edward C. Carterette
and **Morton P. Friedman**

ERRATUM

On the title page, the editors' affiliation should read:

Department of Psychology
University of California, Los Angeles
Los Angeles, California

Friedman and Carterette: *Cognitive Ecology*
ISBN: 0-12-161966-4

Cognitive Ecology

Edited by

Morton P. Friedman

Edward C. Carterette

Department of Psychology
University of Southern California, Los Angeles
Los Angeles, California

AP

Academic Press

San Diego New York Boston
London Sydney Tokyo Toronto

This book is printed on acid-free paper. ∞

Academic Press, Inc.
A Division of Harcourt Brace & Company
525 B Street, Suite 1900, San Diego, California 92101-4495

United Kingdom Edition published by
Academic Press Limited
24-28 Oval Road, London NW1 7DX

Library of Congress Cataloging-in-Publication Data

Cognitive ecology / edited by Morton P. Friedman, Edward C. Carterette.
p. cm. -- (Handbok of perception and cognition (2nd ed.))
Includes bibliographical references and index.
ISBN 0-12-161966-4 (alk. paper)
1. Cognition and culture. 2. Human information processing. 3. Perception. 4. Senses and Sensation. I. Friedman, Morton P. II. Carterette, Edward C. III. Series.
BF311.C5515 1995
153--dc20 95-20313
CIP

PRINTED IN THE UNITED STATES OF AMERICA
95 96 97 98 99 00 BB 9 8 7 6 5 4 3 2 1

Contents

Mind and Culture

The Arts

3 *Confluence and Divergence in Empirical Aesthetics, Philosophy, and Mainstream Psychology*

Gerald C. Cupchik and Andrew S. Winston

4 *Music Perception and Cognition*

Roger A. Kendall and Edward C. Carterette

6 *The Perception of Motion Pictures*

Julian Hochberg and Virginia Brooks

7 *The Art of the Puzzler*

Arthur Schulman

Sensory Evaluation

8 *Flavor*

Harry T. Lawless

Contributors

Numbers in parentheses indicate the pages on which the authors' contributions begin.

Virginia Brooks (205)
Department of Film
Brooklyn College, CUNY
Brooklyn, New York 11210

Edward C. Carterette (87)
Department of Psychology
University of California, Los Angeles
Los Angeles, California 90024

Gerald C. Cupchik (61)
Life Sciences Division,
Scarborough College
University of Toronto
Scarborough, Ontario
Canada M1C 1A4

Julian Hochberg (151, 205)
Department of Psychology
Columbia University
New York, New York 10027

Roger A. Kendall (87)
Department of Ethnomusicology
University of California, Los Angeles
Los Angeles, California 90024

Harry T. Lawless (325)
Department of Food Science,
New York State College
of Agriculture and Life Sciences
Cornell University
Ithaca, New York 14850

Eleanor Rosch (5)
Department of Psychology
University of California, Berkeley
Berkeley, California 94720

Arthur Schulman (293)
Department of Psychology
University of Virginia
Charlottesville, Virginia 22903

Jaan Valsiner (29)
Department of Psychology
University of North Carolina
at Chapel Hill
Chapel Hill, North Carolina 27599

Andrew S. Winston (61)
University of Guelph
Guelph, Ontario
Canada N1G 2W1

Foreword

The problem of perception and cognition is in understanding how the organism transforms, organizes, stores, and uses information arising from the world in sense data or memory. With this definition of perception and cognition in mind, this handbook is designed to bring together the essential aspects of this very large, diverse, and scattered literature and to give a précis of the state of knowledge in every area of perception and cognition. The work is aimed at the psychologist and the cognitive scientist in particular, and at the natural scientist in general. Topics are covered in comprehensive surveys in which fundamental facts and concepts are presented, and important leads to journals and monographs of the specialized literature are provided. Perception and cognition are considered in the widest sense. Therefore, the work will treat a wide range of experimental and theoretical work.

The *Handbook of Perception and Cognition* should serve as a basic source and reference work for all in the arts or sciences, indeed for all who are interested in human perception, action, and cognition.

Edward C. Carterette and Morton P. Friedman

Preface

Cognitive Ecology is concerned with relationships between the traditional academic fields of Perception and Cognition and some broader issues of human activity.

The volume is organized into three sections: The first section considers issues of human mind and culture. The second section deals with cognitive psychology and the arts. The third section deals with sensory evaluation.

Mind and Culture

In Chapter 1, Eleanor Rosch considers the relation of mind to its environment. She discusses trends in cognitive psychology that suggest alternatives to the traditional dualistic view "in which mind and environment are treated not as separate objects or topics but as codefining poles of experiences and actions (p. 23)."

Much of mainstream Cognitive Psychology is based on models that ignore cultural issues. Jaan Valsiner's (Chapter 2) "Cultural Organization of Cognitive Functions" examines the problems that this narrow view creates for our understanding of human nature and considers attempts to integrate concepts of culture and cognitive functions.

The Arts

Gerald Cupchik and Andrew Winston (Chapter 3) outline major philosophical themes in aesthetics and discuss how these themes have guided psychological research. They also explore relationships between "mainstream analyses of perception and cognition in everyday life and conceptualizations of aesthetic process (p. 62)."

Roger Kendall and Edward Carterette's "Music Perception and Cognition" (Chapter 4) approaches the topic "from an ecological perspective as the interrelationship between composer, performer, and listener (p. 88)."

Julian Hochberg's "Perception of Pictures and Pictorial Art" (Chapter 5) discusses applications of cognitive psychology to an understanding of how we perceive and understand pictures, and conversely, what we can learn from picture perception about human cognition. Similarly, Chapter 6, "The Perception of Motion Pictures" by Julian Hochberg and Virginia Brooks, analyzes the perception and cognition of motion pictures in terms of principles of cognitive psychology.

The fourth chapter in this section, "The Art of the Puzzler" (Chapter 7) by Arthur Schulman, deals with the analysis of word puzzles (e.g., crosswords, anagrams) in terms of coginitive psychological principles.

Sensory Evaluation

The final chapter in this volume, "Flavor" (Chapter 8) by Harry Lawless, offers a comprehensive analysis of how our senses interact to produce human flavor perception.

Mind and Culture

CHAPTER 1

The Environment of Minds:

Toward a Noetic and Hedonic Ecology

Eleanor Rosch

Ecology is the study of the interrelation of living creatures with their environment. For ecological psychology the focus is the relation of mind to environment. This, of necessity, raises issues fundamental to any psychology: How are we to understand the nature and relation of the subject of experience to its object, of perceiver to perception, knower to known, motivated creature to the object of its desires, and actor to the field of action? These are not simply philosophical questions; they can profoundly affect how research is done, perhaps even how lives are lived.

There is a standard classical portrait of the relation of mind to environment in much of modern psychology, cognitive science, biology, and common sense. It may be called the Cartesian vision. Inside the mind is the thinking, attending, perceiving, feeling, desiring, reasoning, choosing subject; outside is the rest of the world that acts both as stimulus for the mind and as the recipient of its eventual actions. The hallmark of this view is that the mind, with its basic mechanisms, and the world, with its information and its potential for the survival or destruction of that mind, are inherently different and separate entities and should be treated as such by theorist and experimenter.

If the mind (however that is operationally defined) is conceived as different and separate from environmental context, then it makes sense to seek to study its basic mechanisms in as controlled and decontextualized a laboratory setting as possible. Until roughly the 1970s, such methods were the rule in the study of learning,

Cognitive Ecology

memory, categorization, perception, and other fields of psychology. One current trend is to move away from exclusive reliance on artificial stimuli and tasks toward more general sampling of real-world behavior settings (Neisser, 1984). Learning may be studied with natural stimuli in homes, schools, forests, and other species-specific habitats; memory studies may deal with stories, scripts, general knowledge, and autobiographies; and perception research may explore the complex optical structures that are created as animals move through their environments. Such a trend has been described as a move to more ecologically valid research, and portions of the resultant body of work are referred to as ecological psychology (Neisser, 1976, 1984, 1987).

There may, however, be a further meaning to the shift to ecological psychology; the Cartesian vision itself can be challenged. In present psychology and cognitive science there would seem to be two basic versions of the Cartesian model: one (corresponding to the ancient *faculty of cognition*) in which the knower is separate from the world known (and must perform elaborate indirect acts to acquire knowledge), the other (corresponding to the *faculty of conation*) in which the goal-directed, motivated desirer is separate and independent from his or her desired outcomes.

The cognitive model of the separate knower is that of basic information-processing psychology (some classic examples are Anderson, 1985, and Lindsay & Norman, 1977). The assumption is that there is information in the world that the information processor must recover. Information in a raw form enters via the sense organs, which the processor then decodes. In most versions the processor builds up a cognitive representation of the world with which the incoming information interacts to produce an interpretation. The system's behavior tests this interpretation, producing feedback that may lead to modification of the cognitive representation. (Social psychologists have referred to this as the model of human's as lay scientists—Fiske & Taylor, 1984; Tetlock, 1993).

The conative or affective model is the rational model of classical economic decision theory (see, for example, Fishburn, 1970; Sen, 1987). Outcomes in the world possess varying degrees of positive or negative value for an actor with varying probabilities of occurrence. Actors know what they want and rationally combine utility with probability in making decisions. Nonconscious desires function in the same manner (they just do not happen to be conscious).

In both models, the mind (the knower, desirer, doer) is separate from its ecological context of the known, desired, and acted upon. This tends to be taken for granted both in science and folk psychology as an incontrovertible fact of nature. However, Clifford Geertz (1975) contested this from anthropological observation:

> The Western conception of the person as a bounded, unique, more or less integrated motivational and cognitive universe, a dynamic center of awareness, emotion, judgment and action organized into a distinctive whole and set contrastively against a social and natural background is, however incorrigible it may seem to us, a rather peculiar idea within the context of the world's cultures. (p. 48)

Can it be that the duality of mind and environment, which may seem so obvious, is actually a metaphysical assumption? We need not depend on the contrastive vision of anthropology alone. From within psychology and cognitive science there have been a variety of explicit and implicit challenges to the Cartesian vision both of the separate knower and the separate motivated actor. The remainder of this chapter will explore some of these bodies of research.

I. THE HISTORICAL BACKGROUND: BEHAVIORISM

"Behaviorism . . . is wrong theoretically, wrong technically, wrong morally and wrong politically" (Rowan quoted in Blackman, 1980, p. 99). In present-day psychology, behaviorism tends to be almost as unfashionable as Aristotle and perhaps as little read. In the straw man version of behaviorist models, living systems are portrayed as simple mechanisms in which single stimuli give rise mechanically to simple responses, both organism and mind being confined to the black box in which experimentation is impossible and theorizing outlawed. Such a portrait is unjust both to the complexity of past models and to the theoretical and experimental sophistication of the surviving current practitioners. In this account I will emphasize work of the present groups that call themselves radical behaviorists, trace their lineage from Skinner's operant conditioning (Skinner, 1938), and designate their enterprise as the experimental analysis of behavior (Catania & Harnad, 1988; Hineline, 1980). They claim that all of mental life, including language, thought, and emotion, can be encompassed by the methods of behavior analysis.

Behaviorism is of theoretical interest to our present topic for many reasons beyond its historical parentage of contemporary experimental psychology. In the first place, both methodologically and substantively, the environment is inherently a part of any behaviorist analysis because environment (reinforcers) and behavior mutually define each other. Second, the units of analysis are necessarily actions rather than postulated structures. Third, at least in contemporary behavior analysis, the experimenter may be explicitly included as part of the field of analysis.

In response to the accusation that behaviorism is mechanistic, modern behavior analysts would argue that what they deal with is not a two-term, mechanical, associationistic relationship between a "stimulus" and a "response" but rather a functional or contextual portrait of predictable correlational regularities in environment–organism activities. The basic two-term unit is the *operant,* which is a relationship between a behavior and its *subsequent* environmental state (its consequence). Defining that subsequent state as an operant stimulus (a positive or negative reinforcer) depends on the relevant consequent behavior (whether it is observed to increase or decrease). An environmental change that was totally unrelatable to any behavioral change would not be a viable unit in behavior analysis. Similarly a behavior unrelatable to any environmental change would not be a studiable operant response. Thus operant response and stimulus mutually define one another. In the words of Morris (1988),

> From a contextual perspective, stimulus functions and response functions develop historically and exist simultaneously with respect to one another, and are defined in terms of each other in context. As such, stimulus functions have no more "control" over behavior than do response functions—the two are interdependently and mutually defining. (pp. 307–308)

This is interestingly reminiscent of Dewey's (1896) pragmatist argument that the proper basic unit of psychological analysis should be neither an isolated action nor even an interaction between two independently defined entities, but rather a functional unity of action and its consequence, which he called a *transaction*.

In experimental behavior analysis, circularity in such a definition, or at least an unmanageable appearance of circularity, is at times avoided by the introduction of a third party, the experimenter, who is also in an operant relationship with the organism he is studying (Hineline, 1992). The manipulations that he introduces into the experiment are operant responses that only continue to be emitted when reinforced by the behaviors of his subjects—as in the well-known cartoon from the *Columbia Jester* in which one rat says to another, "Boy do we have this guy conditioned. Every time I press the bar down he drops a pellet in."

More complex behavior–environment relations, an understanding of context, an explanation of language, and an interpretation of internal behaviors such as thoughts and emotions all depend in behavior analysis on the introduction of discriminant and conditional functions that expand the two-term relation in the basic operant to a many-term contingency relation (Hineline, 1992; Sidman, 1986). For example, if a rat learns to press a bar to get food, that is a two-term operant; if food is obtained from a bar press only if the bar is marked with a square and not if marked with a circle, this is a three-term discriminant relation; if the positive and negative role of the discriminant square and circle function only in the presence of a red light but are reversed in the presence of a green light, that is a four-term contingency relation; and so on for five and more terms. Nonhuman as well as human organisms can learn such contingencies to considerable depth and can also learn to chain contingencies in elaborate temporal sequences. Such contingencies can create equivalence classes of operant stimuli that behavior analysts claim are the basis of categorization, symbol using, and thus of language (Sidman, 1986).

What is called linguistic reference is, in behavior analytic terms, a complex set of contingent, discriminant, operant actions learned within the reinforcement contingencies of a linguistic community (Hineline, 1980). The contemporary behavior analyst's riposte to Chomsky's (1959) classic attack on Skinner's (1957) *Verbal Behavior* is to designate the cognitive or grammatical analysis of language as structural and the behavior analytic as functional, two cooperative and (ideally) converging modes of analysis (Catania, 1973). Supposedly internal or private states are considered learned responses that have their origins in public, primarily linguistic, reinforcement contingencies (Hineline, 1992). This is what enables much of the practical work in cognitive therapy derived from behavior analysis to be carried out (Wegner & Pennebaker, 1993). Through all of this complexity, ideally the interdependence of definition of environmental context and behavior are to be maintained.

The unit of analysis consists of particular behaviors, activities, or behavior–environment contingency links—never enduring traits, structures, or persons. This is particularly striking in contexts where folk and other psychologies would normally speak substantively. Thus in behavior modification, one would not speak of a trainer rewarding, let us say, a retarded child for tying his shoes, but of particular actions of the trainer reinforcing particular overt (and perhaps covert internal) actions of the child. Likewise, one would not speak of a patient being depressed but of the functional relationship of depressive ruminations to the self-reinforcement schedules that maintain these painful patterns. Much of the success of behavior–modification techniques appears to depend on the ability of the psychologist to segment the flow of phenomena into novel but, in retrospect, ingeniously appropriate behavioral units and manipulable contingencies that promote change (Kendall, 1982; Kohlenberg & Tsai, 1991; Wegner & Pennebaker, 1993; Wolpe, 1978). The emphasis on acts as the basic psychological unit is another way in which behavior analysis is reminiscent of pragmatism (James, 1890, 1907; Rorty, 1982)—as well as prescient, perhaps, of trends to be described later in this chapter.

Critiques of behaviorism, of course, are legion. The pervasive sense of many of them is that behavior analysis, for all its claims, is simply too restrictive a format to deal with the full range of human and animal life (i.e., it throws the baby out with the bath water—see for example Lerner, 1976). Often unrecognized is the extent to which behaviorism provides the intellectual underpinnings of modern experimental psychology. It may thus be provocative to see that the core definitions and assumptions of behavior analysis violate the commonsense metaphysic of a separation of mind, organism, and environment. Whatever its limitations, behavior analysis offers the background for one form of an ecological model of mind.

II. NOETIC ECOLOGY: THE KNOWER AND KNOWLEDGE

A. Perception

Perception is a major and primary form of knowing the world in virtually all philosophical and psychological systems (for non-Western examples, see Radhakrishnan & Moore, 1957). The bulk of perception research in experimental psychology has consisted of psychophysics, which seeks laws relating judgments about sensory events to physically measurable properties of proximal stimuli, and sensory psychophysiology, which seeks to relate such events to activities within the nervous system. The aspect of the environment at issue for both fields is the energy impinging on a sensory organ. However, as Brunswik (1956) pointed out, the major task of an organism is to interact appropriately with distal objects in its environment.

When organism, mind, and environment are treated as separate independent entities, perception of distal objects is necessarily indirect. An example is the formulation of the "problem of depth perception" in information-processing models: the physical environment, describable by the laws of physics, consists of space in which objects are located at varying distances from the eye of the perceiver. Light reflected from those objects strikes the eye simultaneously; it bears with it no inherent information about the distance it has traveled. The problem for the perceiver is to recover the information about depth in the environment from the cues provided by that light. To do this the perceiver must combine, perhaps through a process of unconscious inferential reasoning, raw data from the sense organs with the cognitive representation of the environment that has been built up from past learning.

All of this was anathema to that giant of the ecological approach to perception, J. J. Gibson (1950, 1966, 1979), who spoke of *direct* perception. Gibson's approach can be interpreted in relatively narrow or broad terms. Generally, *direct perception* is taken as the claim that there are information invariants available to the senses that portray the environment—that is, that information invariants sufficient to specify all objects and events in the organism's environment can be directly extracted from the flux of information available in its sensory arrays (Shepard, 1984). In the example of depth perception, information from texture gradients in the optical array and disparity of retinal images are directly available. Many research programs have been engaged in the search for the multitudes of perceptual invariances (Garner, 1974; MacLeod & Pick, 1974).

Gibson's ecological approach, however, was arguably broader than the search for higher order perceptual invariances alone.

> It is often neglected that the words *animal* and *environment* make an inseparable pair. Each term implies the other. No animal could exist without an environment surrounding it. Equally, although not so obvious, an environment implies an animal . . . to be surrounded. (1979, p. 8)

This is no empty verbal definition; the environment for Gibson is not the world as described by physics, and the eye of the animal is not a decontextualized mirror of a physical universe. Gibson developed a new way to describe the world, ecological optics, in which light, space, motion, and other abstract properties were necessarily designated in organismically relevant and dependent ways and in which the perceiving organism was necessarily described in environmentally relative and dependent ways. For example, in ecological optics the perceived environment does not contain the empty space of physics, which is an abstraction and not visible, but rather a gradient of features of the ground (grasses, stones, plants, objects) of decreasing optical size and increasing optical density which *is* visible; it contains not the abstract time of physics but various rates of perceived change.

Perception of oneself and one's environment, so defined, are inseparable.

> The optical information to specify the self, including the head, body, arms, and hands, accompanies the optical information to specify the environment. . . . The one could

> not exist without the other. . . . The supposedly separate realms of the subjective and the objective are actually only poles of attention. The dualism of observer and environment is unnecessary. The information for the perception of "here" is of the same kind as the information for the perception of "there," and a continuous layout of surfaces extends from the one to the other. (Gibson, 1979, p. 116)

An important aspect of Gibson's description of the organism–environment world is the concept of *affordances,* the functions that the perceived world afford the animal. For example, the ground affords support, enclosures afford shelter, elongated objects afford pounding and striking. Because it is an organismically meaningful world that is directly perceived, these affordances are the information that constitutes perception. Affordance may seem a somewhat mysterious attribute unless it is remembered that perception for Gibson is inseparable from appropriate functional action in the environment:

> The point to remember is that the visual control of the hands is inseparably connected with the visual perception of objects. The act of throwing complements the perception of a throwable object. The transporting of things is part and parcel of seeing them as portable or not. (Gibson, 1979, p. 235)

Might one accuse Gibson of a return to mentalism? Is his replacement of a physical description of the environment with a functional ecological one a covert way of placing the environment inside the organism's head? Gibson replies:

> The movement of the hands do not consist of responses to stimuli. . . . Is the only alternative to think of the hands as instruments of the mind? Piaget, for example, sometimes seems to imply that the hands are tools of a child's intelligence. But this is like saying that the hand is a tool of an inner child. . . . This is surely an error. The alternative is not a return to mentalism. We should think of the hands as neither triggered nor commanded but *controlled.* (Gibson, 1979, p. 235).

At his most innovative, Gibson is trying to develop an analysis of perception and action that captures just that level of functional description *between* the subject of perception and its object, between the mental and physical, between action and perception—a level of description that Gibson would call truly ecological.

B. Attention

Attention is inextricably bound with knowledge. William James (1890) distinguished two qualitatively different aspects of attention: (1) a clear nucleus or focus of attention, and (2) a *fringe* to that experience. The focus was what is normally meant by attention or awareness—that which is revealed by the searchlight in the contemporary searchlight metaphor for attention. The fringe for James included many different types of experience, such as: (1) feelings of familiarity, (2) feelings of knowing, such as tip-of-the-tongue-experiences, (3) feelings of relation between objects or ideas, (4) feelings of action tendency, as in intentions, (5) feelings of expectancy, (6) feelings of rightness or being on the right track. To this list we might add other examples of one's sense of one's own metaknowledge, such as metaknowledge of one's memory or one's abilities. Perhaps the most pervasive fringe feeling is that of meaningfulness, that one knows the larger context of any given moment of focal attention although that context is not part of the content of attention. Might the experience of the fringe represent the awareness aspect of ecological context?

After a century of neglect, interest in James's distinction between focal and fringe attention is slowly returning, spearheaded by Mangan's (1991) innovative dissertation. In Mangan's account of the fringe, the feeling of rightness serves as a summary or condensation of nonconscious knowledge structures (and is a phenomenological instantiation of the connectionist concept of goodness-of-fit—see Section IIC). Like chunking of information, the fringe is an effective way to deal with the limited capacity of focal awareness. The fringe experience is a way to present, in summary form, the contexts of relation that give meaning to discrete items present in the focus of awareness. Because the fringe experience is nonverbal and ungraspable by focal attention, Mangan suggested it is the basis of "ineffable" aesthetic and mystical experiences.

Galin (1992, 1993) extended Mangan's account, suggesting that focal attention has to do with featural specification, whereas the fringe indicates a broad-based self-monitoring system. Self-monitoring may concern knowledge (such as feelings of knowing and familiarity), action (such as intentions and sense of agency), and the fate of goals (monitored by the emotion system). By such a taxonomy, various clinical phenomena such as blind sight, agnosia, and hallucinations may be understood as malfunctions of various parts of the self-monitoring system.

Fringe experience is relevant to the ecology of mind in three ways. One aspect is the issue of knowledge of context. If mind is seen as something separate from its environment, it would be natural to have isolated atoms of experience strung together temporarily with no necessary connection to context. However, if mind and environment are viewed as analytically inseparable, one would expect awareness to mirror this connectedness in some way—perhaps as in the fringe experience.

A second aspect is the issue of capacity. Attention is normally viewed either explicitly, or more recently implicitly, as a limited-capacity system (Galin, 1993; Kahneman, 1973; Navon, 1984; Parasuraman & Davies, 1984—although for some radical exceptions see Hirst, Spelke, Reaves, Caharack, & Neisser, 1980; Neisser, 1976). This may be because only focal attention is normally investigated. A mind that is defined literally as part of its environment (the subjective pole of attention in a subject–object field) should have much broader attentional capacities than a mind defined as separate. Many of the anomalies of attention and consciousness research, such as blind sight and the other agnosias, are cases that violate the standard limited-capacity conception. Investigation of fringe phenomena may serve to expand, or perhaps undermine, models of attentional limits.

The third issue is that of agency. The concept of the fringe may serve to solve some puzzles about action and agency. When mind and environment are seen as separate, action is as mysterious as perception. One tends to fall either into a deterministic account in which stimuli cause responses (e.g., move the hand) or a free will account in which mind moves the hand. As we have seen, in Gibson's account, a third alternative is possible: seeing the action of the hand as *under the control* of the organism–environment field. James's category of the fringe might serve as the experiential correlate of being embedded in such a field.

C. Connectionism

In current cognitive science, there is a class of computer models known as connectionist (parallel distributed processing) systems specifically designed to violate certain key assumptions in the representational information-processing models we have been citing (Martindale, 1991; McClelland & Rumelhart, 1986; Rumelhart & McClelland, 1986). Connectionism arose as an alternative to the dominant approach of symbol manipulation as the basis of artificial intelligence. Connectionist systems consist of simple, neuronlike, unintelligent components which, when appropriately connected, develop intelligent global properties. Processes in the system function in parallel and are nonlocalized. Thus representation of a particular entity, such as a category, is a state of activation defined over the entire memory system rather than an invariant component of memory retrieved from a particular location. Perception and knowledge are states of activation of the system; there is no question of invariant objects.

There are two major classes of learning methods currently being explored in connectionist systems. The first is learning by correlation (Rosenbaum, 1989). If two nodes are active together, their connection is strengthened; otherwise it is diminished. (Hebb, 1949, postulated this principle as the basis for neural learning.) Some nodes of the system may be treated as sensory receptors and presented with a pattern. Links between nodes that are active together are increased, and when the system is again presented with the pattern, it recognizes it in the sense that it falls into the unique global state or internal configuration that was correlated with the presentation of that pattern. The second major learning method is copying, known as *backpropagation*. In this technique, changes in the neuronal connections inside the network (called *hidden units*) are assigned so as to minimize the difference between the network's response and what is expected of it (McClelland & Rumelhart, 1986; Rumelhart & McClelland, 1986). Here learning resembles someone trying to imitate an instructor.

Connectionist systems are excellent demonstrations of how complex global effects can appear as emergent properties out of the organization of simple components. And they demonstrate how such a system can function without cognitive representations. It is not obvious, however, that they present any new understanding regarding the relation of a system with its environment. Information tends to be introjected directly into the system already in its system-appropriate form. In relation to the environment, these systems may resemble nothing so much as the brain-in-a-vat riddle beloved of philosophers. There is often a good deal of *ad hoc* adjustment necessary to get any particular system to perform its intended task, and when successful the system is generally sufficiently complex that one cannot tell how it did so—just as with humans. One important Gibsonian aspect of these systems is that they demonstrate forcefully that a perceptual system does not require representations of the environment in order to operate. Beyond that, as a mode of research, connectionist systems appear to possess the virtues and vices of any other artificial-intelligence simulation.

III. HEDONIC ECOLOGY: DESIRES AND THINGS DESIRED

It is a commonplace that organisms and minds are goal directed—whether goals are conceptualized as survival, reproduction, the satisfaction of drives, the fulfillment of human desires, or the social enactment of values. The standard assumption of folk psychology, psychology, and cognitive science is that desires or goals are inside the organism or mind and that the objects of those desires are in the environment—those two being separate and independent realms. However, just as Gibson found it necessary to conceptualize the perceived environment as a world of organism-dependent ecological optics (not the world of physics), so the world of the desiring organism may require the development of its own hedonic ecological description.

The standard objectivist view of desires is perhaps most clearly delineated in classical economic decision theory (Fishburn, 1970; Kahneman & Varey, 1991; Sen, 1982, 1987; Von Neumann & Morgenstern, 1947). Utility is assigned to tangible and objectively identifiable aspects of a decision maker's environment (usually material assets) and is measured by choice behavior. It is assumed that decision makers know their present and future utilities and rationally combine utility with probability of outcome to make choices. Gain consists of obtaining objects of utility, loss of losing them, a symmetrical function.

All aspects of this model are being challenged by psychological research.

A. Neutral Reference Points for Gain and Loss

A self-monitoring, goal-directed system needs to keep continual track of the fate of its goals, and to monitor whether it is gaining or losing. In an objectivist view, gains and losses are external (e.g., a raise in salary is a gain, a reduction a loss). But suppose such a raise is accompanied by the information that all co-workers received a higher increase? It may well then be coded as a loss. In a more phenomenological description of gain and loss, gains and losses would appear to be measured from a neutral reference point (what is seen as the norm, the way things are), which is psychological and can be highly volatile, that is, subject to momentary shifts depending on interpretation (Kahneman & Tversky, 1979; Kahneman & Varey, 1991; Tversky & Kahneman, 1981). Thus one rule of a hedonic ecology may be that the hedonic ecological environment is perceived in terms of neutral norm(s) with reference to which possible changes of state are gains and losses.

Whether something is categorized as a gain or loss can have far-reaching impact on the way it is then processed. Although people are normally risk averse for gains (they opt for smaller sure gains rather than larger risky ones), they are risk taking for losses (as owners of casinos and racetracks know, even ordinary people will risk a great deal to attempt recovery of a sure loss) (Kahneman & Tversky, 1979, 1984; Tversky & Kahneman, 1981). A special case of loss aversion may be the *endowment effect* (Knetsch, Thaler, & Kahneman, 1987; Thaler, 1980); as soon as anything is defined as one's possession or even associated with oneself, it increases in value and one will show monetary commitment and physiological arousal at the possibility of loss. Note that possessions are a somewhat anomalous category (as William James, 1890, pointed out)—if one is dividing the world into independent subjects and objects, possessions (by definition almost) join subject and object; they partake of both.

Loss aversion has been used to account for several other social psychological phenomena. For example, people are generally conservative about making changes. If the attributes of a potential new situation are coded as gains and attributes of the old as losses, loss aversion predicts that people will have a great tendency to stay where they are (Kahneman & Tversky, 1984; Thaler, 1980). Judgments of fairness are another domain that can be elucidated by understanding the principles of neutral reference points and loss aversion (Kahneman, Knetsch, & Thaler, 1986a, 1986b). For example, survey respondents deem it unfair for an employer to lower an employee's salary due to the availability of cheaper labor, although it would be fair for him to hire a new employee at the lower wage.

B. Habituation

Once a situation or hedonic object has become the norm or neutral reference point, it changes its hedonic value. This is the well-known phenomenon of habituation; responses to a repeated stimulus decrease or cease (Appley, 1971; Helson, 1964; Solomon, 1980; Thompson & Spencer, 1966). Organisms respond to change and contrast, not to steady states. This gives rise to the paradoxical phenomenon that if one wishes to maximize a particular state (the experience, for example, of green or of pleasure), one must introduce an optimal amount of its opposite (red or displeasure) (Kahneman & Varey, 1991). Pleasures to which we have habituated (Scitovsky, 1976, calls them comforts) no longer produce enjoyment by their presence but do create pain by their absence—as many an intrepid vacationer has discovered regarding the comforts of home. (Note that this is technically the definition of an addiction.) Alas for humankind, gains tend to become new neutral reference points and pleasures habituate far more readily than their negative counterparts. A state of health is seldom compared joyously to a former ailment, but an illness is almost inevitably seen as a loss from former health. The phenomenon of mourning may also be understood in these terms—people retain as reference point the lost relationship or situation. All of these phenomena argue that, at a minimum, a hedonic ecology will require an account of the mutability of objects of desire.

C. Comparison

In an objectivist utility theory, satisfaction with one's life, grades, income, housing, friends, or physical condition should correspond to objective circumstances and be measurable by external indicators. One ought to be able to judge the extent of one's own satisfaction without reference to the satisfaction of others. It should be unproblematic to judge whether the present is better than the past with respect to some attribute. And one should, indeed must, be able to predict what states of the world one will find satisfying in the future. Research calls all of these assumptions into question.

A basic finding of well-being research is that objective circumstances and actual achievements are poor predictors of satisfaction in any domain across populations (Argyle, 1987; Campbell, 1981; Diener, 1984, Duncan, 1975; Easterlin, 1974). Instead, satisfaction appears mainly determined by a process of comparison. One tradition of research has emphasized the role of social comparisons—that is, comparisons of one's present state with those of other persons within one's reference group. It is social comparison that can give rise to the phenomenon of *relative deprivation* (e.g., a wealthy person may feel impoverished by comparison with wealthier associates) (Crosby, 1982; Festinger, 1954; Stouffer, 1949). Another tradition emphasizes comparisons to an adaptation level, which is often determined by an individual's personal history (Brickman & Campbell, 1971; Helson, 1964). More recently, these two standards of comparison have been shown to interact; in public domains, such as income, subjects appear to rate their own satisfaction by comparison to others, but in private domains, such as love life, they appear to rate others in comparison to themselves (Fox & Kahneman, 1992).

Yet a third tradition calls into question subjects' ability to perform any kind of comparison in an objectivist fashion, even the ability to compare one's own past and present states. Judgment of any state—whether present, past, one's own, or another's—is an ad hoc construction determined by the categories, framing, and reference points evoked by the particular task in its particular context (Schwarz, Bless, & Wanke, 1992; Schwarz & Strack, 1991a; Strack, Argyle, & Schwarz, 1991). For example, which time period is categorized as the present and which as the past is easily manipulated by slight differences in the wording of survey questions (Schwarz & Strack, 1991b) (e.g., mention of a temporal boundary, such as graduating from college, can reverse experimental effects) (Schwarz, Bless, & Wanke, 1992; Strack, Schwarz, & Nebel, 1987). The positive or negative valence of recalled life events either increases or decreases the valence of ratings of current satisfaction depending upon how the time periods are categorized (Schwarz & Bless, 1992). Furthermore, the direction of comparison has a large effect; when one compares a current problematic situation with the past, one is likely to conclude that things are getting worse, whereas when comparing a past problem with the present that things are getting better (Dunning & Parpal, 1989; Schwarz et al., 1992). There are many other clever demonstrations of the mutability of reference points and temporal judgments (see Schwarz et al., 1992, and Schwarz & Strack, 1991a, for reviews). The conclusion seems unavoidable that judgments of subjective well-being can be conceived neither as a read-out of a stable inner state nor as a comparison of such a state with either an objectively or subjectively represented reference point. Rather, the experimenter and his tasks are a part of the on-line experimental context, and subjects' judgments are immediately generated actions shaped by the shifting conditions of that context. Inner and outer, past and present, satisfied and dissatisfied, are relative to each other and to context and have no demonstrable intrinsic stable referent.

The ability to predict one's future states of satisfaction has been little studied, perhaps because an unchallenged cornerstone of rational decision theory is that one knows what one wants (Elster, 1984). Currently, at least one program of research is investigating such judgments. In laboratory settings, students' predictions of their enjoyment of eating particular foods over time were found unrelated to their ratings of enjoyment at the actual time of eating (Kahneman & Snell, 1990; Snell, 1991). Such a finding may be surprising and yet accord well with experience—most people can probably readily recall life experiences where events occurred as expected but affective reactions to them did not.

D. Categorizations

1. Framing of Events

Desired objects or states of affairs are categorized as parts of larger units or frames. Tversky and Kahneman (1981) speak of the framing of decisions, the framing of probabilities, and the framing of outcomes. The framing of outcomes may be the most intuitively comprehensible; the following are some examples. Financial decisions may be made on the basis of *accounts* for particular items rather than on the basis of total wealth; thus, in the context of a large financial expenditure, such as purchase of a home, consumers are likely to spend large additional sums on accessories that they would not purchase under normal circumstances. In experiments, subjects are less likely to say they would replace a lost ticket to the theater than that they would buy the ticket, having just lost the identical amount of money. And having already paid for a particular consumer item that one then finds unacceptable (e.g., a wretched desert), one is likely to suffer through consuming it anyway, an activity called *sunk costs* (Thaler, 1980). These and many other examples (Kahneman, Slovic, & Tversky, 1982; Tversky & Kahneman, 1981) demonstrate that the perceived units and forces of the hedonic environment are not the same as those of a neutrally conceived environment.

2. Summarization of Events and Loss of Temporal Extension

Hedonic tone and emotional content of events fluctuate from moment to moment (Varey & Kahneman, 1992). In a rational model of utility, overall rating of pleasantness for an event should be a summation over time of the ratings of individual parts. Longer pleasant and aversive events should be rated as more hedonically extreme than shorter events. However, subjects' evaluations of film clips showed little relation to on-line ratings; rather, global judgments of pleasantness appeared a function only of the most extreme point of pleasantness in the film and the ending point (Fredrickson & Kahneman, 1993). Importance of the ending point of an event has paradoxical implications. Laboratory subjects prefer a longer unpleasant event that terminates gradually (a hand in unpleasantly cold water which is slowly warmed) to a shorter version that ends abruptly (Kahneman, Fredrickson, Schreiber, & Redelmeier, 1993). This finding was replicated with a painful procedure in a medical setting and is of potential practical importance (Redelmeier & Kahneman, 1993).

In an extreme form, summary judgments of the pleasantness or unpleasantness, and hence desirability, of an event may have no relation at all to the hedonic experience of the event itself—the judgment may be based on memories that have become dissociated from the event. Such dissociation could result from a general bias to reconstruct memories as pleasant (Fiske & Taylor, 1984; Matlin & Strang, 1978); social pressures (e.g., one is supposed to enjoy a visit to Disneyland) (Festinger, 1957; Schlenker, 1980; Sutton, 1989); or the use of selected positively valenced photographs or other mementos as cues (Chalfen, 1981; Milgram, 1977; Sontag, 1973). Intriguingly, the desirability or fearfulness of an anticipated event may be based on imagining how one would feel at hearing the news (e.g., that one has just won a lottery or tested HIV-positive) rather than on a realistic appraisal of one's future state. And some events may be anticipated in terms of the memories they will leave—such as vacations. At the very least it is clear that temporal order in the hedonic environment bears little resemblance to the abstract time of physics but operates on its own (yet to be formulated) laws.

E. Probability

A rational economic decision maker computes probability in a normatively correct fashion (such as Bayes's theorem) and combines it with utility to inform his choices. There is accumulating evidence, however, that people do not use prior odds or base rates in judging probability; rather they base judgments on a collection of heuristics such as representativeness and availability (Kahneman et al., 1982; Kahneman & Tversky, 1973; Nisbett & Ross, 1980). In a hedonic context, certainty is preferred to probability, giving rise to what is called the *pseudo-certainty effect* (Tversky & Kahneman, 1981). All insurance coverage is partial insurance against disaster but people prefer that it be framed as complete coverage for a clearly specified subset of possibilities. For example, if there are two strains of flu virus, subjects find a vaccine that gives complete immunity to strain A and no immunity to strain B considerably more attractive than a vaccine that gives 50% immunity against both (Kunreuther, 1978; Tversky & Kahneman, 1981). The hedonic environment may have its own laws for the computation of probabilities.

F. Counterfactuals

One of the most unique, perhaps startling, aspects of the ecological environment for mind, whether noetic or hedonic, is that a large part of it is counterfactual (Kahneman & Miller, 1986; Kahneman & Tversky, 1982). That is, most cognitive and emotional judgments take place in a context where what is judged to be the actual state of affairs is compared to what could be or could have been. For example, imagine that Mr. A and Mr. B both arrive late at the airport. A's plane left on time and is long since gone. However, B's plane was also delayed; it only just departed. Which man will experience more distress? Note that the difference between their situations is not actual but counterfactual; yet the alternative world in which B caught his plane seems *closer* to the actual. The judgment that a particular event *almost* occurred has a peculiar status (Kahneman & Varey, 1990). In one sense it is an objective judgment with intersubjective reliability; however, it does not correspond to anything that happened, and it may have no formal consequences (A and B have identical immediate practical plights).

Let us take another example (Kahneman & Tversky, 1982). Mr. T leaves his office 20 minutes early one day and is killed in an automobile accident. Despite the fleshing out of the incident with additional narrative details, subjects agree that when Mr. T's relations "undo" the accident, they will think, "If only he hadn't left early that day," rather than changing any of the other events of the story that would have removed him from the time and place of the accident. Normal, routine situations are the most salient counterfactual comparisons to violations of routine.

We live in worlds of counterfactuals. Perception is of discontinuity and change, both contrasts to the background state of the world that would prevail in their absence (Appley, 1971). To say that an object has a particular orientation is meaningful only within a spatial reference frame in which other orientations are a possibility (Garner, 1974). Narrative is of unusual events made newsworthy because in the background routines and scripts of daily life they did not occur (Clark & Haviland, 1977). Children must learn that these background routines do not constitute a narrative (Snow, 1991; Eisenberg, 1985). Questions (at least *why* questions) can normally be asked only about a violation of routine, not about what is conceived as the normal state of the world (Lehnert, 1978). Focal attention is drawn by change and violations of expectation, by things made interesting because they could be expected to be otherwise (Berlyne, 1960). In fact, the very concept of selective attention is a counterfactual one; we say that *x* was *attended to* because we have reason to construct a counterfactual world in which something else was attended instead (that is, in which we had some other perception or thought). Any act of categorization consists not only of placing an item in its appropriate category but of discriminating it from the other possible categories of which it is *not* a member (Rosch, 1994).

Emotion occurs primarily in a counterfactual context. It requires comparison of a present state of the hedonic world with a possible past or future (Lazarus, 1991). Fear sees an imminent change of state for the worse, hope a potential change for the better. Regret sees a past that could have been better by one's own action. Surprise expected a very different state than what occurred. Envy sees a cognitively available counterfactual world in which one has what is now another's.

Each of the five principles already enumerated that distinguish the hedonic environment from an environment of objectivist utilities have a strong counterfactual aspect. When a neutral reference point is compared with a state of the world to determine whether something is a loss or gain, one of the terms of the comparison is necessarily a counterfactual. The same holds true for other comparisons—comparing one's present hedonic state to others, comparing past and present, or predicting future satisfaction. Categorizations necessarily involve the choice between possible alternatives. It is an issue of debate whether representations that summarize an event entirely erase memories or exist against a background (perhaps a *fringe*) memory for the original affect, retrievable in appropriate contexts. And the computation of probabilities, by whatever heuristic, must involve the envisioning of counterfactuals.

Given the ubiquitousness of counterfactuals in the hedonic environment, do we know anything about the laws and logic of counterfactual thinking? Kahneman and Miller (1986) proposed a model to account for the construction and use of norms in which they suggested that a counterfactual possibility will appear close to reality to the extent that it can be reached by altering the features of reality that are conceived as the most mutable. The concept of mutability appeals to the intuition that some modifications of reality seem more natural than others, whereas some attributes seem particularly resistant to change.

Several rules are suggested that determine mutability:

1. Exceptions are more mutable than routines. We have already seen the application of this rule in the example of Mr. T who left his office early.

2. Ideals are less mutable than nonideals. When asked to change the outcome of a card game or tennis match, subjects do so by imagining an improvement of the losing game rather than a deterioration of the winning one.

3. Reliable knowledge is less mutable than unreliable. People are more confident in predicting the score on a short IQ test from a long one than vice versa (Tversky & Kahneman, 1982).

4. Causes are less mutable than effects. A child may be described as "big for his age" but not as "young for his size" and as overachieving rather than undertalented (Tversky & Kahneman, 1982).

5. The actions of the focal or attended actor in a situation are more mutable than those of a background actor. If subjects are told that a near collision between two cars was averted and asked how, preventive actions tend to be attributed to the actor from whose point of view the incident had been narrated. This aspect of counterfactual reasoning could lead to the well-documented tendency to blame victims for their plight (Lerner & Miller, 1978). If focus of attention is on the victim, then actions of the perpetrator will tend to be presupposed, whereas actions of the victim can easily be imagined to have been otherwise.

Counterfactuals play a large role in explaining affective response. The same undesirable outcome is judged more upsetting when its alternative seems closer. People regret exceptional actions that ended in tragedy more than routine ones. Actions taken that lead to losses are regretted more than opportunities missed (Kahneman & Tversky, 1982). Victims are awarded more compensation when struck down performing unusual actions than routine ones and more still for tragedies that came close to being avoided (Miller & McFarland, 1986).

Causal reasoning is another domain heavily dependent on counterfactual reasoning. In the first place, we tend to request a causal account (ask *why* questions) only for exceptions to background routines ("Why didn't you eat your lunch?" but not "Why did you eat your lunch?"). In fact, simply asking a question conveys the presupposition that the queried fact is nonnormative and questionable (Lehnert, 1978). In the second place, default values of a situation cannot be offered as a cause. One cannot say that the fire in the wastebasket was caused by the oxygen in the air unless it is normative for oxygen to be absent from that location (Hart & Honore, 1959). Finally, causal questions for the same situation will receive legitimately different answers depending on the counterfactual norm that is evoked and used as a standard of comparison. A man may blame an attack of indigestion on eating parsnips whereas his doctor blames it on the man's ulcers—the man is comparing the situation to the other instances when, despite ulcers, he had no attack; the doctor is comparing the man to other patients without ulcers (Hart & Honore, 1959).

In summary, the ecological environment of goal-directed, acting minds is not a neutral world describable by physics but a noetic and hedonic environment consisting of affordances, normative reference points, counterfactual possible worlds, and a continual emotively toned process of comparison. The ecological environment is always defined and judged with respect to a goal-directed mind. But what of that mind? What we have been suggesting is that the definition of mind is as dependent on the noetic and hedonic environment as the environment is on the goal-directed mind. This claim is addressed further (hopefully clarified) in the following section.

IV. THE SELF

There is a certain naive sense of oneself—perceptions, thoughts, memories, body, desires, plans, feelings, accomplishments, failures, agency, habits, sense of integrity—that phenomenologically we tend to take as an actual self. It is in relation to this self that one has personal goals and the emotions that are associated with the fate of those personal goals. Who but the self is insulted when insulted, pleased when praised, or afraid when threatened? Perhaps it is this sense of a self that makes it natural to postulate that one has a mind that is separate from one's ecological context.

Historically, the phenomena of the self have been approached from two markedly different perspectives, from the point of view of the life sciences (psychology, biology, cognitive science) and through the reflective disciplines of contemplation and meditation. At present there tends to be a marked split between the sciences of mind and mind as it is handled in daily life (Rosch, 1991). Just as no one would think to take seriously claims of cognitive science (such as that consciousness is epiphenomenal) in daily life, so the cognitive scientist does not find it necessary to develop a method for the serious study of lived experience. It is thus fortunate to have history provide us with converging operations for the study of the self.

What kind of self is implied by the phenomenological, emotive sense of self?

> Such a self has to be lasting, for if it perished every moment one would not be so concerned about what was going to happen to it the next moment; it would not be one's "self" anymore. Again it has to be single. If one had no separate identity, why should one worry about what happened to one's "self" any more than one worried about anyone else's? It has to be independent or there would be no sense in saying "I did this" or "I have that." . . . We all act as if we had lasting, separate, and independent selves that it is our constant preoccupation to protect and foster. It is an unthinking habit that most of us would normally be most unlikely to question or explain. (Gyamtso, 1986, pp. 20–21)

Both the scientific and contemplative traditions *do* question and analyze. Do they identify a self with properties like those above? In the classic words of David Hume:

> For my part, when I enter most intimately into what I call *myself,* I always stumble on some particular perception or other, of heat or cold, light or shade, love or hatred, pain or pleasure. I never can catch *myself* at any time without a perception, and never can observe anything but the perception. (Hume, 1964, I:iv)

Various sciences of mind study the self. The self-concept and self-attribution processes are studied by social, cognitive, personality, and cross-cultural psychology (Gergen, 1991). How the self-concept is conceived and measured varies considerably with the orientation of the researcher and with prevailing fashions about what a concept is believed to be (Neisser, 1988). However, all are studying self-*concepts,* not a lasting, unitary, independent self. Another trend is to study the self as narrative—

the narratives that people store in memory and tell to others (Bruner, 1990). This self obeys the laws of narrative that are taught to children from the time they begin learning to speak (Miller & Sperry, 1988; Snow, 1991). Again, this is an account of self-concepts and self-construction, not a lasting, unitary, independent self.

Psychoanalysis, from its inception, divided the mind into separate regions and structures and unseated the notion of a rational governing self. One contemporary school of psychoanalysis, object relations theory, actually views the self as a society of interacting schemas (Klein, 1987; Turkle, 1988). These are basically interpersonal schemas in which forms or aspects of oneself are envisioned in relation to forms or aspects of others. Thus representations of the self are inherently multiple and relational. Some social psychology has also turned to the concept of a *relational self*, suggesting that one has a different self in different interactional contexts (Gergen, 1991). Such trends in psychoanalysis and social psychology are sometimes said to accord well with what, in the humanities, is called "the breakdown of the self in the post modern period" (Gergen, 1991). Note that this is a more macrodefinition of relationality and self than the ecological analysis we have been suggesting here.

In cognitive science, mind may be modeled in different ways. Cognitivist accounts reduce mental functions to symbols that are manipulated by rules. (This is the tradition in which experiential aspects of mind tend to be considered epiphenomenal—Churchland, 1986; Jackendoff, 1987.) A particular program or model may include a self or ego as a supervisor function, but it tends to be an *ad hoc* construction and is certainly not a lasting, unitary, independent self. Society-of-mind models divide the supervisory functions into many separate processes or selves (Minsky, 1986). Connectionist models have no supervisory structures as such; higher order functions are emergents from the states into which the system settles in its functioning (McClelland & Rumelhart, 1986). There may be a new form of cognitive science, which Varela, Thompson, and Rosch (1991) called *enaction*. This approach explicitly addresses the issue of the self and of failures to find an actual self in either the scientific, phenomenological (Heidegger, 1962; Merleau-Ponty, 1962), or contemplative traditions. In the enactive view, self and world are simultaneously enacted moment by moment on the basis of the history of such enactions.

It is often tempting to see the lack of a real self in the theories of psychology and cognitive science as contradictory to the way humans actually experience their minds. It thus becomes interesting to note parallels between the cognitive sciences and the contemplative and meditative traditions (Rosch, 1994). Meditation methods of concentration, mindfulness, relaxation, and examination reputedly can calm and sharpen the mind such that it can serve as an instrument for its own observation. Reports of meditators parallel some aspect of cognitive theories:

1. An analytic component. Meditators typically report noticing that what they had taken as single experiences (e.g., being angry all day) are actually composed of many different, changeable, smaller experiences.

2. Experience of emergence. Higher order aspects of experience (such as judgments, emotions, self-concepts) are reported to be seen as apparent and functional without necessarily being substantively real.

3. Interdependence of subject and object in experience. The subject of experience is reportedly seen not to be a thing separate from the experience itself.

The latter point is particularly important for the present argument. It is analyzed further in one of the contemplative traditions (the Buddhist Abhidharma—Rabten, 1981; Varela et al., 1991). Each moment of experience is said to take the form of a particular consciousness that has a particular object to which it is tied by particular relations. For example, a moment of "seeing consciousness" is composed of a seer (the subject) who sees (the relation) a sight (the object). These components, however, are neither analytically nor experientially separable; they arise codependently. Some meditators report that they do actually come to see this in a way that is very personally meaningful. They further report that *seeing* it involves a switch from the usual subject–object mode of perception to a more comprehensive sense of knowing that is quite reminiscent of descriptions of the *fringe* experience. Thus lived experience might be said to fall *between* what is ordinarily thought of as the mind and what is ordinarily thought of as the world—or, perhaps, to include both.

V. CONCLUSIONS

We have asked about the relation of mind to its environment. The classic answer both in folk psychology and science is a dualism in which mind and environment are seen as separate objects that can be studied as such. However, this may be a metaphysic of our culture rather than an incontrovertible fact of nature. We have outlined a number of powerful trends in psychology and cognitive science that argue that mind and environment must be treated as a unity.

In behaviorism, the parent of modern experimental psychology, use of the *operant* as a basic unit of analysis automatically tied behavior and its environmental outcome together into a single unit in which behavior and environment are mutually defining. In the study of perception, J. J. Gibson developed an ecological optics to replace physical optics; according to ecological optics the environment is defined and described as it is perceived by a goal-directed, acting organism. In a Gibsonian analysis, perceiver and perceived environment are two poles of the same perception, never separable. In attention research, rejuvenation of William James's concept of *the fringe* suggests a model in which each act of mind in focal attention is accompanied by a broader form of knowing that includes the context of the experience.

The study of desires, decisions, and satisfaction suggests that the objects of such acts constitute a hedonic environment quite different from the world as described by physics or objectivist utility theory. Subject and object are codefining in continual acts of comparison, many of which involve reference to counterfactual rather than actual worlds. Finally, study of the self—which might have been expected to unify the knowing, perceiving, attending, desiring, and acting subject of experience into something enduring and separate from its environment—upon analysis reveals instead the changing play of subject–object relations.

These trends suggest a new field, a deep ecology of mind, in which mind and environment are treated not as separate objects or topics but as codefining poles of experiences and actions. Knowing, desiring, and acting may not be analytically independent; for a goal-oriented organism the noetic and hedonic environment are the same. Given the difficulties that an egocentric viewpoint provides both individuals and the world, it might be hoped that such a change might enliven not only the cognitive sciences but, perhaps thereby, also modes of human behavior.

References

Anderson, J. R. (1985). *Cognitive psychology and its implications* (2nd ed.). New York: W. H. Freeman.

Appley, M. H. (Ed.). (1971). *Adaptation-level theory: A symposium*. New York: Academic Press.

Argyle, M. (1987). *The psychology of happiness*. London: Methuen.

Berlyne, D. E. (1960). *Conflict, arousal and curiosity*. New York; McGraw-Hill.

Blackman, D. E. (1980). Images of man in contemporary behaviorism. In A. J. Chapman & D. M. Jones (Ed.), *Models of man* (pp. 99–112). London: The British Psychological Society.

Brickman, P., & Campbell, E. T. (1971). Hedonic relativism and planning the good society. In M. H. Appley (Ed.), *Adaptation-level theory: A symposium*. New York: Academic Press.

Bruner, J. S. (1990). *Acts of meaning*. Cambridge, MA: Harvard University Press.

Brunswik, E. (1956). *Perception and the representative design of psychological experiments* (2nd ed.). Berkeley: University of California Press.

Campbell, A. (1981). *The sense of well-being in America: Recent patterns and trends*. New York. McGraw-Hill.

Catania, A. C. (1973). The psychologies of structure, function, and development. *American Psychologist, 28*, 434–442.

Catania, C., & Harnad, S. (Eds.). (1988). *The selection of behavior: The operant behaviorism of B. F. Skinner: Comments and consequences*. New York: Cambridge University Press.

Chalfen, R. (1981). Redundant imagery: Some observations on the use of snapshots in American culture. *Journal of American Culture, 4*, 106–113.

Chomsky, N. (1959) Review of B. F. Skinner's *Verbal Behavior. Language, 35*, 26–58.

Churchland, P. S. (1986). *Neurophilosophy*. Cambridge MA: MIT Press, A Bradford Book.

Clark, H. H., & Haviland, W. E. (1977). Comprehension and the given-new contract. In R. O. Freedle (Ed.), *Discourse production and comprehension* (pp. 1–40). Norwood, NJ: Ablex.

Crosby, F. (1982). *Relative deprivation and the working woman*. New York: Oxford University Press.

Dewey, J. (1896). The reflex arc concept in psychology. *Psychological Review, 3*, 357–370.

Diener, E. (1984). Subjective well-being. *Psychological Bulletin, 95*, 542–575.

Duncan, O. (1975). Does money buy satisfaction? *Social Indicators Research, 2*, 267–274.

Dunning, D., & Parpal, M. (1989). Mental addition and subtraction in counterfactual reasoning: On assessing the impact of personal action and life events. *Journal of Personality and Social Psychology, 57*, 5–15.

Easterlin, R. A. (1974). Does economic growth improve the human lot? Some empirical evidence. In P. A. David & M. W. Reder (Eds.), *Nations and households in economic growth*. New York: Academic Press.

Eisenberg, A. (1985). Learning to describe past experiences in conversation. *Discourse Processes, 8,* 177–204.

Elster, J. (1984). *Ulysses and the sirens: Studies in rationality and irrationality* (2nd ed.). Cambridge, UK: Cambridge University Press.

Festinger, L. (1954). A theory of social comparison process. *Human Relations, 7,* 230–243.

Festinger, L. (1957). *A theory of cognitive dissonance.* Evanston, IL: Row & Peterson.

Fishburn, P. (1970). *Utility theory for decision making.* New York: Wiley.

Fiske, S. T., & Taylor, S. E. (1984). *Social cognition.* Reading, MA: Addison-Wesley.

Fox, C. R., & Kahneman, D. (1992). Correlations, causes and heuristics in surveys of life satisfaction. *Social Indicators Research, 27,* 221–234.

Fredrickson, B. L., & Kahneman, D. (1993). Duration neglect in retrospective evaluations of affective episodes. *Journal of Personality and Social Psychology, 65,* 45–55.

Galin, D. (1992). Theoretical reflections on awareness, monitoring, and self in relation to anosognosia. *Consciousness and Cognition, 1,* 152–162.

Galin, D. (1993). *The structure of awareness: Contemporary application of William James' forgotten concept of "the fringe."* Unpublished manuscript, University of California at San Francisco, San Francisco, CA.

Garner, W. R. (1974). *The processing of information and structure.* Potomac, MD: Lawrence Erlbaum.

Geertz, C. (1975). On the nature of anthropological understanding. *American Scientist, 63,* 47–53.

Gergen, K. J. (1991). *The saturated self: Dilemmas of identity in contemporary life.* New York: Basic Books.

Gibson, J. J. (1950). *The perception of the visual world.* Boston: Houghton-Mifflin.

Gibson, J. J. (1966). *The senses considered as perceptual systems.* Boston: Houghton-Mifflin.

Gibson, J. J. (1986). *The ecological approach to visual perception.* Potomac, MD: Lawrence Erlbaum. (First published in 1979).

Gyamtso, K. T. (1986). *Progressive stages of meditation on emptiness.* Oxford: Longchen Foundation.

Hart, H. L., & Honore, A. M. (1959). *Causation and the law.* London: Oxford University Press.

Hebb, D. O. (1949). *The organization of behavior: A neuropsychological theory.* New York: Wiley.

Heidegger M. (1962). *Being and time.* New York: Harper & Row.

Helson, H. (1964). *Adaptation level theory.* New York: Harper & Row.

Hineline, P. N. (1980). The language of behavior analysis: Its community, its functions, and its limitations. *Behaviorism, 8,* 67–86.

Hineline, P. N. (1992). A self-interpretive behavior analysis. *American Psychologist, 47,* 1274–1299.

Hirst, W., Spelke, E., Reaves, C. C., Caharack, G., & Neisser, U. (1980). Dividing attention without alternation or automaticity. *Journal of Experimental Psychology: General, 109,* 98–117.

Hume, D. (1964). *A treatise of human nature.* (L. A. Selby-Bigge, Ed.). Oxford: Clarendon Press.

Jackendoff, R. (1987). *Consciousness and the computational mind.* Cambridge, MA: MIT Press, A Bradford Book.

James, W. (1950). *The principles of psychology.* New York: Dover. (Original work published in 1890)

James, W. (1907). *Pragmatism.* New York: New American Library.

Kahneman, D. (1973). *Attention and effort.* Englewood Cliffs, NJ: Prentice-Hall.

Kahneman, D., Fredrickson, B. L., Schreiber, C. A., & Redelmeier, D. A. (1993). When more pain is preferred to less: Adding a better end. *Psychological Science, 4,* 401–405.

Kahneman, D., Knetsch, J. L., & Thaler, R. H. (1986a). Fairness and the assumptions of economics. *Journal of Business, 59,* s285–s300.

Kahneman, D., Knetsch, J. L., & Thaler, R. H. (1986b). Fairness as a constraint on profit seeking: Entitlements in the market. *The American Economic Review, 76,* 728–741.

Kahneman, D., & Miller, D. T. (1986). Norm theory: Comparing reality to its alternatives. *Psychological Review, 93,* 136–153.

Kahneman, D., Slovic, P., & Tversky, A. (Eds.). (1982). *Judgment under uncertainty: Heuristics and biases.* New York: Cambridge University Press.

Kahneman, D., & Snell, J. (1990). Predicting utility. In R. M. Hogarth (Ed.), *Insights in decision making: A tribute to Hillel J. Einhorn* (pp. 295–310). Chicago: University of Chicago Press.

Kahneman, D., & Tversky, A. (1973). On the psychology of prediction. *Psychology Review, 80,* 237–251.

Kahneman, D., & Tversky, A. (1979). Prospect theory: An analysis of decision under risk. *Econometrica, 47,* 2, 263–291.

Kahneman, D., & Tversky, A. (1982). The simulation heuristic. In D. Kahneman, P. Slovic, & A. Tversky (Eds.), *Judgment under uncertainty: Heuristics and biases* (pp. 201–208). New York: Cambridge University Press.

Kahneman, D., & Tversky, A. (1984). Choices, values and frames. *American Psychologist, 39,* 341–350.

Kahneman, D., & Varey, C. (1990). Propensities and counterfactuals: The loser that almost won. *Journal of Personality and Social Psychology, 6,* 1101–1110.

Kahneman, D., & Varey, C. (1991). Notes on the psychology of utility. In J. Elster & J. E. Roemer (Ed.), *Interpersonal comparisons of well-being* (pp. 127–153). Cambridge, UK: Cambridge University Press.

Kendall, P. C. (Ed.). (1982). *Advances in cognitive-behavioral research and therapy* (Vol. 1). San Diego, CA: Academic Press.

Klein, J. (1987). *Our need for others and its roots in infancy.* London: Tavistock Publications.

Knetch, J. L., Thaler, R. H., & Kahneman, D. (1991). The endowment effect, loss aversion, and status quo bias. *Journal of Economic Perspectives, 5,* 193–206.

Kohlenberg, R. J., & Tsai, M. (1991). *Functional analytic psychotherapy.* New York: Plenum.

Kunreuther, H. (1978). *Disaster insurance protection: Public policy lessons.* New York: Wiley.

Lazarus, R. S. (1991). *Emotion and adaptation.* Oxford: Oxford University Press.

Lehnert, W. (1978). *The process of question answering.* Hillsdale, NJ: Erlbaum.

Lerner, R. M. (1976). *Concepts and theories of human development.* Reading, MA: Addison-Wesley.

Lerner, M. J., & Miller, E. T. (1978). Just world research and the attribution process: Looking back and ahead. *Psychological Bulletin, 85,* 1030–1051.

Lindsay, P. H., & Norman, D. A. (1977). *Human information processing: An introduction to psychology* (2nd ed.). New York: Academic Press.

MacLeod, R. B., & Pick, H. L. (Eds.). (1974). *Perception: Essays in honor of James J. Gibson.* Ithaca, NY: Cornell University Press.

Mangan, B. (1991). *Meaning and the structure of consciousness: An essay in psycho-aesthetics.* Unpublished doctoral dissertation, University of California at Berkeley, Berkeley, CA. (Microfilms No. 92033636).

Martindale, C. (1991). *Cognitive psychology: A neural network approach.* Pacific Grove, CA: Brooks-Cole.

Matlin, M., & Strang, D. (1978). *The pollyanna principle.* Cambridge, MA: Schenkman.

McClelland, J. L., & Rumelhart, E. E. (Eds.). (1986). *Parallel distributed processing. Vol. 1: Foundations.* Cambridge, MA: MIT Press.

Merleau-Ponty, M. (1962). *Phenomenology of perception.* (C. Smith, Trans.). London: Routledge & Kegan Paul.

Milgram, S. (1977, January). The image-freezing machine. *Psychology Today,* 50–54 & 108.

Miller, D. T., & McFarland, C. (1986). Counterfactual thinking and victim compensation: A test of norm theory. *Personality and Social Psychology Bulletin, 12,* 513–519.

Miller, P., & Sperry, C. (1988). Early talk about the past. *Journal of Child Language, 15,* 293–315.

Minsky, M. (1986). *The society of mind.* New York: Simon & Schuster.

Morris, E. K. (1988). Contextualism: The world view of behavior analysis. *Journal of Experimental Child Psychology, 46,* 289–323.

Navon, D. (1984). Resources: A theoretical soup stone. *Psychological Review, 91,* 216–234.

Neisser, U. (1976). *Cognition and reality.* New York: W. H. Freeman.

Neisser, U. (1984). *Toward an ecological cognitive science.* Emory Cognition Project Report #1. Atlanta, GA: Emory University.

Neisser, U. (1987). *Concepts and conceptual development: Ecological and intellectual factors in categorization.* Cambridge: Cambridge University Press.

Neisser, U. (1988). *Five kinds of self knowledge.* (Emory Cognition Project working paper #14). Atlanta, GA: Emory University.

Nisbett, R. E., & Ross, L. (1980). *Human inference: Strategies and shortcomings of social judgment.* Englewood Cliffs, NJ: Prentice-Hall.

Parasuraman, R., & Davies, E. C. (Eds.). (1984). *Varieties of attention.* Orlando: Academic Press.

Page, A. N. (1968). *Utility theory: A book of readings.* New York: Wiley.

Rabten, G. (1981). *The mind and its functions.* Mt. Pelverin, Switzerland: Tharpa Choeling.

Radhakrishnan, S., & Moore, C. A. (1957). *A sourcebook in Indian philosophy.* Princeton, NJ: Princeton University Press

Redelmeier, D. A., & Kahneman, D. (1993). On the importance of a good ending: Patient's real-time and retrospective evaluations of the pain of colonoscopy. Berkeley, CA: University of California.

Rorty, R. (1982). *Consequences of pragmatism.* Minneapolis: University of Minnesota Press.

Rosch, E. (1991 August). *Outside and inside the self: Convergence of cognitive science and the mindfulness tradition.* Paper presented at the meeting of The American Psychological Society, San Francisco, CA.

Rosch, E. (1994). *The original psychology.* Unpublished manuscript, University of California, Berkeley, CA.

Rosch, E. (1994). Categorization. In V. S. Ramachandran (Ed.) *The encyclopedia of human behavior.* San Diego, CA: Academic Press.

Rosenbaum, I. (1989). *Readings in neurocomputing.* Cambridge, MA: MIT Press.

Rumelhart, E. E., & McClelland, J. L. (Eds.). (1986). *Parallel distributed processing* (Vol. 1,) Cambridge, MA: MIT Press.

Schlenker, B. R. (1980). *Impression management.* Monterey, CA: Brooks-Cole.

Schwarz, N., & Bless, H. (1992). Constructing reality and its alternatives: Assimilation and contrast effects in social judgment. In L. L. Martin & A. Tesser (Eds.), *The construction of social judgment* (pp. 217–245). Hillsdale, NJ: Erlbaum.

Schwarz, N., Bless, H., & Wanke, M. (1992 May). *Subjective assessments of change: Lessons from social cognition research.* Paper presented at the East-West meeting of the European Association of Experimental Social Psychology, Munster, Germany.

Schwarz, N., & Strack, F. (1991a). Evaluating one's life: A judgment model of subjective well being. In F. Strack, M. Argyle, & N. Schwarz (Eds.), *Subjective well being: An interdisciplinary perspective* (pp. 27–47). Oxford: Pergamon.

Schwarz, N., & Strack, F. (1991b). Context effects in attitude surveys: Applying cognitive theory to social research. In W. Stroebe & M. Hewstone (Eds.), *European review of social psychology* (Vol. 2, pp. 31–50). Chichester: Wiley.

Scitovsky, T. (1976). *The joyless economy.* Oxford: Oxford University Press.

Sen, A. (1982). *Choice, welfare, and measurement.* Oxford: Basil Blackwell.

Sen, A. (1987). *On ethics and economics.* Oxford: Basil Blackwell.

Shepard, R. N. (1984). Ecological constraints on internal representation: Resonant kinematics of perceiving, imagining, thinking, and dreaming. *Psychological Review, 91,* 417–447.

Sidman, M. (1986). Functional analysis of emergent verbal classes. In T. Thompson & M. D. Zeiler (Eds.), *Analysis and integration of behavioral units* (pp. 213–245). Hillsdale, NJ: Erlbaum.

Skinner, B. F. (1938). *The behavior of organisms: An experimental analysis.* New York: Appleton-Century.

Skinner, B. F. (1957). *Verbal behavior.* New York: Appleton-Century-Crofts.

Snell, J. S. (1991). *A bias in hedonic self prediction.* Unpublished doctoral dissertation, University of California, Berkeley, CA.

Snow, C. (1991). Building memories. In D. Cicchetti & M. Beeghly (Eds.), *The self in transition* (pp. 213–242). Chicago: University of Chicago Press.

Solomon, R. L. (1980). The opponent-process theory of acquired motivation: The costs of pleasure and the benefits of pain. *American Psychologist, 35*(7), 691–712.

Sontag, S. (1973). *On photography.* New York: Delta.

Stouffer, S. A. (Ed.). (1949). *The American soldier: Adjustment during wartime life* (Vol. I). Princeton, NJ: Princeton University Press.

Strack, F., Argyle, M., & Schwarz, N. (Eds.). (1991). *Subjective well-being: An interdisciplinary perspective.* Oxford: Pergamon.

Strack, F., Schwarz, N., & Nebel, A. (1987, March). *Thinking about your life: Affective and evaluative consequences.* Paper presented at the conference on "Ruminations, self-relevant cognitions, and stress." Memphis State University, Memphis, TN.

Sutton, R. I. (1989, April). *Feeling about a Disneyland visit: Photography and the reconstruction of bygone emotions.* Paper presented at a meeting of the National Academy of Management, Anaheim, CA.

Tetlock, P. E. (1993). *Behavior, society and interpersonal conflict.* Oxford: Oxford University Press.

Thaler, R. (1980). Toward a positive theory of consumer choice. *Journal of Economic Behavior and Organization, 1,* 39–60.

Thompson, R. F., & Spencer, W. A. (1966). Habituation: a model phenomenon for the study of neuronal substrates of behavior. *Psychological Review, 73,* 16–43.

Turkle, S. (1988). Artificial intelligence and psychoanalysis: A new alliance. *Daedelus* (Winter), 241–269.

Tversky, A., & Kahneman, D. (1981). The framing of decisions and the psychology of choice. *Science, 211,* 453–458.

Tversky, A., & Kahneman, D. (1982). Causal schemas in judgments under uncertainty. In D. Kahneman, P. Slovic, & A. Tversky (Eds.), *Judgment under uncertainty: Heuristics and biases* (pp. 117–128). New York: Cambridge University Press.

Tversky, A., & Kahneman, D. (1986). Rational choice and the framing of decisions. *Journal of Business, 59,* s251–278.

Varela, F. J., Thompson, E., & Rosch, E. (1991). *The embodied mind: Cognitive science and human experience.* Cambridge, MA: MIT Press.

Varey, C., & Kahneman, D. (1992). Experiences extended across time: Evaluation of moments and episodes. *Journal of Behavior Decision Making, 5,* 169–185.

Von Neumann, J., & Morgenstern, O. (1947). *Theory of games and economic behavior.* Princeton, NJ: Princeton University Press.

Wegner, D. M., & Pennebaker, J. W. (1993). *Handbook of mental control.* Englewood Cliffs NJ: Prentice-Hall.

Wolpe, J. (1978). Cognition and causation in human behavior and its therapy. *American Psychologist, 33,* 437–446.

CHAPTER 2

Cultural Organization of Cognitive Functions

Jaan Valsiner

I. INTRODUCTION

The issue of culture has had a varied fate in the realm of psychology—its relevance has been episodically denied, while at other times culture has been recognized as central for understanding human psychology. However, that latter recognition has usually remained unproductive, because the research practices of psychology have been built upon scientific models for which culture inclusiveness is a threat, rather than an asset (see Danziger, 1990; Valsiner, 1994).

Recently, different groups of investigators have started to talk about the field of culture-linked psychological issues, often in conjunction with interest in mental functions (Cole, 1992; Cole, 1995; Markus & Kitayama, 1991; Shweder, 1990, 1992; Shweder & Sullivan, 1993). This orientation towards bring different notions of culture into the social sciences can be observed in different disciplines—sociology, ethnology, psychology, and so on (e.g., Boesch, 1983, 1991; Bourdieu, 1973, 1985; Eckensberger, 1991; Ivic, 1978, 1988; Kon, 1988; Krewer, 1990, 1992; Lyra & Rossetti-Ferreira, 1995; Moscovici, 1981, 1982, 1988; Ratner, 1991; Rio & Alvarez, 1990, 1992; Rogoff, 1990, 1992, 1993; Tulviste, 1991). This effort seems to gain momentum, hence it is in a phase of constructing (rather than possessing) its view upon the complex psychological phenomena. However, complex psychological phenomena have always been difficult for psychologists to han-

Cognitive Ecology

dle, and it remains to be seen how psychologists of the 1990s can address the task that previous decades have failed to accomplish. Reviews of recent efforts to summarize empirical finding on the social foundations of cognition (e.g., Levine, Resnick, & Higgins, 1993) have demonstrated how the discipline suffers within its theoretical stagnation and repetition of old research practices, dressed in new verbiage.

In the context of the history of science, the focus on culture in psychology is better seen not as a new development in the discipline, but rather as a restoration of the interests that were characteristic of human sciences in the nineteenth century (see Jahoda, 1990, 1993, 1995). Wilhelm Wundt's version of *Völkerpsychologie* (which served as both an opponent and proponent of different twentieth-century tendencies in psychology) grew out of that general cultural-psychological orientation (Van Hoorn & Verhaeve, 1980). Before the revitalization of American psychology by behaviorism, cognitively oriented psychology of the turn of this century benefited from regular exchange of information between ethnologists, sociologists, and psychologists (e.g., Simmel, 1906, 1908; Thurnwald, 1922; Wertheimer, 1912). Culturally situated perspectives on human mental processes have a long tradition in science (see Jahoda, 1990).

Obviously, any cultural perspective on cognition has to face psychology's traditional conceptual confusion. In this chapter, I relate three ill-defined idea complexes —culture, cognition, and ecology. All three terms have traditionally been generic "family names" for different research orientations that have at times proclaimed their revolutionary status (Gardner, 1985), even if such claims can be viewed as rhetorical (Gergen, 1989; Valsiner, 1991). Rhetorics in science are productive for constructing socially shared understanding (and misunderstanding), or at least a particular perspective on reality.

II. SHARED MYSTERIES OF CULTURE AND COGNITION

A. The Concept of Culture from an Anthropological Standpoint

It may be a surprise to find out that culture—the central concept of cultural (or social) anthropology—has been a troublesome theoretical entity. In that discipline, cultures have traditionally been viewed as homogeneous classes where all persons "share" the culture that may be said to consist of strictly defined and relatively stable rules, beliefs, or folk models. Culture has been usually viewed as either fully "shared" by all successfully socialized members of the given social group (i.e., thus treating cultures as homogeneous groups), or as shared by different *degree* (in likeness to a fuzzy set) by persons who occupy different positions in their status, role, and social relationships with one another (see Linton, 1936; Swartz, 1982; Wallace, 1970).

1. Past Views on Culture

Anthropology has borrowed from psychology regarding cultural transmission. Thus culture has been defined as a "mass" of learned and transmitted motor reactions, habits, techniques, ideas, and values—and the behavior they induce" (Kroeber, 1948, p. 8). The holistic focus on phenomena is likewise not lost in anthropology—culture has been defined as an integrated whole in which systemic interconnections between physiological drives and their transformations through social institutions were the core of the concept (Malinowski, 1944). Within that perspective,

> culture is an integral composed of partly autonomous, partly coordinated institutions. It is integrated on a series of principles such as the community of blood through procreation; the contiguity in space related to cooperation; the specialization in activities; and last but not least, the use of power in political organization. Each culture owes its completeness and self-sufficiency to the fact that it satisfies the whole range of basic, instrumental and integrative needs. (Malinowski, 1944, p. 40)

2. Anthropology's Flirtation with Cognitive Science

In the 1980s and 1990s, anthropologists' difficulties with the concept of culture are amplified. The consensual beliefs of previous decades that culture is a relatively homogeneous and stable entity (shared by all of its "members"—i.e., persons), is being eroded by a number of critical tendencies in the social sciences (see Strauss, 1992). In contemporary anthropology there exist three major kinds of views on "culture" in anthropology (see D'Andrade, 1984, pp. 115–116):

1. *Culture is seen as knowledge:* it is the accumulation of information (irrespective of the extent to which that information is shared between "members of the culture").

2. *Culture is seen as consisting of core conceptual structures that provide basis for intersubjectively shared representation of the world in which the persons live.* This perspective does not emphasize the moment of accumulation (of information), but is rather a set of rules that makes it possible for persons to arrive at shared understandings.

3. *Culture is construction of conceptual structures* by activities of persons.

Undoubtedly these explanations of culture indicate appropriation of concepts brought to anthropologists' social discourse by the popularity of the "cognitive revolution." If we carefully examine all the three cognitive views on culture, we can reach an understanding that much of anthropology tries to study culture *through persons* (individual members of the culture) without taking the persons *as persons* into account. In this respect, the anthropology seems to be under the influence of the "cognitive revolution" in psychology (and other disciplines). This modern state of affairs seems to repeat other episodes in anthropology's history, where psychoanalytic or behavioristic ideas were appropriated from psychology. Yet the conceptual problem remains unclear: How can personal, culturally constructed knowledge be "shared" interpersonally? The concept of culture has become a target of ambivalent attitudes in contemporary anthropology (Moore, 1994). Its static connotations lead anthropologists to reject it, yet all of cultural anthropology is based on the axiomatically set relevance of the ontological view of the culture.

B. Psychology's Inferential Fallacies and Discourse about Culture

Psychology's conceptual apparatus has been largely essentialistic. Thus, statements about direct linear causation ("*X* causes *Y*") have been rampant, and probably on the increase in recent decades (see Gigerenzer et al., 1989). Thus, talk about "cultural effects" upon "cognition" (or vice versa: "effects" of "cognitive models" upon "culture," or "accounting for variance" by way of "variables") abounds in the literature, yet we have rather few insights into the actual processes that are involved in such implied causal effects.

1. Fascination with Variables: Constructing Black Boxes

Psychology's usual logic of inference has proceeded from outcomes (products of some psychological processes) to finding plausible causes for these outcomes. The latter are usually viewed as static essences (or at least as terms referring to *black boxes,* which include processes that psychologists consensually agree not to tear open—such as "learning"—see Goodyear, 1995). The black box construction is further facilitated by the inference from difference between phenomena to the hypothetical causal agent that stands behind that difference.

For example, an investigator discovers a difference between two samples from different geographical locations (and their corresponding index features) on some measure of outcomes (e.g., a test composed of accumulated responses of subjects) and claims that "culture" has had an "effect" that explains the demonstrated difference. Of course, culture (however defined) cannot have an "effect" upon the marks made on response sheets directly, but it can effect subjects'(at most) *the process* of generating their responses. Furthermore, it can effect the subjects' process of personal development up to the very moment of test-response generating. However, in both cases the effective entity is actually some (usually unspecified) process that leads to the outcomes, rather than a black box label designating that process (e.g., "problem solving" as describing the process of response generating), or some essentialistic posited entity (culture, intelligence, etc.). Psychology has provided a framework for appropriate relabeling of these causal attributions, while keeping conceptualization of the actual processes from direct analysis.

For psychology, the need to define culture more specifically remains an important task (see Fogel, 1993; Mercer, 1993). Culture is most often utilized in psychology within its cross-cultural subbranch. There it is usually viewed as if it were "transmitted" from one generation to the next through teaching and learning (see Berry, Poortinga, Segall, & Dasen, 1992). Again, the use of psychology's traditional terms keeps the processes involved in both (i.e., in the actual reality of teaching and learning) closed for further investigation. Instead of analyzing the actual process of cultural transmission in all its complexity, cross-cultural psychology has explained away the need for such investigation. By (obviously undisputable) common-language claims that "children learn culture" we do not clarify the processes involved in the actual cultural transmission. What we do affirm via such discourse is contrasts with other possible common-language claims (e.g., "'culture is inborn"), thus restating the old nature–nurture controversy in the domain of culture and cognition.

2. How Is Psychological Argumentation Different from Common Sense?

Psychology has been a hostage of the commonsense language of psychologists (as laypersons). Hence, it is perhaps appropriate to start the analysis of culture in cognition from the psychologist-as-researcher level (rather than analyzing his or her subjects). In this respect, all of psychology's understanding of human mental functions is based on language-encoded reflection, which introduces static moments into the realm of constant flow of life events.

Psychology's reliance upon a common language is ambivalent. On the one hand, the richness of common language allows psychology a basis for sophisticated understanding (see Siegfried, 1994). On the other hand, it limits psychological expertise to those aspects that may be the historical and cultural particulars of the investigator. Psychology, embedded in a common language, needs to transcend the boundaries of that constraint system (Valsiner, 1985); otherwise, it may remain part of the collective culture (of the persons acting in the role of "psychologists"), and as such create different kinds of socially desirable myths under the halo effect of science (Sarbin, 1990). Collective-cultural definitions of *culture* are obstacles to using it as a scientific concept.

3. Culture in Psychology: From Index Variables to Constructive Processes

As has been pointed out thus far, culture has been treated as an illusionary independent variable in psychology (although it has merely been an index variable). The usual research tradition is to use it as a label that is appropriately applicable to a sample of subjects (e.g., the comparisons of German vs. Zulu cultures). Hence there has been little interest in explicating the meaning of culture in systemic (organizational) terms, and with the focus upon particular persons. Psychology has become quasi-sociology in its knowledge construction: inference is made from samples to populations (rather than from single to generic case), data accumulation follows implicitly the ideals of "majority rule" (e.g., in metanalyses of data), and systemic-structural organization of the phenomena is not part of investigative focus.

Nevertheless, psychology provides a starting point for a consideration of culture as an inherently heterogeneous organizational form. Social psychology's focus on interpersonal relationships can reconcile the heterogeneous reality with investigators' folk models (see Pepitone, 1986). The classic contributions to social psychology in the area of construction of social norms (Sherif, 1936) or the construction and reconstruction of intra- (and inter-) group relations (Sherif, Harvey, White, Hood, & Sherif, 1961) can be seen as psychology's contribution to finding a solution to the problem of culture. Culture from that point of view is constantly in the process of construction in social relationships (hence inherently and normatively structurally heterogeneous). Therefore *culture cannot be "shared" between persons outside of social communication* (other than by way of a metaphoric illusion), and can be located only in the psychological (intra- and extramental) sphere of persons and their environments. The latter relationship is also the target of cognitive investigations.

C. Cognition: Process Situated within Contexts

Despite the wide proliferation of the modern catchword *cognitive science,* what is meant by cognition (or even more so—by the frequent use of the plural—cognition*s*) in psychology remains a mystery. Interest in intrapsychological processes is not new to psychology, as the rich (but forgotten and stigmatized) traditions of introspectionist methodology have demonstrated.

Basically, approaches in contemporary psychology that bear the label "cognitive" are a hybrid of psychologists' historical interest in the intricacies of the intramental psychological processes, and their (behavioristically and socially well-trained) avoidance of the inevitable feature of those processes (i.e., their direct accessibility via introspection only—see Danziger, 1990). It is therefore not surprising that cognition is a widely used but loosely defined concept (see also Skinner, 1985), which goes only half-way towards making sense of the psychological functioning of human and nonhuman organisms.

Cognitive-psychological issues were high profile in turn-of-the-century psychology, as the more fundamental issues of consciousness, free will, and dynamics of thought processes were actively discussed (Baldwin, 1906; James, 1890, 1904; Morgan, 1892). The reappearance of mental psychology under "cognitive revolution" since 1956 has focussed extensively on empirical work, largely underestimating the greater theoretical unsolved issues (Valsiner, 1991). Nevertheless, the so-called cognitive science movement remains a historically particular reaction to the excesses of behaviorism in psychology, rather than a new breakthrough in modern science.

1. Mental Practices in Cognitive Studies

Let us consider some usual efforts to explain what cognitive perspectives in our modern science entail. Putting aside most of the cognitivists' efforts to define their mental products, a rather refreshing view by an intelligent outsider in the doings of cognitive scientists seems to be appropriate:

> In cognitive science the usual procedure is to isolate some psychological phenomenon, make a theoretical model of the postulated mental processes, and then test the model, by computer simulation, to make sure it works as the author thought it would. If it fits at least some of the psychological facts it is then thought to be a useful model.

The fact that it is rather unlikely to be the correct one seems to disturb nobody. (Crick, 1988, pp. 149–150)

Crick seems to capture a very general problem of psychology at large—the belief that a *consensually sufficient* fit of theoretical models to reality is also a real fit between the two. A biological scientist looking at psychology as an outsider certainly can see that serious limitation, which has rendered psychology rich in published papers but poor in actual understanding. Modern-day cognitive psychology is likewise filled with self-glorifying slogans (e.g., "If cognitive science does not exist then it is necessary to invent it"—Johnson-Laird, 1980, p. 71) and history constructions (Gardner, 1985). Yet it remains a curious invention (see Kessen, 1981), which shrugs off theory, forms exclusive social clubs, and takes excessive interest in talking about context. Paradoxically, it turns out that for integrating cultural and mental processes into its general knowhow, psychology may have to forget about the vicissitudes of "cognitive revolution" and the slogan of interdisciplinary integration with anthropology. Culture may become a central concept of the psychology of mental processes, not an artificially imported term from anthropology.

2. Cognition and Context

It has become widespread to talk of cognition as *situation in* context (Butterworth, 1993; Lave, 1988; Rogoff & Lave, 1984; for criticism of that tradition see Vera & Simon, 1993a, 1993b), the latter being often seen as culturally organized. This emphasis replicates the more general focus on persons-within-environments (Lave, 1988), and leads to the issue of what is that curious mental process that is posited to be situated. On the one hand—given the uniqueness of every personal experience—situatedness of mental processes is an axiomatic given. On the other hand, however, the person who experiences these context-situated events is an integrated whole (self), who constantly goes beyond the situatedness of one's mental phenomena in the direction of transcontextual transfer and utilization of newly constructed cognitive tools in totally new context. It can be said that the emphasis on the situatedness of cognitive functions in context in contemporary research creates a major need for theoretical reconstruction of all psychology.

D. Context-Bound Cognition and Ecological Approaches

Modern cognitive discourse at times becomes linked with ecological discourse (Hoffman & Nead, 1983; Neisser, 1982, 1987; Turvey & Carello, 1980; Turvey, Shaw, Reed, & Mace, 1981) in ways of opposing the mentalistic constructions. Yet the emphasis on "direct perception" may have been generalized beyond proportions and has begun to overlook the relevance of higher mental functions. Currently a more balanced view is in practice (see Reed, 1995), which recognizes the relevance of both direct perceptual and indirect symbolic-informational inputs in the processes of knowing (Reed, 1992).

The ecological approach to cognition seems to stem from strictly functionalist roots, and hence is built on the concept of adaptation (see Shettleworth, 1993). However, as is pointed out in the present chapter (see Section III.A.), the concept of adaptation itself can be viewed differently. The ecological approaches to cognition tend to borrow the conceptual model of *adaptation-as-fit* (contrast this with the relevance of misfit—Valsiner & Cairns, 1992). Cognitive processes may be emergent mechanisms that increase the lack of immediate fit between the mentally "processing" agent (person) and the given context. It is the irreversible nature of time that guarantees the processual primacy of the adaptation-as-construction notion. A cultural-ecological perspective necessarily adopts a systemic causality perspective—cognitive functions are generated by a process entailing a causal system, parts of which are organismic, and their corresponding parts may be located in the environment. For instance, the complementary processes of internalization and externalization (see Lawrence & Valsiner, 1993), or Piaget's system of assimilation and accommodation, constitute instances of systemic causality. Systemic causality notions are of course well known in developmental biology (see Weiss, 1978).

E. Language Constraints on the Understanding of Flow of Mental Processes

A great conceptual difficulty in accounting for the irreversible time in explicating process mechanisms has been prominent in psychology (see Van Geert, 1988). Psychology lacks terminology to describe (and less so—explain) dynamic change as it takes place. This deficiency was well expressed at the beginning of this century in the organismic philosophy of Henri Bergson, whose emphasis on irreversibility of time has proven to be profound (Prigogine, 1987).

Bergson's focus of interest was primarily on the world of human psychology, as his "intuitivistic" bases in understanding the world philosophically started from the highly personal act of introspection, the acceptance of which has been difficult for most of our modern cognitive psychology. He was in many ways close to William James's cognitivist perspective on "stream of consciousness" (James, 1890). Bergson's emphasis on semiotic mediation of human consciousness has all the features of the latter versions of sociocultural perspectives (Rommetveit, 1979; Vygotsky, 1934; Werner & Kaplan, 1963). Language use allows consciousness to separate the flow of experience into units with the help of symbols (Bergson, 1910, p. 128). At the level of symbol use, persons discover their "impersonal residue" of feelings communicable across persons in society. Language performs a dual function: on the one hand, it generates self-reflexive stability (thus eliminating the real "flow" of irreversible personal experience by translating it into symbols reflecting stability). On the other hand, thanks to its stability-constructing role, language makes it possible for human consciousness to transcend the present here-and-now and reconstruct memories of the past (as well as transfer reflections of the present to a new context).

It should be clear that all theoretical constructions of culture–cognition relationships are canalized by the constraints of language on conceptualization. This constitutes a severe handicap for psychology as a discipline, because its borrowing of everyday concepts and transferring them to scientific terms usually carries unintended connotational baggage that later delimits investigators' understanding of relevant substantive issues. The inheritance of such common-language connotations surfaces in our contemporary efforts to create theoretical links of culture and mental processes.

III. CONTEMPORARY ATTEMPTS TO UNDERSTAND CULTURE WITHIN MENTAL PROCESSES

Contemporary psychology provides an increasing flow of both theoretical and empirical efforts to clarify the relation between culture and mental processes. This chapter does not attempt to provide an overview of all the empirical directions within which culture is being considered as a catchword. Instead, I will concentrate on the analysis of those few theoretical stances that can be viewed as of fundamental relevance.

Obviously, a number of existing perspectives (see Bronfenbrenner, 1979, 1989, 1993; Bronfenbrenner & Crouter, 1983; Fuhrer, 1993) have been widely known in the pertinent literature. Here I concentrate directly on the analysis of culture–activity and culture–semiotic mediation-related approaches, while recognizing the central relevance of cultural ecological thought. Much of that perspective has been prepared by classic thinkers like Ernst Cassirer (see Cassirer, 1929), Lev Vygotsky (Van der Veer & Valsiner, 1994; Vygotsky, 1934; Vygotsky & Luria, 1930/1993) and Karl Bühler (Bühler, 1934/1990). In an interesting parallel, a number of relevant ideas from literary science (e.g., the semiotic analyses of Mikhail Bakhtin—see Bakhtin, 1975) have had a substantive impact on the development of cultural focus in contemporary psychology of mental processes.

A. Cultural Voices in (and around) the Mind

James Wertsch's (1985a, 1985b, 1991) emphasis on the semiotic mediation of thinking persons in the area of culture and cognition is of central relevance for understanding culture in mental processes. Wertsch began from Vygotsky's cultural-historical viewpoint (see Wertsch, 1979, 1983), adding to it the activity theoretic perspective (Leontiev, 1981). He set up the *dynamic process of situation redefinition* as the primary means by which persons involved in joint activity guide one another's development.

Interaction partners are constantly in some relation of intersubjectivity—they share a similar situation definition—which they transcend by the process of situation redefinition (Wertsch, 1984, pp. 7–13). Communication about the situation definition (and redefinition) takes place by semiotic means, and the structure of the activities involved guides that communication (Wertsch, Minick & Arns, 1984).

More recently, Wertsch integrated his semiotically mediated activity approach with the wider sociolinguistic context (see Wertsch, 1985a; Wertsch & Stone, 1985) that has been characterized by the dynamic world view of Mikhail Bakhtin's literary theory (Bakhtin, 1981). Wertsch has combined that focus also with Lotman's semiotic approach to communication, and Tulviste's view on the heterogeneity of thinking processes (Tulviste, 1991; Wertsch, 1991).

The activity framing remains in the background of Wertsch's accounts, but the main focus becomes the level of utterance as appropriate for analysis of the dialogue of "voices" in and around the mind. Wertsch utilizes Bakhtin's focus on dialogicality and makes it work in his system, where the analysis of "voices" affords the revealing of complexity of messages (Wertsch, 1990, 1991, 1995). The result is a consistent return to the study of ambivalences embedded in communicative messages—in the form of "polyphony of voices" or "heteroglossia" (Wertsch, 1985c, pp. 62–68). Different "voices" can be seen in the utterances in ways that "interanimate" or dominate each other in the act of speaking in situated activity contexts. Bakhtin's legacy allows Wertsch to advance his theory of communication into the realm of conceptualizing *processual relations* between the components in a dialogue (i.e., different "voices"). For example, "privileging" in relation between "voices" (i.e., the "foregrounding" of voice X while voice Y is simultaneously being "backgrounded") becomes a central issue for Wertschian analysis (Wertsch, 1990, pp. 119–122; Wertsch & Minick, 1990, p. 85).

This dynamic differentiation by way of "privileging" makes Wertsch's theory not only appreciative of the fluidity of mental phenomena, but also their social-institutional and historical situatedness. At the level of analysis of utterances as units, he proceeds to analyze the process of takeover ("appropriation") of different speech genres and social languages by the narrative constructors. This process involves joint construction by partners, and "tactics of resistance" (Wertsch & O'Connor, 1992; Wertsch & Smolka, 1993). This mosaic of strategic voices leads to the construction of unique narratives.

Wertsch's theoretical perspective has been developed further in the context of analyses of voices explicated in educational settings (Pino, 1994; Smolka, 1993, 1994; Smolka, Goés & Pino, 1995). In a series of empirical studies conducted at the State University of Campinas (UNICAMP) in Brazil in the area of education, different dynamic forms of interplay between voices have been demonstrated in the classroom activities (Goés, 1994; Smolka, 1990, Smolka, Fontana, & Laplane, 1994). The perspective of analyzing cultural voices within persons' situations in activity contexts is linked with the emphasis on appropriation.

B. Dialogical Nature of Mental Processes: Ivana Markova and the Prague School

In a way that transcends Wertsch–Bakhtin's multivoiced emphasis, Ivana Markova's persistent focus on the dialectical analysis of the mental processes is of great relevance for understanding culture in the mind. She transcends the Cartesian dualism by way of taking Hegel's dialectical scheme and applying it to processes of human cognitive functioning (Markova, 1987a, 1987b). Similar to the coconstructionist approaches, Markova's intellectual roots go back also to J. M. Baldwin and G. H. Mead (Markova, 1990a, 1990b; Markova & Foppa, 1991).

Markova also relies on the traditions of the Prague Linguistic Circle of the 1920s and 1930s (see Markova, 1993a), particularly on the notion of dialogical quality of speech. Markova avoids the theoretical temptation to create another black box explanation of cognitive functions (e.g., by relabeling them dialogical), and instead proceeds to analyze dialogic events via use of three step units (i.e., dialogic events A1, B1, A2—where A2 is the integrative reflection upon the events A1 and B1). By using the three-step unit (in which the third step includes dialectical integration of the previously constructed opposites), Markova brings the notion of dialectics home to the reality of analysis of dialogues (both interpersonal and intrapersonal or intramental ones). Within her analytic scheme, the concentration on the emergence of A2 on the basis of coconstructed A1 and B1 highlights the location of novelty construction.

Markova's dialogical approach integrates the cognitive and language functions in the process of analysis. She builds upon the rich continental European tradition of looking at language as the main vehicle of mental organization of psychological functions (see also Markova, 1993b). With both Wertsch and Markova, psychological functions are best analyzed as processes of a semiotic kind that take place in activity contexts, but transcend those by virtue of heteroglossia and dialogicality of the cognitive functions.

C. Barbara Rogoff's Ethnography of Participatory Appropriation

Barbara Rogoff has been an active promoter of the cultural nature of situated cognition over the recent decade (Rogoff, 1982, 1986, 1990, 1992, 1993; Rogoff & Lave, 1984). Her own focus is mostly ethnographical, which allows her to take into account complex intricacies of culture-embedded mentality in action. Although wary of abstract theory, Rogoff's explication of her ideas includes a clear ideological and theoretical stance that needs analysis.

Rogoff is consistent in her emphasis on context-linkage of all developmental processes all of the time (Rogoff, 1990, 1992, 1993). She provides an explicit solution to the problem of the context—it is the "sociocultural activity" that involves "active participation of people in socially constituted practices" (Rogoff,

1990, p. 14) that acts in the role of the unit of analysis. Within that unit, persons are involved in interactive problem solving rather than in lengthy intrapersonal contemplations or soul-searching.

Rogoff's viewpoint is explicitly developmental. She emphasizes the need to study developmental processes (rather than products). The central focus on everyday events as problem-solving settings is characteristic for Rogoff's perspective. In these events, microgenetic (specific problem solving) and ontogenetic aspects of development become united in Rogoff's version of the notion of "zone of proximal development." The active (but not always persistent) guidance by the "social others" of the developing person is complemented by the person's own constructive role in one's own development. The child is always an *active* apprentice who *participates* in the *socially guided activity settings.*

The use of the metaphor of apprenticeship allows Rogoff to emphasize the active-but-subordinate role of the developing child in his or her own ontogeny. The crucial link between central concepts of Rogoff's theory is the dynamically coordinated interdependence between guidance and apprenticeship. This interdependence makes it possible for Rogoff's theory to overcome the individualistic ethos of most of the existing psychological theories, which include neither the systemic (i.e., coordination of parts of the system and its environment) nor the dynamic notions into their conceptual texture.

On the side of interindividual processes of communication, Rogoff builds her ideas on the foundation of the notion of intersubjectivity. All the culturally organized settings for adults–child communication set up the frameworks for intersubjectivity—not only in situations where the adults integrate children into their own activities (and tolerate children's active intrusions), but also in conditions where children in many settings are not supposed to directly partake in the activities.

For Rogoff, the concept of *appropriation* of culture (see Rogoff, 1993, p. 139–141) seems preferred for her construction of activity-centered theory of mental development. That concept (widely used in present-day sociocultural thinking) is largely (and explicitly—see Rogoff, 1993, pp. 126–127) based on the ideas expressed by John Dewey:

> The development within the young of the attitudes and dispositions necessary to the continuous and progressive life of a society cannot take place by direct conveyance of beliefs, emotions, and knowledge. It takes place through the intermediary of the environment. The environment consists of the *sum total of conditions* which are concerned in the execution of the activity characteristic of a living being. The social environment *consists of all the activities* of fellow beings that are *bound up* in the carrying on of the activities of any one of its members. It is truly educative in its effect in the degree in which an individual shares or participates in some conjoint activity. *By doing his share in the associated activity, the individual appropriates the purpose which actuates it,* [italics added] becoming familiar with its methods and subject matters, acquires needed skill, and is saturated with its emotional spirit. (Dewey, 1980, p. 26)

Dewey's emphasis on the associationistic fluidity of the environment (e.g., references to "sum total" and "all of activities," and to the indeterminate "bounding" of the latter) continues in our modern-day socio-mental psychology at large. Rogoff's theoretical construction is a good example. She is interested in overcoming the Cartesian dualism between individual psychological functions and their sociogenetic origins by way of appropriation.

Rogoff's view of participatory appropriation entails acceptance of transformation of cultural forms into novel states (both by persons and by groups: "participatory appropriation involves individuals changing through their own adjustments and understanding of the sociocultural activity" (Rogoff, 1993, p. 141). This emphasis on the creative (constructive) open-endedness of the appropriation process is particularly relevant, because it turns her theoretical system into a generative developmental framework. Culture in that framework is never a given entity, but is constantly in the process of change by way of participating individuals who are involved in the appropriation process.

D. Lutz Eckensberger's Ecological Approach for Analyzing Culture in the Mind

Eckensberger's theoretical work stems from the rich tradition of German psychology, which has historically taken mind–environment relationships seriously. Within the German post-World War II psychology, the theoretical perspective on culture in psychology that has been advanced in Saarbrücken (see Boesch, 1983, 1989, 1991) and elsewhere (Trommsdorff, 1993a, 1993b) has produced relevant results at the intersection of action theory and cross-cultural psychology. The historical roots of this perspective are in the richness of continental European psychological thought, and are similar to some directions within the coconstructionist perspective.

Originally phrased within the cross-cultural psychology perspective (see Eckensberger, 1979; Eckensberger & Kornadt, 1977; Eckensberger & Burgard, 1983; Eckensberger, Krewer, & Kasper, 1984), Eckensberger has moved on to construct a sophisticated dynamic theoretical perspective of cultural psychology of human action and thinking (Eckensberger, 1990, 1992). First (and foremost), the emphasis on *persons' goal-directed actions* and their *emerging reflexive abstraction* are the cornerstone of Eckensberger's cultural view of mental processes. The agency notion is carefully retained, and it is seen to both construct "action barriers" and negotiate those in the domain of action and reflexive abstraction (Eckensberger, 1992, 1995). Three levels of action are distinguished by Eckensberger: (1) primary (actions applied directly to the world); (2) secondary (reflections and regulations that are applied to actions themselves), and (3) tertiary actions (self-oriented reflection and contemplation). By concentrating upon the emergence of higher levels of actions in the mental sphere, Eckensberger manages to solve the complex problem that has boggled the minds of scientists in the last two centuries (see Daston, 1982)—that of integration of intentionality into a cognitive-psychological scheme.

The "witchhunt for dualisms" that we can see in some North American conceptualizations of mental processes is substituted by an effort to analyze the developmental unity of intra- and interpsychical phenomena.

E. Michael Cole and the Laboratory of Comparative Human Cognition

Cole's work in the area of culture and cognition spans more than two decades (Cole, 1975, 1981, 1990, 1991, 1992, 1995; Cole & Bruner, 1971; Laboratory of Comparative Human Cognition, 1979, 1983). Starting from indebtedness to the cultural-historical school of psychology (of Alexander Luria and Lev Vygotsky), Cole has built a consistent research program of cultural psychology (Cole, 1990, 1992).

For Cole, the main mechanism by which culture and person are related is that of *mutual interweaving* (see Cole, 1992, p. 26—for the use of the metaphor of "intermingling of threads from two ropes"—those of biological "modules" and cultural contexts). This interweaving reflects the general process in which "the culture becomes individual and the individuals create their culture" (Laboratory of Comparative Human Cognition, 1983, p. 349)—or, in other terms, the culture and cognition are *mutually constituted*. The locus of this mutual constituting process is in the concrete activities that are carried out in everyday life (see Cole, 1985; Newman, Griffin, & Cole, 1989). The cognitive processes that are established within such contexts can be transferred to other contexts under social facilitation of such transfer.

Cole's emphasis on socially organized transfer between context leads him to use the concept of appropriation (Newman et al., 1989, pp. 62–65). The culture provides a range of cultural mediating devices (tools or signs) to the developing child in specific activity contexts; the child actively takes over (appropriates) those cultural means, reconstructing those in the process of activity.

F. Carl Ratner's Sociohistorical Perspective

In the context of different approaches in culturally oriented cognitive thought, Carl Ratner's work (see Ratner, 1991) is particularly notable. Ratner follows the footsteps of Lev Vygotsky, taking a perspective that differs from Cole and relates to Bronfenbrenner's "nested system" model.

Ratner's sociohistorical perspective requires a treatment of the issue of psychological universals. Given his context-oriented viewpoint, he advocated the study of concrete phenomena before abstracting commonalities. Ratner distinguished so-called false and true universals. The former are the result of either a priori beliefs in universality of some psychological function, or selective generalization on the basis of data that do not represent the cultural variability of the world. In contrast, Ratner's "true universals" are described in conjunction with variability, which needs to be explained (Ratner, 1991, p. 122). Where Vygotskian bases of Ratner's perspective end, Bronfenbrenner's start—at the conceptualization of the multilevel structure of society and the ways in which that structure guides human development. Ratner goes beyond the usual concentration of sociogenetic researchers on the microsocial phenomena.

G. Richard Shweder's "Cultural Psychology"

Starting from an anthropological background, Shweder's claims of recent years have undoubtedly pointed to the need to consider culture in psychology as a primary constituting factor of the self (Shweder, 1984; 1992, 1995; Shweder & Much, 1987; Shweder & Sullivan, 1993). Shweder's perspective fits the North American cultural anthropology. He can be observed to create a narrative about the history of a discipline that itself is in the process of creation (see Shweder & Sullivan, 1993).

Shweder (1977, 1980) started from the study of the concept of person (see also Shweder & Bourne, 1984), and moral reasoning and its contexts. He recognizes the heterogeneity and culture inclusiveness of moral reasoning by human beings Shweder, Mahapatra, & Miller, 1987; Shweder & Much, 1987). Shweder makes an effort to explain cultural psychology's aims as

> to imaginatively conceive of subject-dependent objects (intentional worlds) and object-dependent subjects (intentional persons) interpenetrating each other's identities or setting the conditions for each other's existence and development, while jointly undergoing change through social interaction. (Shweder, 1990, p. 25)

The personal minds (object-dependent persons) construct mental and affective order out of chaos of everyday events—hence, an illusory view of reality is constructed by persons, but on the basis of the culture (Shweder, 1980, p. 77). The latter facet led Shweder to reject the reconstructed versions of Piaget's ideas that circulated among American cognitive child psychologists. The specific processes by which (in Shweder's theoretical system) mental and cultural processes are working has so far remained unexplicated, and would constitute a next interesting stage in the development of this integration effort of cultural anthropology and psychology.

H. The Perspective of Coconstructionism

In a step forward from simple constructionist claims, the coconstructionist viewpoints generally emphasize both person's mental construction and the social suggestions by other persons (Wozniak, 1986, 1993). This perspective has been of importance in the area of moral cognition (Kurtines, Alvarez, & Azmitia, 1990), but is main elaboration has taken place in conjunction with viewing children's development within their culturally organized environments (Fogel, 1993; Valsiner, 1987, 1988a, 1988b, 1989b; 1995; Winegar & Valsiner, 1992a, 1992b).

The latter direction within the coconstructivist perspective is based on the historical roots of the work of James Mark Baldwin, Jean Piaget, Lev Vygotsky, and William Stern (Valsiner, 1994). It can be described as *sociogenetic personology,* because it preserves the uniqueness of individual persons *within* their interdependence with the culture. It is exactly in the person–culture relationship that the micro–macro problem finds its solution here: persons are microlevel ("personal-culture") parts of the macrolevel entity ("collective culture"—see Valsiner, 1989a), with the relationship between the two levels describable as that of bidirectional culture transformation. The person is a joint constructor of one's own personal culture, as well as (from a subdominant position) of the collective culture.

Obviously, not every construction within the personal culture is of relevance for the collective culture, and vice versa. Both levels can be seen as *temporarily* and *conditionally* open for acceptance of inputs from the other, with every input submitted to an analysis and new synthesis at the level of the recipient. The adequate model of causality that is applicable to this relationship is that of catalyzed systemic kind (see Valsiner, 1989a, p. 30). A causal entity is always a system of integrated parts, the joint functioning of which leads to some outcomes (while each part individually does not lead to the outcome). Furthermore, the actual production of the outcome is made possible by some "catalysts"—aspects of the given context that either enable the causal system to produce the outcome, or block that from happening.

I. Conclusions: Multivoicedness in Contemporary Theorizing

It should become clear that the various contemporary theoretical perspectives that treat culture–mental processes relations are basically attempting to grasp the very same (but in time vanishing) phenomenon—construction of semiotic mediating devices that, once constructed, change the flow of the very same mental (and action) processes from which they emerged. Thus some of the approaches outlined here concentrate on the study of cultural–mental phenomena (e.g., Wertsch, Markova, Eckensberger, Shweder), others put the emphasis on persons' actions in activity contexts (e.g., Cole, Ratner, Rogoff). Nevertheless, both directions of theory building are aimed at the same goal—integrating culture into our understanding of human mental processes and their development. The achievements of modern cognitive science have been insufficient for reaching that goal.

Here is the interesting paradox: culture (semiotic mediation) makes the mind free to wander in directions where usual research practices of cognitive psychology fail to follow it. Hence the central issue of establishing a viable research program that looks at culture in (and around) the mind requires methodological innovations.

IV. PATHWAYS TO METHODOLOGY

Much of the existing empirical investigative effort in viewing topics of culture and cognition has been hampered by research conventions that have rendered the systemic analysis of cultural and mental processes unresearchable. First, the preponderance of associativistic summative inference strategies (e.g., analysis of variance models superimposed on the psychological phenomena) have obscured our understanding. Secondly, the "quantophrenic" obsessions of psychologists (see Sorokin, 1956) have further obscured the investigators' field of vision. Psychologists' deep belief in the act of quantification as a guarantee of scientific nature of their discipline has led to a situation where the structured nature of most psychological phenomena is eliminated in the very beginning of the research process. The result is a discipline that may leave (to outsiders) an impression of a "hard science," yet remains a consistently far from adequate representation of the phenomena it claims to study.

A. What Kind of Methodology Is Needed?

If culture–mental processes relationships are to be viewed in ways that do not eliminate the first, relatively strict expectations for methodology are specifiable. First, the emphasis on person's *independent dependence* (see Winegar, Renninger, & Valsiner, 1989) on culture needs to be recognized. Secondly, methodology has to take into account the *emergent processes*.

Both of these methodological cornerstones have been overlooked in psychology. Traditionally, psychology has been oriented toward describing psychological phenomena the form of which has already emerged in the course of development. It has tried to superimpose upon the highly variable (developing) phenomena categorical systems that are prestructured by the investigators by way of "metacontracts" (Elbers, 1986). A similar need to rethink methodological adequacy has been set forth in cases of cultural practice theory (Laboratory of Comparative Human Cognition, 1983, p. 348). Adequate methodology for culture-inclusive psychology of mental processes needs to consider the investigator as an active coconstructor of the data, and of knowledge at large.

B. Developmental Perspective as the Methodological Norm

It becomes clear that methodological imperatives of traditional psychology do not fit the goals of making sense of the sociogenetic process (Lightfoot & Folds-Bennett, 1992). Therefore, the developmental perspective on culture–cognition relations implies that the person be observed in the process of movement from "chaotic" or "fluid" (and hence—categorically difficult to describe) phenomena to the emergence of clear forms (see Basov, 1991). It is exactly the unclassifiable phenomena we observe in child development or any human "dialogical" encounter with others or with one's own self, that may need to be the target of empirical analysis—yet at present, these phenomena are consistently being overlooked and eliminated from the research process (see Valsiner, 1993, 1994).

It becomes obvious that the emergence of novel psychological phenomena is embedded in the fuzzy nature of a person(s)'s concrete interaction with a context. That process of emergence can be approached through microgenetic research strategies (Catán, 1986; Siegler & Crowley, 1991). In the history of sociogenetic thought, Lev Vygotsky's "method of double stimulation" (see Valsiner, 1989a, ch. 3) constituted an effort to directly observe (and trigger) the emergence of novel psychological phenomena.

The function of experimentation in the evocation of the emergence of novelty makes it close to the concerns of hermeneutic methodologists. An issue arises in the hermeneutic process of knowledge construction in experimenter–subject relationships, where it may be the moments of sudden mutual divergence of communication where the relevant phenomena are discovered (see Hermans, 1991; Hermans & Bonarius, 1991a, 1991b; Hermans & Kempen, 1995; Hermans, Kempen, & Van Loon, 1992). The person who takes the role of a "subject" in the research process is constantly creating novelty on the basis of the previous state of relationship with the world. In the process of research, the roles of "subject" and "investigator" are constantly in the process of modification, based on their dialogic integration (see Markova, 1993b; Santamaria, 1994). This stance denies the possibility of an "experimenter's control over the experiment" in its exaggerated form, and looks at the negotiation of mutual misunderstandings in the research process as places where experimental manipulations actually work.

C. Levels of Complexity and Units of Analysis

1. The Individual-Socioecological Frame of Reference

The issue of an investigative frame of reference is central for building adequate methodology. In order to capture the mutual interdependence (of cultural and mental processes), the individual-socioecological frame of reference (Valsiner, 1987, 1989a) constitutes an adequate solution to the problem. That frame entails the use of units of analysis that focus on the dynamic interdependence between person and the environment, while explicitly considering the goal-oriented canalizing functions of the cultural world. The Agent–Context interdependence is regulated by cultural constraints at multiple levels, from actions to intrapersonal ideations to social processes.

Some exemplification of this solution to the question of analysis units could be considered. First, the "traditional" unit of analysis of culture in cognition can be seen in the case of "cultural features" (prototypes, "folk models," etc.), which are "taken over" (appropriated) by the developing persons, who "make" those (features, prototypes, models) "their own." In this description, an unidirectional influence of the culture upon the persons is inherent.

In contrast, the alternative unit of analysis would treat such "cultural features" as part of the canalization efforts whereby the social agents at any given time attempt to guide persons' acting and thinking and feeling in some goal-oriented direction. Such cultural input may set the background that organizes the appearance of specific forms of agent–context interaction in the foreground. No "influence" in any direct sense takes place, yet the process of person–context relations is becoming cultural.

2. Focus on Functioning Structures

Psychology has usually fluctuated between static (structural) and dynamic (functional) descriptive efforts, failing to recognize the unity of those two. Hence, structuralistic foci on culture–cognition relations are functionalistic in their construction, and vice versa (functional aspects of mental existence lead to the construction of new structural phenomena). The emphasis here is on function*ing* (rather than function*al*) structures—the former can be ones that function, without being functional, under conditions of challenges in person–environment transaction (see Bornstein, 1995; Kojima, 1995).

3. Processes and Outcomes of Adaptation

Much of argumentation about functions of cognitive structures has been centered on the issue of adaptation. Two basic views can be discerned in coverage of adaptation: (1) adaptation as conformity with the given environmental demands; and (2) adaptation as a process that transcends the immediate environmental demands.

In its first meaning adaptation has been viewed as a direct reaction to the conditions that are causing change—either "positive" (by way of giving rise to new variations) or "negative" (elimination of misfitting emerged variations). The folk model of static fit between the organism (here-and-now) and the demands is maintained under that construction of adaptation. For instance, sociobiological post factum explanations of different organizational forms of psychological functions (e.g., MacDonald, 1988) are based on a demonstration of such a fit having emerged by selection processes.

In contrast, proponents of a constructive (organismic) position in cognitive development (e.g., see Bergson, 1911, p. 63), have called for seeing adaptation in the process of emergence of novel mechanisms in ways *coordinated with* context demands (but not "molded" or "shaped" by those). Thus, in psychological development the psychological functions develop new organizational forms that make it possible for them to encounter new *possible* conditions in the future (as opposed to the idea of "fitting in" with the environmental demands at the present). The adaptations are organic (systemic) growths, oriented towards a set of future possibilities (which, as those do not exist in present, cannot be precisely defined).

In case of creative adaptation, the organizational forms that emerge in adaptation go beyond the "fit with" the present state of the survival conditions, and set the basis for facing the challenges of the possible future demands. Thus, the "goodness-of-misfit" characterizes the process of development (see Valsiner & Cairns, 1992). Hence, there is no need to prove that a particular set of cognitive processes fits exactly with the demands of the given context. Just the opposite—by locating cognitive functions that seem to be "surplus" (given the currently immediately available demands), we have discovered a developmental constructive adaptation for the possible future needs. This view on adaptation would set up the traditional "competence–performance" issue of cognitive psychology in a new light, as it is expected that surplus competence (relative to performance demands) is exactly the constructed adaptational form in the domain of cognition.

D. Analytic Levels and Parsimony in Investigation

Most of cultural-cognitive discourse in modern psychology has been struggling with the problem of analytic levels of the phenomena most useful for knowledge construction, and with the issue of relations between these levels. The empirically observable phenomena (from which data are derived) are accessible only at specific microlevel (afforded by some sampling of events that take place in irreversible time). Yet, the integration of cognitive functions does not take place in these immediate ("situated activity") contexts, but somehow transcends them. In other terms, the actual psychological development—based on the microlevel experiential foundation—takes place in ways that cannot be immediately accessed by researchers (i.e., at some macrolevel, in contrast to the microlevel, however defined).

All issues of relationship of the levels regarding human mental functions in collective-cultural contexts are thus systemic (i.e., can be analyzed within individual-socioecological reference frame). There exists limited indeterminacy between the micro- and macrolevels: Under some conditions (at some time) the input from one to the other leads to a reconstruction of the latter (with feedback to the former), but a very similar input at another time (and conditions) is ineffective.

A simple solution to the micro- and macrolevels problem in psychology would be an outright a priori denial of the specificity of different levels. Even if the latter may be observable as if they are different, a psychologist may claim that one level of complexity can be actually reduced to another. Such reductionism has been widespread in different psychologies—usually it entails reduction of the more complex levels of analysis to more elementary ones (e.g., physiological, biological, or genetic reductionism). However, it is possible to observe also cases of reduction of psychological functions to more complex levels (e.g., sociological reductionism—viewing of psychological functions as mere reflections of discourse, rhetorics, or "collective representations"). Finally, psychologies have been active in proliferating horizontal reductionism—translating explanations in one theoretical language into another, at the same level of complexity (e.g., translation of behaviorist explanations into cognitive ones, see Valsiner, 1991).

In the history of comparative psychology of species, similar kinds of reductionism can be traced. Intellectual capacities of Homo sapiens have at times been viewed either as not different from those of the nearest other species, or the capabilities of those other species have been elevated to the level of humans (Glickman, 1985; Romanes, 1988). All these controversies about evaluation of differences of levels led C. Lloyd Morgan to the formulation of his famous "canon":

> In no case may we interpret an action as the outcome of the exercise of a higher psychical faculty if it can be interpreted as the outcome of the exercise of one which stands lower in the psychological scale. (Morgan, 1894, p. 53)

Morgan's law of parsimony has often been used as a legitimization for downward-oriented reductionism, which of course is not adequate. Morgan introduced the law in the context of argument with any kind of reductionism, and in contrast with the commonsense interpretations of animal intellect that glossed over the issues for investigation. Morgan had a clear-cut emphasis on development of psychological functions towards higher levels of complexity, and in order to be able to provide an adequate account of the emergence of the new levels, a certain rule of thumb (in the form of the "canon") was necessary. A developmental reformulation would restore the focus of the "canon" on the process of emergence of psychological levels of different complexity: we cannot interpret a phenomenon as generated by a function that has not yet emerged (but may be in the process of doing so), but only by way of currently developed functions that can generate transformation of the organism to another state (Valsiner, 1990).

V. GENERAL CONCLUSIONS: DYNAMIC INTERDEPENDENCE OF CULTURE AND COGNITION

If the texture of psychological knowledge outlined above is taken seriously, if follows that both culture and cognition constitute mutually interdependent processes. The (self) reflective human being is constantly involved in the construction of mental experiences (cognitive processes) that are made possible by the semiotic activity of constructing and using signs (cultural process). The latter process feeds constantly into the former (aside from providing externalized products as outcomes), and the former is the basis for further cultural construction. It is at this mutual interdependence of the cultural and mental processes that person and culture constantly (and unavoidably) "meet."

In the realm of methodology solutions for integration of culture into the study of mental (cognitive) processes can emerge. Semiotically mediated psychological functions need to be analyzed in their flexibility, open-endedness, and generativity. The development perspective is therefore the rule (rather than exception) for further efforts to unite culture and cognition. Furthermore, return to general abstract theory building, which was widespread in the psychology of the past, is a way to arrive at general knowledge that transcends differences in psychologists' local dialects of descriptive terminology. In other terms, psychology of culture in cognition (and vice versa) can make use of the most remarkable characteristic of our ever-wandering minds:

> Occasionally, it hits me how marvelous a creature Man is. Among all his numerous talents, the one that fascinates me most is his ability to create abstract notions out of something which he does not really understand. (Toda, 1975, p. 314)

Cognitive science's efforts to understand mental processes in the flow of contexts seem to be a good specimen of such marvel created by Homo sapiens. It makes investigation of human cultural mentality both complicated, and rewarding—when the real psychological phenomena are studied with the goal of creating general knowledge.

References

Bakhtin, M. M. (1975). *Voprosy literatury i estetiki. Problems of literature and aesthetics* Moscow: Khudozhestvennaya Literatura.

Bakhtin, M. M. (1981). *The dialogic imagination.* Austin, TX: University of Texas Press.

Baldwin, J. M. (1906). *Thought and things: A study of the development and meaning of thought, or genetic logic: Vol. 1. Functional logic, or genetic theory of knowledge.* London: Swan Sonnenschein & Co.

Basov, M. (1991). The organization of processes of behavior. In J. Valsiner & R. Van der Veer (Eds.), *Structuring of conduct in activity settings: The forgotten contributions of Mikhail Basov. (Part 1). Soviet Psychology, 29,* 5, 14–83.

Bergson, H. (1910). *Time and free will.* London: George Allen & Unwin.

Bergson, H. (1911). *Creative evolution.* New York: Henry Holt & Co. [English translation of Bergson, 1907]

Berry, J. W., Poortinga, Y. H., Segall, M. H., & Dasen, P. R. (1992). *Cross-cultural psychology: Research and applications.* Cambridge: Cambridge University Press.

Boesch, E. E. (1983). *Das Magische und das Schöne: zur Symbolik von Objekten und Handlungen.* [On the magical and good: On symbolics of objects and actions] Stuttgart: Frommann.

Boesch, E. E. (1989). Cultural psychology in action-theoretical perspective. In Ç. Kagitçibasi (Ed.), *Growth and progress in cross-cultural psychology* (pp. 41–51). Lisse: Swets & Zeitlinger.

Boesch, E. E. (1991). *Symbolic action theory and cultural psychology.* New York: Springer.

Bornstein, M. H. (1995). Form and function: Implications for studies of culture and human development. *Culture & Psychology, 1,* 1, 123–137.

Bourdieu, P. (1973). Cultural reproduction and social reproduction. In R. Brown (Ed.), *Knowledge, education, and cultural change* (pp. 71–112). London: Tavistock Publications.

Bourdieu, P. (1985). The social space and the genesis of groups. *Social Science Information, 24,* 2, 195–220.

Bronfenbrenner, U. (1979). *The ecology of human development.* Cambridge, MA: Harvard University Press.

Bronfenbrenner, U. (1989). Ecological systems theory. *Annals of Child Development, 6,* 185–246.

Bronfenbrenner, U. (1993). The ecology of cognitive development: Research models and fugitive findings. In R. H. Wozniak & K. W. Fischer (Eds.), *Development in context* (pp. 3–44). Hillsdale, NJ: Erlbaum.

Bronfenbrenner, U., & Crouter, A. C. (1983). The evolution of environmental models in developmental research. In W. Kessen (Ed.), *Handbook of child psychology: Vol. 1. History, theory, and methods* (pp. 357–414). New York: Wiley.

Bühler, K. (1990). *Theory of language: The representational function of language.* Amsterdam: John Benjamins.

Butterworth, G. (1993). Context and cognition in models of cognitive growth. In P. Light & G. Butterworth (Eds.), *Context and cognition: Ways of learning and knowing* (pp. 1–13). London: Erlbaum.

Cassirer, E. (1929). *Philosophie der symbolischen Formen* [Philosophy of symbolic forms]. *Vol. 3. Phänomenologie der Erkenntnis* [phenomonology of Cognition]. Berlin: Bruno Cassirer Verlag.

Catán, L. (1986). The dynamic display of process: Historical development and contemporary uses of the microgenetic method. *Human Development, 29,* 252–263.

Cole, M. (1975). An ethnographic psychology of cognition. In R. W. Brislin, S. Bochner, & W. Lonner (Eds.), *Cross-cultural perspectives on learning* (pp. 157–175). New York: Wiley.

Cole, M. (1981). Society, mind, and development. In F. S. Kessel & A. W. Siegel (Eds.), *The child and other cultural inventions* (pp. 89–123). New York: Praeger.

Cole, M. (1985). The zone of proximal development: Where culture and cognition create each other. In J. V. Wertsch (Ed.), *Culture, communication, and cognition: Vygotskian perspectives* (pp. 146–161). Cambridge: Cambridge University Press.

Cole, M. (1990). Cultural psychology: A once and future discipline? In J. Berman (Ed.), *Nebraska Symposium on Motivation* (Vol. 37, pp. 279–336). Lincoln: University of Nebraska Press.

Cole, M. (1991). On putting Humpty Dumpty together again: A discussion of the papers on the socialization of children's cognition and emotion. *Merrill-Palmer Quarterly, 37, 1,* 199–208.

Cole, M. (1992). Context, modularity and the cultural constitution of development. In L. T. Winegar & J. Valsiner (Eds.), *Children's development within social context. Vol. 2. Research and methodology* (pp. 5–31). Hillsdale, NJ: Erlbaum.

Cole, M. (1995). Culture and cognitive development: From cross-cultural research to creating systems of cultural mediation. *Culture & Psychology. 1,* 1, 25–54.

Cole, M., & Bruner, J. S. (1971). Cultural differences and inferences about psychological processes. *American Psychologist, 26,* 867–876.

Crick, F. (1988). *What mad pursuit: A personal view of scientific discovery.* London: Penguin Books.

D'Andrade, R. (1984). Cultural meaning systems. In R. A. Shweder & R. A. LeVine (Eds.), *Culture theory: Essays on mind, self and emotion* (pp. 88–119). Cambridge: Cambridge University Press.

Danziger, K. (1990). *Constructing the subject.* Cambridge: Cambridge University Press.

Daston, L. J. (1982). The theory of will versus the science of mind. In W. R. Woodward & M. G. Ash (Eds.), *The problematic science: Psychology in nineteenth-century thought* (pp. 88–115). New York: Praeger.

Dewey, J. (1980). *The middle works, 1899–1924 (Vol. 9).* Carbondale: Southern Illinois University Press.

Eckensberger, L. (1979). A metamethodological evaluation of psychological theories from a cross-cultural perspective. In L. H. Eckensberger, W. J. Lonner, & Y. H. Poortinga (Eds.), *Cross cultural contributions to psychology* (pp. 255–275). Lisse: Swets & Zeitlinger.

Eckensberger, L. H. (1990). From cross-cultural psychology to cultural psychology. *The Quarterly Newsletter of the Laboratory of Comparative Human Cognition, 12,* 1, 37–52.

Eckensberger, L. H. (1991). Moralische Urteile als handlungsleitende normative Regelsysteme im Spiegel der kulturvergleichenden Forschung. In A. Thomas (Ed.), *Einführung in der kulturvergleichende Psychologie* [Introduction to Cultural-comparative psychology]. Göttingen: Hogrefe.

Eckensberger, L. H. (1992). Agency, action and culture: Three basic concepts for psychology, in general and for cross-cultural psychology, in specific. *Arbeiten der Fachrichtung Psychologie, Universität des Saarlandes.* No. 165. Saarbrücken.

Eckensberger, L. H. (1995). Activity or action: Two different roads towards an integration of culture into psychology? *Culture & Psychology, 1,* 1, 67–80.

Eckensberger, L. H., & Burgard, P. (1983). The cross-cultural assessment of normative concepts: Some considerations on the affinity between methodological approaches and preferred theories. In S. H. Irvine & J. W. Berry (Eds.), *Human assessment and cultural factors* (pp. 459–480). New York: Plenum.

Eckensberger, L. H., & Kornadt, H. J. (1977). The mutual relevance of the cross-cultural and the ecological perspective in psychology. In H. McGurk (Ed.), *Ecological factors in human development* (pp. 219–227). Amsterdam: North-Holland.

Eckensberger, L. H., Krewer, B., & Kasper, E. (1984). Simulation of cultural change by cross-cultural

research: some metamethodological considerations. In P. Baltes (Ed.) *Life-span developmental psychology: Historical and generational effects* (pp. 73–107). New York: Academic Press.
Elbers, E. (1986). Interaction and instruction in the conservation experiment. *European Journal of Psychology of Education, 1,* 1, 77–89.
Fogel, A. (1993). *Developing through relationships.* Chicago: University of Chicago Press.
Fuhrer, U. (1993). Living in our own footprints—and those of others: Cultivation as transaction. *Schweizerische Zeitschrift für Psychologie, 52,* 2, 130–137.
Gardner, H. (1985). *The mind's new science: A history of the cognitive revolution.* New York: Basic Books.
Gergen, K. J. (1989). Social psychology and the wrong revolution. *European Journal of Social Psychology, 19,* 463–484.
Gigerenzer, G., Swijtink, Z., Porter, T., Daston, L., Beatty, J., & Krüger, L. (1989). *The empire of chance.* Cambridge: Cambridge University Press.
Glickman, S. E. (1985). Some thoughts on the evolution of comparative psychology. In S. Koch & D. E. Leary (Eds.), *A century of psychology as science* (pp. 738–782). New York: McGraw-Hill.
Góes, M. C. R. (1994). The modes of participation of others in the functioning of the subject. In N. Mercer & C. Coll (Eds.), *Explorations in socio-cultural studies.* Vol. 3. *Teaching, learning, and interaction* (pp. 123–128). Madrid: Fundación Infancia y Aprendizaje.
Goodyear, T. (1995). Observation and learning: Theoretical paradoxes of observational learning. In J. Valsiner & H.-G. Voss (Eds.), The structure of learning processes. Norwood, NJ: Ablex.
Hermans, H. J. M. (1991). The person as co-investigator in self-research: valuation theory. *European Journal of Personality, 5,* 217–234.
Hermans, H. J. M., & Bonarius, H. (1991a). The person as co-investigator in personality research. *European Journal of Personality, 5,* 199–216.
Hermans, H. J. M., & Bonarius, H. (1991b). Static laws in a dynamic psychology? *European Journal of Personality, 5,* 245–247.
Hermans, H. J. M., & Kempen, H. J. G. (1995). Body, mind, and culture: Dialogical nature of mediated action. *Culture & Psychology, 1,* 1, 104–114.
Hermans, H. J. M., Kempen, H. J. G., & van Loon, R. J. P. (1992). The dialogical self: Beyond individualism and rationalism. *American Psychologist, 47,* 1, 23–33.
Hoffman, R. R., & Nead, J. M. (1983). General contextualism, ecological science and cognitive research. *Journal of Mind & Behavior, 4,* 4, 507–560.
Holland, D. C., & Quinn, N. (Eds.). (1987). *Cultural models in language and thought.* Cambridge: Cambridge University Press.
Ivic, I. (1978). *Covek kao animal symbolicum* [Man as symbolic animal]. Belgrade: Nolit. [in Serbian].
Ivic, I. (1988, June). *Semiotic systems and their role in ontogenetic mental development.* Paper presented at the 3rd European Conference on Developmental Psychology, Budapest.
Jahoda, G. (1990). Our forgotten ancestors. In R. A. Dienstbier & J. J. Berman (Eds.), *Nebraska Symposium on Motivation 1989: Vol. 37. Cross-cultural perspectives* (pp. 1–40). Lincoln: University of Nebraska Press.
Jahoda, G. (1993). *Crossroads between culture and mind.* Cambridge, MA: Harvard University Press.
Jahoda, G. (1995). The ancestry of a model. *Culture & Psychology, 1,* 1, 11–24.
James, W. (1890). *The principles of psychology* (Vol. 1). New York: Henry Holt.
James, W. (1904). Does 'consciousness' exist? *Journal of Philosophy and Scientific Methods, 1, 18,* 477–491.
Johnson-Laird, P. N. (1980). Mental models in cognitive science. *Cognitive Science, 4,* 71–115.
Kessen, W. (1981). Early settlements in New Cognition. *Cognition, 10,* 167–171
Kojima, H. (1995). Forms and functions as categories of comparison. *Culture & Psychology, 1,* 1, 139–145.
Kon, I. S. (1988). *Rebenok i obshchestvo* [Child and society]. Moscow: Nauka.
Krewer, B. (1990). Psyche and culture—can a culture-free psychology take into account the essential features of the speceis "homo sapiens"? *The Quarterly Newsletter of the Laboratory of Comparative Human Cognition, 12,* 1, 24–36.

Krewer, B. (1992). *Kulturelle Identität und menschliche Selbsterforschung* [Cultural identity and self-search]. Saarbrücken: Breitenbach.

Kroeber, A. L. (1948). *Anthropology.* New York: Harcourt & Brace.

Kurtines, W. M., Alvarez, M., & Azmitia, M. (1990). Science and morality: The role of values in science and the scientific study of moral phenomena. *Psychological Bulletin, 107,* 3, 283–295.

Leontiev, A. N. (1981). The problem of activity in psychology. In J. V. Wertsch (Ed.), *The concept of activity in Soviet psychology* (pp. 37–71). Armonk, NY: Sharpe.

Laboratory of Comparative Human Cognition (1979). What's cultural about cross-cultural cognitive psychology? *Annual Review of Psychology, 30,* 145–172.

Laboratory of Comparative Human Cognition (1983). Culture and cognitive development. In W. Kessen (Ed.), *Handbook of child psychology, Vol. 1. History, theory & methods* (pp. 295–356). New York: Wiley.

Lave, J. (1988). *Cognition in practice.* Chicago: University of Chicago Press.

Lawrence, J. A., & Valsiner, J. (1993). Conceptual roots of internalization: From transmission to transformation. *Human Development, 36,* 150–167.

Levine, J. M., Resnick, L. B., & Higgins, E. T. (1993). Social foundations of cognition. *Annual Review of Psychology, 44,* 585–612.

Lightfoot, C., & Folds-Bennett, T. (1992). Description and explanation in developmental research: Separate agendas. In J. Asendorpf & J. Valsiner (Eds.), *Stability and change in development: A study of methodological reasoning* (pp. 207–228). Newbury Park, CA: Sage.

Linton, R. (1936). *The study of man.* New York: Appleton-Century-Crofts.

Lyra, M. C., & Rossetti-Ferreira, M. C. (1995). Transformation and construction in social interaction: A new perspective of analysis of the mother–infant dyad. In J. Valsiner (Ed.), *Child development in culturally structured environments* (Vol. 3, pp. 57–87). Norwood, NJ: Ablex.

MacDonald, K. B. (Ed.). (1988). *Sociobiological perspectives on human development.* New York: Springer.

Malinowski, B. (1944). *A scientific theory of culture.* Chapel Hill, NC: University of North Carolina Press.

Markova, I. (1987a). The concepts of the universal in the Cartesian and Hegelian frameworks: Consequences for psychology. In A. Costall & A. Still (Eds.), *Cognitive psychology in question* (pp. 213–233). Brighton: Harvester Press.

Markova, I. (1987b). On the interaction of opposites in psychological processes. *Journal for the Theory of Social Behaviour, 17,* 279–299.

Markova, I. (1990a). On three principles of human social development. In G. Butterworth & P. Bryant (Eds.), *Causes of development* (pp. 186–211). Hillsdale, NJ: Erlbaum.

Markova, I. (1990b). The development of self-consciousness: Baldwin, Mead, and Vygotsky. In J. E. Faulconer & R. N. Williams (Eds.), *Reconsidering psychology: Perspectives from continental philosophy* (pp. 151–174). Pittsburgh: Duquesne University Press.

Markova, I. (1993a). On the structure and dialogicity in Prague semiotics. In A. H. Wold (Ed.), *The dialogic alternative: Towards a theory of language and mind* (pp. 45–63). Oslo: Scandinavian University Press.

Markova, I. (1993b). Sociogenesis of language: Perspectives on dialogism and on activity theory. In W. de Graaf & R. Maier (Eds.), *Sociogenesis reexamined.* New York: Springer.

Markova, I., & Foppa, K. (Eds.). (1991). *Asymmetries in dialogue.* Hemel Hempstead: Harvester Press.

Markus, H. R., & Kitayama, S. (1991). Culture and the self: Implications for cognition, emotion and motivation. *Psychological Review, 98,* 2, 224–253.

Mercer, N. (1993). Culture, context and the construction of knowledge in the classroom, In P. Light & G. Butterworth (Eds.), *Context and cognition: Ways of learning and knowing* (pp. 28–46). London: Erlbaum.

Moore, D. C. (1994). Anthropology is dead, long live anthro(a)pology: Poststructuralism, literary studies, and anthropology's "nervous present." *Journal of Anthropological Research, 50,* 345–365.

Morgan, C. L. (1892). The law of psychogenesis. *Mind, 1,* 72–93.

Morgan, C. L. (1894). *An introduction to comparative psychology.* London: Walter Scott Ltd.

Moscovici, S. (1981). On social representations. In J. P. Forgas (Ed.), *Social cognition* (pp. 181–209). London: Academic Press.

Moscovici, S. (1982). The coming era of representations. In J.-P. Codol & J. -P. Leyens (Eds.), *Cognitive analysis of social behavior* (pp. 115–150). The Hague: Martinus Nijhoff.

Moscovici, S. (1988). Crisis of communication and crisis of explanation. In W. Schönpflug (Ed.), *Bericht über der 36, Kongress der Deutschen Gesellschaft für Psychologie in Berlin* (Vol. 2, pp. 94–109). Göttingen: Hogrefe.

Neisser, U. (1982). *Memory observed: Remembering in natural contexts.* San Francisco: W. H. Freeman.

Neisser, U. (1987). *Concepts and conceptual development.* Cambridge: Cambridge University Press.

Newman, D., Griffin, P., & Cole, M. (1989). *The construction zone: Working for cognitive change in school.* Cambridge: Cambridge University Press.

Pepitone, A. (1986). Culture and the cognitive paradigm in social psychology. *Australian Journal of Psychology, 38,* 3, 245–256.

Pino, A. (1994). Public and private categories in an analysis of internalization. In J. Wertsch & J. D. Ramirez (Eds.), *Explorations in socio-cultural studies.* Vol. 2. *Literacy and other forms of mediated action* (pp. 33–42). Madrid: Fundación Infancia y Aprendizaje.

Prigogine, I. (1987). Exploring complexity. *European Journal of Operational Research, 30,* 97–103.

Ratner, C. (1991). *Vygotsky's sociohistorical psychology and its contemporary applications.* New York: Plenum.

Reed, E. S. (1992). Knowers talking about the known: Ecological realism as a philosophy of science. *Synthese, 92,* 9–23.

Reed, E. S. (1993). The intention to use a specific affordance: A conceptual framework for psychology. In R. H. Wozniak & K. W. Fischer (Eds.), *Development in context* (pp. 45–76). Hillsdale, NJ: Erlbaum.

Reed, E. S. (1995). The ecological approach to language development: A radical solution to Chomsky's and Quine's problems. *Language & Communication, 15,* 1, 1–29.

Rio, P. del, & Alvarez, A. (1990). Puntos de convergencia en la psicologia histórico-cultural de lengua española. *Infancia y Aprendizaje, 51–52,* 15–40.

Rio, P. del, & Alvarez, A. (1992). Tres pies al gato: significado, sentido y cultura cotidiana en la educación. *Infancia y Aprendizaje,* 59–60.

Rogoff, B. (1982). Integrating context and cognitive development. In M. Lamb & A. Brown (Eds.), *Advances in developmental psychology* (Vol. 2, pp. 125–170). Hillsdale, NJ: Erlbaum.

Rogoff, B. (1986). Adult assistance of children's learning. In T. E. Raphael (Ed.), *The contexts of school-based literacy* (pp. 27–40). New York: Random House.

Rogoff, B. (1990). *Apprenticeship in thinking.* New York: Oxford University Press.

Rogoff, B. (1992). Three ways of relating person and culture: Thoughts sparked by Valsiner. *Human Development, 35,* 5, 316–320.

Rogoff, B. (1993). Children's guided participation and participatory appropriation in sociocultural activity. In R. Wozniak & K. Fischer (Eds.), *Development in context* (pp. 121–153). Hillsdale, NJ: Erlbaum.

Rogoff, B., & Lave, J. (Eds.). (1984). *Everyday cognition: Its development in social context.* Cambridge, MA: Harvard University Press.

Romanes, G. J. (1888). *Mental evolution in man: Origin of human faculty.* London: Kegan Paul, Trench, & Co.

Rommetveit, R. (1979). On common codes and dynamic residuals in human communication. In R. Rommetveit & R. Blakar (Eds.), *Studies of language, thought and verbal communication* (pp. 163–175). London: Academic Press.

Santamaria, A. (1994). The experimental situation as a communicative situation. In N. Mercer & C. Coll (Eds.), *Explorations in socio-cultural studies.* Vol. 3. *Teaching, learning, and interaction* (pp. 65–71). Madrid: Fundación Infancia y Aprendizaje.

Sarbin, T. R. (1990). Metaphors of unwanted conduct: A historical sketch. In D. E. Leary (Ed.), *Metaphors in the history of psychology* (pp. 300–330). Cambridge: Cambridge University Press.

Sherif, M. (1936). *The psychology of social norms.* New York: Harper & Brothers.

Sherif, M., Harvey, O. J., White, B. J., Hood, W. R., & Sherif, C. W. (1961). *Intergroup conflict and cooperation: The Robbers Cave experiment.* Norman, OK: The University Book Exchange.

Shettleworth, S. J. (1993). Where is the comparison in comparative cognition? *Psychological Science, 4,* 3, 179–184.

Shweder, R. (1980). Rethinking culture and personality theory. Part III. From genesis and typology to hermeneutics and dynamics. *Ethos, 8,* 1, 60–94.

Shweder, R. (1984). Anthropology's romantic rebellion against the enlightenment, or there is more to thinking than reason and evidence. In R. Shweder & R. A. LeVine (Eds.), *Culture theory* (pp. 27–66). Cambridge: Cambridge University Press.

Shweder, R. (1990). Cultural psychology—what is it? In J. W. Stigler, R. A. Shweder, & G. Herdt (Eds.), *Cultural psychology* (pp. 1–43). Cambridge: Cambridge University Press.

Shweder, R. (1992). Ghost busters in anthropology. In R. D'Andrade & C. Strauss (Eds.), *Human motives and cultural models* (pp. 45–57). Cambridge: Cambridge University Press.

Shweder, R. A. (1995). The confessions of a methodological individualist. *Culture & Psychology, 1,* 1, 115–122.

Shweder, R. A., & Much, N. (1987). Determinations of meaning: Discourse and moral socialization. In W. M. Kurtines & J. L. Gewirtz (Eds.), *Moral development through social interaction* (pp. 197–244). New York: Wiley.

Shweder, R., & Bourne, E. (1984). Does the concept of person vary cross-culturally? In R. Shweder & R. A. LeVine (Eds.), *Culture theory* (pp. 158–199). Cambridge: Cambridge University Press.

Shweder, R., Mahapatra, M., & Miller, J. G. (1987). Culture and moral development. In J. Kagan & S. Lamb (Eds.), *The emergence of morality in young children* (pp. 1–83). Chicago: University of Chicago Press.

Shweder, R., & Sullivan, M. (1993). Cultural psychology: Who needs it? *Annual Review of Psychology, 44,* 497–523.

Siegfried, J. (Ed.). (1994). *The status of common sense in psychology.* Norwood, NJ: Ablex.

Siegler, R. S., & Crowley, K. (1991). The microgenetic method: A direct means for studying cognitive development. *American Psychologist, 46,* 6, 606–620.

Simmel, G. (1906). The sociology of secrecy and of secret societies. *American Journal of Sociology, 11,* 4, 441–498.

Simmel, G. (1908). Vom Wesen der Kultur. *Österreichische Rundschau, 15,* 36–42.

Skinner, B. F. (1985). Cognitive science and behaviorism. *British Journal of Psychology, 76,* 291–301.

Smolka, A. L. B. (1990, October). *School interactions: An analysis of speech events in a Brazilian public school setting.* Paper presented at the Boston University Conference on Language Development, Boston, MA.

Smolka, A. L. B. (1992, September). *Towards the co-constructive methodology in the study of human development.* Paper presented at the First Conference on Socio-Cultural Studies, Madrid.

Smolka, A. L. B. (1993). A dinamica discursive no ato de escrever: relações or alidade escitura. On A. L. B. Smolka & M. C. Goés (Eds.), *A linguagem e o outro no espaco escolar: Vygotsky e a construção do conhecimento* [Language and the other in educational space: Vygotsky and reconstruction of consciousness] Campinas: Papyrus.

Smolka, A. L. B. (1994). Discourse practices and the issue of internalization. In A. Rosa & J. Valsiner (Eds.), *Explorations in Socio-cultural studies.* Vol. 1, *Historical and theoretical discourse* (pp. 75–83). Madrid: Fundacion Infancia y Aprendizaje.

Smolka, A. L. B., Goes, M. C., & Pino, A. (1995). The constitution of the subject: A persistent question. In J. Wertsch & B. Rogoff (Eds.), *Sociocultural studies of the mind.* Cambridge: Cambridge University Press.

Smolka, A. L. B., Fontana, R., & Laplane, A. (1994). The collective process of knowledge construction: Voices within voices. In A. Rosa & J. Valsiner (Eds.), *Explorations in socio-cultural studies.* Vol. 1. *Historical and theoretical discourse* (pp. 109–118). Madrid: Fundacion Infancia y Aprendizaje.

Smolka, A. L. B., Góes, M. C., & Pino, A. (1995). The constitution of the subject: A persistent question. In J. Wertsch & B. Rogoff (Eds.), *Sociocultural studies of the mind*. Cambridge: Cambridge University Press.

Sorokin, P. (1956). *Fads and foibles in modern sociology and related sciences.* Chicago: Regner.

Strauss, C. (1992). Models and motives. In R. D'Andrade & C. Strauss (Eds.), *Human motives and cultural models* (pp. 1–20). Cambridge: Cambridge University Press.

Swartz, M. J. (1982). Cultural sharing and cultural theory: Some findings of a five-society study. *American Anthropologist, 84,* 2, 314–338.

Thurnwald, R. (1922). Psychologie des primitiven Menschen. In G. Kafka (Ed.), *Handbuch der vergleichenden Psychologie, Vol. 1, Part 2.* München: Ernst Reinhardt.

Toda, M. (1975). Time and structure of human cognition. In J. T. Fraser & N. Lawrence (Eds.), *The study of time II* (pp. 314–324). Berlin: Springer.

Trommsdorff, G. (1993a). Kindheit im Kulturvergleich. In M. Markefka & B. Nauck (Eds.), *Handbuch der Kindheitsforschung* [Handbook of childhood research] (pp. 45–65). Neuwied: Lucherhand.

Trommsdorff, G. (1993b). Kulturvergleich von Emotionen beim prosozialen Handeln. In H. Mandl, M. Dreher, & H.-J. Kornadt (Eds.), *Entwicklung und Denken im kulturellen Kontext* [Development and thinking in cultural context] (pp. 3–25). Göttingen: Hogrefe.

Tulviste, P. (1991). *The cultural-historical development of verbal thinking.* Commack, NY: Nova Science Publishers.

Turvey, M. T., & Carello, C. (1980). Cognition: The view from ecological realism. *Cognition, 10,* 313–321.

Turvey, M. T., Shaw, R., Reed, E., & Mace, W. (1981). Ecological laws of perceiving and acting: A reply to Fodor and Pylyshin. *Cognition, 10,* 139–195.

Valsiner, J. (1985). Common sense and psychological theories: The historical nature of logical necessity. *Scandinavian Journal of Psychology, 26,* 97–109.

Valsiner, J. (1987). *Culture and the development of children's action.* Chichester: Wiley.

Valsiner, J. (Ed.). (1988a). *Child development within culturally structured environments: Vol. 1. Parental cognition and adult-child interaction.* Norwood, NJ: Ablex.

Valsiner, J. (Ed.). (1988b). *Child development within culturally structured environments: Vol. 2. Social co-construction and environmental guidance of development.* Norwood, NJ: Ablex.

Valsiner, J. (1989a). *Human development and culture.* Lexington, MA: D. C. Heath.

Valsiner, J. (Ed.) (1989b). *Cultural context and child development.* Toronto-Göttingen-Bern: C. J. Hogrefe and H. Huber.

Valsiner, J. (1991). Construction of the mental: From the "cognitive revolution" to the study of development. *Theory & Psychology, 1,* 4, 477–494.

Valsiner, J. (1993, July). *Irreversibility of time and the construction of historical developmental psychology.* Paper presented at the XII Biennial Meetings of the International Society for the Study of Behavioural Development, Recife, Pernambuco, Brazil.

Valsiner, J. (1994a). Culture and human development: A co-constructivist perspective. In P. van Geert & L. Mos (Eds.), *Annals of theoretical psychology* (Vol. 10, pp. 247–298). New York: Plenum.

Valsiner, J. (Ed.) (1994b) Bi-directional cultural transmission and constructive sociogenesis. In W. de Graaf & R. Maier (Eds.), *Sociogenesis re-examined.* New York: Springer.

Valsiner, J. (Ed.) (1995). *Child development within culturally structured environments. Vol. 3. Comparative-cultural and constructivist approaches.* Norwood, NJ: Ablex.

Valsiner, J. (1996, in press). Constructing the personal through the cultural: Redundant organization of psychological development. In K. A. Renninger & E. Amsel (Eds.), *Change and development.* Hillsdale: Erlbaum.

Valsiner, J., & Cairns, R. B. (1992). Theoretical perspectives on conflict and development. In C. U. Shantz & W. W. Hartup (Eds.), *Conflict in child and adolescent development* (pp. 15–35). Cambridge: Cambridge University Press.

Van der Veer, R., & Valsiner, J. (Eds.). (1994). *The Vygotsky reader.* Oxford: Basil Blackwell.

Van Geert, P. (1988). The concept of transition in developmental theories. In W. J. Baker, L. P. Mos, H. V. Rappard, & H. J. Stam (Eds.), *Recent trends in theoretical psychology* (pp. 225–235). New York: Springer.

Van Hoorn, W., Verhave, T. (1980). Wundt's changing conceptions of a general and theoretical psychology. In W. G. Bringmann & R. D. Tweney (Eds.), *Wundt studies* (pp. 71–113). Toronto: Hogrefe.

Vera, A. H., & Simon, H. A. (1993a). Situated action: A symbolic interpretation. *Cognitive Science, 17,* 7–48.

Vera, A. H., & Simon, H. A. (1993b). Situated action: Reply to reviewers. *Cognitive Science, 17,* 77–86.

Vygotsky, L. S. (1934). *Thinking and speech.* Moscow-Leningrad: Gosudarstvennoe Sotsialno-eknomicheskoe lzdatel'stvo. (in Russian)

Vygotsky, L. S., & Luria, A. R. (1993). *Studies on the history of behavior: Ape, primitive, and child.* Hillsdale, NJ: Erlbaum. (Work originally published in 1930).

Wallace, A. F. C. (1970). *Culture and personality.* New York: Random House.

Weiss, P. A. (1978). Causality: linear or systemic? In G. Miller & Elizabeth Lenneberg (Eds.), *Psychology and biology of language and thought* (pp. 13–26). New York: Academic Press.

Werner, H., & Kaplan, B. (1963). *Symbol formation.* New York: Wiley.

Wertheimer, M. (1912). Über das Denken der Naturvölker. *Zeitschrift für Psychologie, 60,* 321–378.

Wertsch, J. V. (1979). From social interaction to higher psychological processes: A clarification and application of Vygotsky's theory. *Human Development, 22,* 1–22.

Wertsch, J. V. (1983). The role of semiosis in L. S. Vygotsky's theory of human cognition. In B. Bain (Ed.), *The sociogenesis of language and human conduct* (pp. 17–31). New York: Plenum.

Wertsch, J. V. (1984). The zone of proximal development: Some conceptual issues. In B. Rogoff & J. V. Wertsch (Eds.), *Children's learning in the "zone of proximal development"* (pp. 7–17). No. 23. *New Directions for Child Development.* San Francisco: Jossey Bass.

Wertsch, J. V. (1985a). *Vygotsky and the social formation of mind.* Cambridge, MA: Harvard University Press.

Wertsch, J. V. (1985b). Adult–child interaction as a source of self-regulation in children. In S. R. Yussen (Ed.), *The growth of reflection in children* (pp. 69–97). Orlando, FL: Academic Press.

Wertsch, J. V. (1985c). The semiotic mediation of mental life: L. S. Vygotsky and M. M. Bakhtin. In E. Mertz & R. J. Parmentier (Eds.), *Semiotic mediation: Sociocultural and psychological perspectives* (pp. 49–71). Orlando, FL: Academic Press.

Wertsch, J. V. (1990). The voice of rationality in a sociocultural approach to mind. In L. C. Moll (Ed.), *Vygotsky and education* (pp. 111–126). Cambridge, MA: Cambridge University Press.

Wertsch, J. V. (1991). *Voices in the mind.* Cambridge, MA: Harvard University Press.

Wertsch, J. V. (1995). Sociocultural research in the copyright age. *Culture & Psychology, 1,* 1, 81–102.

Wertsch, J. V., & Stone, C. A. (1985). The concept of internalization in Vygotsky's account of the genesis of higher mental functions. In J. V. Wertsch (Ed.), *Culture, communication, and cognition: Vygotskian perspectives* (pp. 162–179). Cambridge, MA: Cambridge University Press.

Wertsch, J. V., Minick, N. (1990). Negotiating sense in the zone of proximal development. In M. Schwebel, C. A. Maher, & N. S. Fagley (Eds.), *Promoting cognitive growth over the life span* (pp. 71–88). Hillsdale, NJ: Erlbaum.

Wertsch, J. V., & O'Connor, K. (1992, September). The cognitive tools of historical representation: A sociocultural analysis. Paper presented at the First Socio-Cultural Studies Conference, Madrid, Spain.

Wertsch, J. V., & Smolka, A. L. B. (1993). Continuing the dialogue: Vygotsky, Bakhtin and Lotman. In H. Daniels (Ed.), *Charting the agenda: Educational activity after Vygotsky* (pp. 69–92). London: Routledge.

Winegar, L. T., & Valsiner, J. (Eds.) (1992a). *Children's development within social context Vol. 1. Metatheory and theory.* Hillsdale, NJ: Erlbaum.

Winegar, L. T., & Valsiner, J. (Eds.). (1992b). *Children's development within social context: Vol. 2. Research and methodology.* Hillsdale, NJ: Erlbaum.

Winegar, L. T., & Valsiner, J. (1992c). Re-contextualizing context: analysis of metadata and some further elaborations. In L. T. Winegar & J. Valsiner (Eds.), *Children's development within social context. Vol. 2. Research and methodology* (pp. 249–266). Hillsdale, NJ: Erlbaum.

Winegar, L. T., Renninger, K. A., & Valsiner, J. (1989). Dependent independence in adult–child relationships. In D. A. Kramer & M. J. Bopp (Eds.), *Movement through form: Transformation in clinical and developmental psychology* (pp. 157–168). New York: Springer.

Wozniak, R. H. (1986). Notes toward a co-constructive theory of the emotion-cognition relationship. In D. J. Bearison & H. Zimiles (Eds.), *Thought and emotion* (pp. 39–64). Hillsdale, NJ: Erlbaum.

Wozniak, R. H. (1993). Co-constructive metatheory for psychology: implications for an analysis of families as specific social contexts for development. In R. Wozniak & K. Fischer (Eds.), *Development in context* (pp. 77–91). Hillsdale, NJ: Erlbaum.

The Arts

CHAPTER 3

Confluence and Divergence in Empirical Aesthetics, Philosophy, and Mainstream Psychology

Gerald C. Cupchik
Andrew S. Winston

Gustav Fechner (1876/1978) hoped that empirical aesthetics would complement, not replace, philosophical aesthetics. But during the early part of this century, North American psychologists became uncomfortable with "speculative" philosophical analyses of art (see Berlyne, 1974) and philosophical discourse in general (see Toulmin & Leary, 1985). From the psychologist's perspective, 2500 years of philosophical debate failed to yield agreement on the nature of aesthetic objects or aesthetic processes. Philosophical conceptions of art based on imitation of the world, expression of emotion, interplay of forms, satisfying aesthetic experience, institutional practices of the art world, or family resemblance appear to have little in common. In the context of such discord, it is not surprising that psychologists would seize on operationalism and might sidestep philosophical debate by declaring that "art is whatever *you* say it is" (Martindale, 1990, p. 17).

Nevertheless, Western philosophical traditions have provided two interrelated themes that reverberate in the psychological aesthetics of the present day: "disinterest" and "unity in diversity." The concept of *disinterest,* and the related concept of *aesthetic distance,* defined aesthetic experience as a separate realm in which everyday concerns and issues external to the object were excluded. The concept of unity in diversity, as a definition of beauty, implied that aesthetic objects are set apart from everyday objects. These themes provide a fundamental tension between theo-

Cognitive Ecology

ries that rely on basic psychological processes to explain aesthetic experience (e.g., see Child, 1978) and those that postulate unique features of aesthetic activity.

This chapter is divided into three major sections. In Section I we will briefly sketch the history of these two philosophical themes, and in Section II we will illustrate how they have guided psychological research in aesthetics. Section III of the chapter will explore the relationship between mainstream analyses of perception and cognition in everyday life and conceptualizations of aesthetic process derived from scientific aesthetics.

I. ORIGINS OF UNITY IN DIVERSITY AND DISINTEREST

For both Plato and Aristotle, the important characteristic of art was its mimetic or imitative aspect. By virtue of mimesis, art was seen as a relatively inferior *techne* or craft in that the artist produced only imitations of objects, while a shoemaker, in contrast, produced a genuine shoe. For Plato, beauty derived from ideal proportions in complex objects and unity, regularity and simplicity in simple objects. Aristotle emphasized order, symmetry, and arrangement in his conception of beauty. The concerns of modern aesthetics are anticipated in Plotinus's (A.D. 205–270) analysis of *beauty* as an independent philosophical topic.

Plotinus rejected the traditional notion that beauty consisted of symmetry or harmony of the parts of a work, and instead proposed that beauty was a singular, *instantly perceivable* quality that is apprehended by a *special faculty*. He emphasized the role of the imagination and expression of the artist in the production of beauty. Moreover, he used the notion of "unity in diversity" in his discussion of how the architect finds beauty when he sees "his inner idea stamped upon the mass of exterior matter, the indivisible exhibited in diversity" (Plotinus ?/1969, p. 58). Before Plotinus, the theme of unity in diversity, as expressed in the pre-Socratic thought of Heraclitus (e.g., see Copplestone, 1946; Kirk, 1954), was used to characterize a fundamental feature of the world and not of beauty in particular. Plotinus had a very substantial impact on Renaissance thought through the translations of Marsilio Ficino in the late 1400s.

The modern use of the word *aesthetics,* and the conception of aesthetics as a separate branch of philosophy, developed rapidly in the eighteenth century. Although Anthony, Earl of Shaftesbury's (1711/1964) aesthetic theory has historical precedence, Baumgarten's (1750) work marks the beginnings of aesthetics as a formal discipline. In the mid-1700s there was an explosion of scholarly production on aesthetics, art history, and art criticism, such as the influential works of Moses Mendelssohn, Johann Winkelmann, Denis Diderot, and Gotthold Lessing (see Barasch, 1990). It was during this period that the themes of unity in diversity and disinterest developed through both the British and Continental philosophical traditions.

Anthony, Earl of Shaftesbury (1711/1964) was one of the earliest British philosophers to distinguish between "disinterested" enjoyment of something for its own sake, and "interested" enjoyment of anticipated benefits. Thus, one could derive pleasure either from the sheer visual beauty of a mountain lake or from the anticipated ownership and use of the lake. According to Dickie (1988), Shaftesbury saw no necessary conflict between interested and disinterested enjoyment, although the desire to possess might sometimes interfere with the appreciation of beauty.

Shaftesbury's disciple, Francis Hutcheson (1725), articulated a more fully developed psychological theory of the experience of beauty and moved aesthetic theory toward a sharper division between aesthetics and everyday life. To define beauty, Hutcheson (1725) employed the notion of "uniformity amidst variety":

> The Figures which excite in us the Ideas of Beauty seem to be those in which there is Uniformity amidst Variety. . . . what we call Beautiful in Objects, to speak in the Mathematical Style, seems to be in a compound Ratio of Uniformity and Variety: so that where the Uniformity of Bodys is equal, the Beauty is as the Variety; and where the Variety is equal, the Beauty is as the Uniformity. (p. 17)

In Hutcheson's theory, the experience of beauty involves a special sense or faculty (taste), a special criterion, and a special, disinterested pleasure that results.

Kant (1790/1914) also used the concept of disinterested satisfaction in which the existence (or nonexistence) of the object, our subjective appetites (e.g., thirst), or any other practical concerns played no role. In contrast to Hutcheson, Kant rejected the notion of a separate faculty of taste that operated subsequent to the ordinary cognitive faculties. For Kant, the experience of beauty involved the same cognitive faculties as used in the experience of ordinary objects, but with beautiful objects these same faculties would operate in a different way.

This difference lies in Kant's difficult concept of "free play" of the cognitive faculties. In ordinary experience, we attempt to understand the object before us in terms of specific concepts (e.g., "dog" or "sunset") and rules in a process termed *determinant judgment*. In the experience of beauty, there is no specific concept and the imagination engages in a cognitive game with flexible rules, searching for a satisfying structure and order in the object and thereby yielding pleasure. This process is "reflective judgment," in which we imagine "the as yet unsensed, or as yet insufficiently sensed, portions of the object and how they are related to what has been sensed so far" (Barker, 1988, p. 82). For Dickie (1988), the important feature of Kant's position is that the beautiful object, when contemplated in this way, is *detached* from the world. Thus Kant's position forms the basis for what came to be called "aesthetic attitude" theories.

Schopenhauer (1818/1969) elaborated Kantian notions in such a way that a genuine theory of "aesthetic attitude" nearly emerges. For Schopenhauer, aesthetic consciousness is a rare state that is achieved with difficulty. Everyday consciousness is in the service of the will, but if the intellect can temporarily subdue the will, a state of objective, disinterested detachment can be achieved in which Platonic ideas can be suitably contemplated. If any trace of "interest" (e.g., the desire to own) intrudes, then this rare state of aesthetic consciousness will be disrupted. Furthermore, the representational or mimetic aspects of a work of art have no place in this special kind of contemplation. Thus, the Greek tradition emphasizing mimesis, previously incorporated into the notion of special aesthetic experience, is now expunged and the separation between aesthetic and everyday processing is complete.

The concept of "psychical distance," introduced by Bullough (1912) in the *British Journal of Psychology* owes much to the Kantian tradition of "disinterest." For Bullough, "distance is obtained by separating the object and its appeal from one's own self, by putting it out of gear with practical needs and ends" (p. 461). He proposed that distance, as a process, had two aspects: an inhibitory process that suppressed everyday cognition and a facilitative process that fostered "elaboration of the experience" in which subtle, hitherto unnoticed features produce a special emotional experience. What is significant about this formulation is that disinterest is now a *variable,* a matter of degree of elaboration, rather than a qualitatively different state.

Beardsley's (see 1958, 1979) theory represents the most fully developed modern notion of aesthetic experience as a distinct process. Beardsley focused on the importance of unity, complexity, and intensity of the work of art, and the unity, complexity, and intensity of the aesthetic experience it produces. He added a further dimension that distinguishes aesthetic experience: there is an intense, *narrowed focus of attention* that is controlled by the work of art. In later versions, Beardsley (1979) termed this notion "object-directedness" and argued that it was an essential feature of aesthetic experience. In addition, he outlined several other "symptoms" of aesthetic experience: active discovery through the challenge of seeking order and intelligibility; a sense of wholeness, a sense of freedom from concerns about past and future, and affect that is slightly detached and slightly distanced. As noted by Dickie (1988), these last two features constitute a modern version of the classic notion of *disinterest.* Unlike earlier theories, Beardsley did not posit a special mental function or faculty that produced disinterest. Nevertheless, his view continues the tradition in which referential features of art and connections between art and outside experiences are seen as detrimental to true aesthetic experience.

Not all contemporary aesthetic theories are predicated on the concept of disinterest. For example, Goodman's (1968) symptoms of the aesthetic (syntactic density, semantic density, syntactic repleteness, exemplification, and multiple reference) are features that stress how works of art function as symbol systems. Wolterstorff (1980) also foregrounds *meaning,* particularly the religious and moral meaning of works of art. For both approaches, the referential features, connections with everyday life, and personal meanings in a work of art would add to rather than detract from the experience.

Despite these alternative viewpoints, psychological aesthetics has frequently incorporated the themes that developed through Shaftesbury, Hutcheson, Kant, and Schopenhauer: aesthetic experience is a special kind of experience in which pleasure is produced by the disinterested contemplation of objects possessing unity in diversity. This philosophical tradition evolved through Bullough, Beardsley, and others into a view of aesthetic experience as a special, psychological process involving the suppression of everyday concerns and focused attention on the object.

II. CONTEMPORARY THEORY AND RESEARCH

A. Unity and Diversity in Psychological Research

The notion of unity in diversity was carried directly into empirical aesthetics by Fechner (1876/1978). He articulated this idea as the "principle of unitary connection of the manifold," according to Berlyne (1974). This principle operated in conjunction with another mechanism, Fechner's "principle of the aesthetic middle." According to Arnheim's (1985) translation, this principle stated that people "tolerate most often and for the longest time a certain medium degree of arousal, which makes them feel neither overstimulated nor dissatisfied by a lack of sufficient occupation" (p. 862). Thus, unity in variety had clear hedonic implications, an idea that Berlyne would exploit 100 years later. Unity without variety would be boring, whereas variety without unity might produce unpleasant feelings of distraction and fragmentation.

Berlyne's (1971, 1974) psychobiology was founded on the idea that bodily states such as arousal and bodily mechanisms, operating within an adaptation theory framework, shape the experience of pleasure. He introduced the terms "arousal-jag" and "arousal-boost" to describe complementary factors that might shape pleasure. The term "arousal-jag" refers to a situation "in which an animal or a human being seeks a temporary rise in arousal for the sake of the pleasurable relief that comes when the rise is reversed" (Berlyne, 1971, p. 136). The term "arousal-boost" refers "to the kind of situation in which a moderate arousal increment is pursued because it is satisfying in itself, regardless of whether it is promptly reversed or not" (p. 136). Variations in arousal level are governed by three classes of stimuli: psychophysical (e.g., stimulus intensity, simultaneous contrast), ecological (e.g., noxious or beneficial learned associations), and collative (e.g., novelty, complexity, and ambiguity) variables.

Berlyne (1974) saw his arousal theory as consistent with the fundamental notion of unity in variety. That is, he viewed variety as the arousal-increasing factor, tied to the collative properties of complexity, novelty, and so on, and unity as the arousal-decreasing factor, tied to order or lawfulness. Berlyne and Boudewijns (1971) designed a series of studies in which unity in variety was studied by varying the similarities and differences in visual patterns. They interpreted the results as indicating that pleasingness and liking for visual patterns was greatest when both similarities (unity) and contrasts (variety) were present. Interestingly, Berlyne did not cast this issue within the larger question of how everyday and aesthetic processing differ.

In his summary of early aesthetics research, Woodworth (1938) noted that research on the Golden Section, including Fechner's (1876/1978) study of the proportions of rectangles, can be understood as investigations of unity in diversity.

A square has unity, but little diversity, whereas a very narrow rectangle has diversity but little unity, according to Woodworth. Research on the Golden Section and debate over the reliability and meaning of the findings has continued up to the present (see Benjafield, 1985; Boselie, 1992; McWhinnie, 1987). Berlyne (1974) cited at least six potential explanations for preference for the golden section. More recently, Boselie (1992) argued that the golden section has no more appeal than the ratio of 1.5.

The issue of unity in diversity has been studied in a variety of other ways. Birkoff's (1933) aesthetic theory posited two factors, order and complexity, which are at least analogous to unity and diversity. He postulated that aesthetic value, *M,* was a function of Order–Complexity, and these variables could be quantified. This formulation served as the inspiration for a variety of later information theory approaches (e.g., Moles, 1968). More recently, Stiny (1978) proposed that the aesthetic value of a form is the ratio of the length of the description of a form (the variety, increasing as the length increases) to the length of the rules for generating the form (the unity, increasing as the length *decreases*). Stiny explicitly traced the groundwork for this approach back to Hutcheson!

Direct interest in unity in variety continues with recent work by Moore (1986), who attempted to specify precisely the conditions necessary for aesthetic unity. The spatial frequency of components in a stimulus pattern was interpreted by Moore as the perceptual equivalent of unity in variety. Such an approach is fundamentally cognitive in orientation, as opposed to the hedonic analysis of unity in variety that characterizes the Fechner–Berlyne tradition.

B. Aesthetic Distance and Aesthetic Experience

Much of the early work on empirical aesthetics dealt with the relationship between simple stimulus dimensions and aesthetic pleasure (see Beebe-Center, 1932; Woodworth, 1938). In this research, and in subsequent theorizing, the notion of aesthetic experience involving distinct processes played little role. The emphasis within the information theory tradition of Birkoff and the later Berlynian tradition was on the explanation of aesthetic preference in terms of basic processes not specific to the aesthetic domain, such as curiosity, arousal, and uncertainty. Within the predominantly behavioristic climate of the 1930s to the 1970s, the question of how to describe the unique features of aesthetic experience was not often asked.

Nevertheless, some notion of aesthetic experience and aesthetic distance continues to play a role in psychological research. For example, Wild and Kuiken (1992) attempted to experimentally manipulate aesthetic distance by providing taped instructions that directed subjects to relax, or to relax and distance themselves from personal issues.

The idea that viewing a work of art is (or should be) "distanced" from practical concerns was integral to the Berlynian tradition through his conceptions of "intrinsic exploratory behavior," as distinct from "extrinsic exploratory behavior" (Berlyne, 1960) and "diversive" as opposed to "specific" exploration (Berlyne,

1971). That is, the philosophical notion that aesthetic experience rests on "disinterested satisfaction" becomes a *psychological* notion that aesthetic experience is intrinsically motivated exploration guided by arousal regulation.

More recently, research and theory in psychological aesthetics has begun to emphasize questions of process. These theories clearly line up along the traditional continuum of philosophical aesthetics: positions that emphasize the discontinuity of aesthetic and everyday processes versus those theories that invoke everyday processes to explain aesthetic experience. Space does not permit a comprehensive review of this literature. Instead, we will illustrate how this theme is played out.

1. Propositions

Neperud (1988) offered a propositional model of aesthetic experiencing based on the conception of visual processing of Palmer (1975), Cooper (1980), and others. The concept of a *proposition,* as borrowed from logic, is the smallest unit of knowledge that stands as an assertion to be judged as true or false. In propositional models, meaning is conceptualized as a network of propositional "nodes" that link concepts and relationships. Neperud characterized "naive experiencing" of art as a style in which we treat the work as an everyday object and attend, usually very quickly, to its most obvious features. In contrast, "disciplined experiencing" treats the work as a special category of object, considers all aspects of the object, and involves a "deeper, more sustained" experience that "brings a more critical stance to the work, rather than expressing personal preference" (Neperud, 1988, p. 274). Thus, Neperud incorporated some aspects of the "aesthetic distance" tradition to define aesthetic experience as distinct from everyday experience.

To date, there have been very few attempts to use propositional models as a guide for empirical research in aesthetics. One example is a recent study by Beyerlein, Beyerlein, and Markley (1991), in which subjects sorted a set of reproductions of artworks into groups that "go together" and were asked to think aloud as they did so. Using protocol data from novices and experts in the visual arts, Beyerlein et al. constructed "concept maps" for representing the propositional knowledge structures implied by the protocols. Their analysis indicated that compared to novices, experts had greater breadth of both explicit and implicit propositions, and as indicated by the number of levels in their concept maps, greater depth in their knowledge structures.

Although a propositional approach may identify unique features of aesthetic processing, such analyses emphasize basic cognitive processes such as categorization (see Corter & Gluck, 1992) and on general cognitive constructs such as prototypes and schemas. These two terms have been used interchangeably, but can be usefully distinguished: *schema* refer to the entire set of propositions that organize the knowledge about a category, whereas a *prototype* is a hypothetical, most typical instance or "best example" of a category. As noted by Lakoff (1987), the concept of schema is closely related to "scripts," "frames," "mental spaces," and "idealized cognitive models." In addition, some discussions of schema and prototype employ the language of propositions and networks of propositions. Lakoff (1987) suggested that theories employing this group of concepts can all be considered as propositional theories. All of these conceptions imply that the categorization and evaluation of artworks depend on the viewer's cognitive model of "works of art" (e.g., Cowan, 1992).

When examined in these terms, the viewing and processing of works of art is simply a special case of everyday cognitive processes. Research on basic visual schemas would then provide an important clue to encounters with artworks. For example, the collection of features that Lorenz (1943) originally described as *das Kinderschema,* commonly referred to as "babyness," are clearly related to nurturance and positive feelings toward mammalian young. Furthermore, the features of this schema affect judgments of how weak, warm, kind, and honest the person is likely to be (see Berry & McArthur, 1986). The use of this schema in art is one of the more obvious factors that contribute to judgments of the *sentimentality* in a work of art (Winston, 1992).

2. Prototypicality

If aesthetic objects and aesthetic experience have no special properties, then it should be possible to account for aesthetic preferences and judgments within a theoretical framework for normal cognitive functioning. Recently, Martindale (1988; Martindale, Moore, & West, 1988) proposed a cognitive model of aesthetic processing that explains preference in terms of prototypicality (i.e., the degree to which a stimulus is a "good example" of a category) and the activation of neural networks. The theory assumes that stimuli (both aesthetic and everyday) are processed by cognitive "nodes" arranged hierarchically and grouped into analyzers at different levels. Nodes that are connected vertically are assumed to have a primarily excitatory effect on each other (e.g., red would excite color), whereas nodes that are laterally adjacent are assumed to have a primarily inhibitory effect on each other (e.g., red would inhibit orange).

The basic prediction of this theory is that aesthetic pleasure and aesthetic preference for a stimulus are a monotonic function of the degree to which the cognitive node for that stimulus has been activated. The activation of a cognitive node (not to be confused with physiological arousal) in turn depends on (1) the strength of the node, (2) the excitatory input from vertically related nodes, and (3) inhibitory input from laterally related nodes. The theory postulates that more prototypical and frequently encountered stimuli are represented by stronger nodes. The balance of excitatory and inhibitory processes can be used to predict that preference will sometimes be a J- or U-shaped function of typicality, rather than a monotonic function. Further assumptions about habituation effects and the relatively faster decay of fatigue over activation are used in predictions concerning novelty and frequency of exposure.

These principles place Martindale's theory in sharp contrast to Berlyne's (1971) hedonic theory in which aesthetic preference is determined by collative variables in an *inverted* U-shaped rather than monotonic function. Martindale's program of research has explicitly pitted these two theories against each other. Martindale et al. (1988) described three studies in which subjects were asked to make preference ratings on a scale of Like–Dislike for exemplars of everyday categories such as birds, colors, trees, animals, and fruits. The exemplars varied in frequency of occurrence in English (novelty), typicality, and frequency of presentation. The effects for typicality were highly significant, accounting for much more variance than the effect of novelty.

The importance of prototypicality was also demonstrated in a series of experiments on color preference (Martindale & Moore, 1988). In these studies, the concept of lateral inhibition of cognitive units was used to predict successive color contrast effects in a priming paradigm. In Martindale and Moore (1990), pure tone pairs were varied in intensity and dissonance. Preference was related to these variables in a monotonic or U-shaped function, rather than an inverted-U function, as predicted by arousal theory. Intensity accounted for much more variance than dissonance, a finding that Martindale and Moore interpreted as inconsistent with Berlyne's emphasis on collative variables.

Martindale, Moore, and Borkum (1990) described four studies of preference for random polygons, with the number of "turns," meaningfulness, color, and color typicality varied. Although Experiment 1 clearly supported Berlyne's inverted-U function between preference and number of turns, the remaining three experiments supported a monotonic function, and color typicality accounted for much more variance than other factors. Three additional experiments used drawings and paintings as stimuli. In all three studies, preference was strongly related to ratings of meaningfulness (by the same subjects). Preference was weakly or negatively related to complexity.

Hekkert and van Wieringen (1990) proposed that the relationship between prototypicality, complexity, and attractiveness would be moderated by the "categorizability" of the visual image. They found evidence that, for abstract paintings, complexity was related to beauty according to an inverted-U function. For representational works, prototypicality showed a linear relationship with judgments of beauty. Thus, prototypicality effects require a literal image.

Martindale did *not* maintain that Berlyne's inverted-U function never holds; he argued that this relationship is only important in cases where variations in the physiological arousal system are evident. Martindale's theory also contains an important disclaimer: the theory is intended to explain only "relatively disinterested, dispassionate aesthetic pleasure" (Martindale et al., 1988, p. 82) and not cases where strong motives or emotions are involved. Thus, the long historical tradition of disinterest continues, but with a twist: disinterest is now something like Beardsley's (1979) detached affect but with the explicit preclusion of strong emotion. This step is an important departure from the Kant-Schopenhauer-Bullough tradition, in which disinterest referred primarily to freedom from any appetitive or practical concerns. In Martindale's theory, this new definition of disinterest involving reduced affect marks off aesthetic experience. By defining aesthetic experience as a calm state, Martindale argued that the same cognitive processes that are important in making calm, detached judgments of laboratory stimuli in basic cognitive research should apply to aesthetic judgments.

Boselie (1991) recently questioned whether these findings do indeed support the importance of prototypicality, and he additionally questioned whether prototypicality and preference are in fact independently measured. Most significantly, Boselie argued that Martindale's studies were not really about aesthetic preference because these preferences were not part of an aesthetic experience. To define aesthetic experience, Boselie cited a list of characteristics, such as "absorption in the object." This part of Boselie's critique is based on the notion of a separate aesthetic experience as defined in the Kant–Schopenhauer tradition.

Whatever the status of the experimental results, the claim that judgments of preference about works of art involve the same processes and determinants as judgments about colors, tones, polygons, and animals is a startling claim, in which the aesthetic world is stripped of any special status. Furthermore, the claim that preference for works of art can largely be explained in terms of typicality and meaningfulness is equally remarkable, and represents a denial of any special qualities, such as Bell's (1913) "significant form," that are central to the aesthetician's vision of how art is judged.

What is missing from Martindale's analysis is that which is the very heart of Danto's (1981) analysis of the "Artworld." In Danto's view, art cannot be defined in terms of an essentialist set of characteristics common to art objects. Instead, art is the use of objects to make statements that are interpretable to Artworld participants who are familiar with an appropriate "art theory." That is, viewers must share some set of rules about the nature and function of art, the role of the artist, and the means for achieving artistic effects (see Winston, 1992). The evolution of such rules is the traditional concern of art historians (e.g., Barasch, 1990; Blunt, 1962). Whether the rules are explicitly codified in the form of a manifesto or implicitly held by naive viewers, the notion that rule violations are essential to the evaluation of art objects implies processes that are not encompassed by such dimensions as meaningfulness and prototypicality.

Part of the difficulty with Martindale's analysis may lie in his nearly exclusive use of undergraduates with little sophistication in art. As recently demonstrated by Winston and Cupchik (1992), the reasons given for their preferences by naive viewers reveal fundamentally different approaches to works of art. Naive viewers emphasized the value of familiarity and meaningfulness in a way consistent with Martindale's results, while experienced (i.e., more than 10 art courses) viewers emphasized the challenge of complex works with less obvious literal meaning. Hekkert and van Wieringen (1990) also suggested that the way in which prototypes affect preferences may depend on expertise.

3. Distinct Aesthetic Experience

In contrast to the analysis of aesthetic experience in terms of basic cognitive processes, some psychologists have carried on the philosophical traditions in which aesthetic experience is fully distinct from everyday experience. For example, Csikszentmihalyi and Robinson (1990) borrowed directly from philosophy and used Beardsley's (1979) list of five fundamental characteristics of aesthetic experience: fixed attention, release from concerns over past and future, detached affect, active discovery, and a sense of wholeness. For Csikszentmihalyi, these features not only characterize aesthetic experience, but are analogous to the features of any experience involving his concept of "flow." Thus the appreciation of a work of art has unique aspects, but much in common with such activities as mountain climbing or chess playing, which are carried out for their own sake. This approach constitutes yet another modern, psychological version of the philosophical concept of disinterest.

Through content analysis of interviews with highly experienced museum curators, Csikszentmihalyi and Robinson (1990) developed a multidimensional description of aesthetic encounters as having perceptual, emotional, intellectual, and communicative aspects. In addition, they identified the critical role of the process of discovery in the encounter with a work of art. The richness of the experience described by their respondents is noteworthy, and stands in contrast to the impoverished versions of aesthetic experience derived from empirical work that minimizes the distinction between the aesthetic and the everyday.

This contrast between the special nature of aesthetic experience as revealed through highly trained viewers, versus the mundane aspects of aesthetic experience as revealed through naive viewers, is a theme that pervades much recent research. For example, work by Lindauer (1990) and his associates, has repeatedly shown no special aesthetic preferences by (minimally) trained viewers, and has even suggested that all viewers are equally comfortable with popular "sofa" art as with "High Art." In contrast, work by Cupchik and Gebotys (1988) and Winston and Cupchik (1992) suggest substantial differences in the nature of aesthetic experience for trained and untrained viewers. With training in visual art, viewers appreciate the challenge posed by artworks, noticing the subtle visual effects associated with style while deemphasizing the literal meaning of artworks. In contrast, naive viewers are more likely to treat a work of art as an everyday object to be identified both in terms of object recognition and in terms of personal relevance, and are more likely to seek pleasant feelings rather than a challenge.

Although theories such as those proposed by Martindale, Neperud, and Berlyne draw on mainstream psychology, there is an overreliance on a single, relatively simple mechanism, such as arousal, or prototypicality, and a lack of integration of processes of attention, perception, and cognition. Our position is that a more diverse set of concepts from mainstream psychology are required to represent the hierarchical, multileveled nature of aesthetic process. In the next section, we sketch such an analysis.

III. AESTHETICS AND MAINSTREAM PSYCHOLOGY

The reconciliation of psychological aesthetics with mainstream psychology is not without certain ironies. From a mainstream perspective, the study of aesthetic process is somewhat exotic, dealing with complex, nonutilitarian objects that defy rigorous control in the laboratory. However, for Gustav Fechner (1876/1978), a founder of both disciplines, aesthetic and everyday processing are continuous. A rapprochement can be achieved by showing that aesthetic activity involves a *transformation* of everyday perceptual, cognitive, and affective processes giving rise to a uniquely structured aesthetic object (i.e. the perceived stimulus; see Dufrenne, 1973) and its correlated experience. One shared problem concerns the ways that physical/sensory and semantic (i.e., referential) information create diversity. Another issue pertains to the ways that unity is fostered by interrelating hierarchical levels of organization in stimuli. It is to these two problems that we now turn.

A. Physical–Sensory and Semantic Information in Everyday and Aesthetic Processing

1. Everyday Processing

Everyday processing is pragmatic and oriented toward the identification of useful objects. Physical–sensory cues stimulate the recognition process. These cues include, for example, edges (Hubel & Weisel, 1962), color, brightness, line ends, tilt, curvature (Treisman, 1985), elongated blobs (Julesz, 1981), spatial frequencies (Graham, 1981), and convex shapes or *geons* (Biederman, 1987). The brain is hardwired to detect these elementary properties out of which are formed the local features of potentially meaningful symbols (e.g., the horizontal bar in an uppercase *A*).

Interestingly, viewers are not aware of having performed these transitory physical–sensory analyses. Gibson (1971) described visual sensations as "a sort of luxury incidental to the serious business of perceiving the world" (p. 31). Continuing on the ecological theme, Hochberg (1986) argued that the physical world imposes constraints on viewers through patterns of information such as the *depth* cues of linear perspective and interposition. The automatic (i.e., overlearned) application of these implicit constraints makes us "conscious of the world, but not conscious of the sensations or the intervening psychological processes" (Hochberg, 1986, p. 287). Treisman (1964) similarly described physical–sensory analyses as *automatically* performed, signaling higher-order semantic meanings (Treisman, 1964).

For the transmission of pictorial (though not aesthetic) information, perceptual "grouping" of elementary features leads to the "emergence" of physical objects (Pomerantz, 1981). This intersection can be expressed in terms of *correlational redundancy* (Garner, 1962), the probability that certain combinations of elements will occur together. An important by-product of this constructive activity is the impression of three-dimensionality within the framework of a two-dimensional pictorial space.

From an attentional viewpoint, hypotheses about what might be present and happening in a scene guide visual scanning processes (Hochberg, 1978, 1979). In terms of a verbal learning perspective, schemas or prototypes from past experience help identify these emergent objects and events. The "depth-of-processing" hypothesis (Lockhart, Craik, & Jacoby, 1976) emphasizes the role of cognitive analyses in identifying a pattern and extracting its meaning. *Enrichment* (Tulving & Madigan, 1970) of the emerging cognitive structure may result from associations that are triggered by the stimulus. Such "elaboration" can occur for sounds as well as sights (Craik & Lockhart, 1972) and has even been demonstrated with odors (Herz & Cupchik, 1992).

In sum, groups of intersecting physical–sensory features trigger semantic meanings in everyday processing. Because this *forward-cuing process* is overlearned and automatically executed, individuals do not invest much attention in physical–sensory analyses per se. As a consequence, they are only consciously aware of the semantic level of analysis where most of the elaborate processing takes place. This semantic processing encompasses object identification, elaborate association, and critical (i.e., comparative) reflection.

2. Aesthetic Processing

a. Physical–Sensory and Semantic Information

In aesthetic processing, both physical–sensory and semantic information are given equal standing. Physical–sensory information is described as possessing both *syntactic* (Berlyne, 1971) and *aesthetic* (Moles, 1968) organization. "Within the same material message, there is a superposition of several distinct sequences of symbols. These symbols are made of the same elements grouped in different ways" (Moles, 1968, p. 129). At a basic level, relations among physical–sensory elements of the medium distributed in space (e.g., dabs of color) define *artistic style*. These elementary symbols group to denote objects, people, settings, and events at the higher order level of semantic organization.

A valuable distinction can be drawn between *first-* and *second-order* physical–sensory properties. The list of *first-order* visual elements is open-ended and varies in accordance with the medium in question. In paintings, these properties include color, tone, texture, use of positive and negative space, and so on. Each of these qualities can be abstracted from the overall configuration and examined independently. Thus, one can examine tonality (dark-light range) separately from texture (coarse-smooth), and so on. These elements can also be manipulated (e.g., by brush, pallete knife, etc.) to produce unique visual effects. In this way, psychologically neutral "elongated blobs" (Julesz, 1981) became meaningful and evocative impressionist "*taches*" (colored dabs of paint).

The role of *second-order* properties, such as symmetry, good figure, rhythm (Garner, 1978), harmony, and contrast is especially important in aesthetics. According to Gestalt psychology (Arnheim, 1986a), the horizontal and vertical axis is dominant so that figures group "symmetrically around a virtual axis" (p. 302). Symmetry is a quality of visual displays that serves as a "landmark" for exploration (Locher & Nodine, 1989) and is perceived preattentively in a single glance (Locher & Wagemans, 1993). Harmony is related to "dynamic symmetry," the weighting and counterweighting of compositional elements about an axis (Osborne, 1986). Research has shown that dynamic symmetry enhances exploration (Locher & Nodine, 1989), particularly for art-trained viewers using short gazes (see Nodine, Kundel, Toto, & Krupinski, 1992; Nodine, Locher, & Krupinksi, 1993).

Semantic information in artworks possesses a denotative (Goodman, 1968), logical, utilitarian character (Moles, 1968) informing about physical and social conditions in the world (Berlyne, 1971). The symbolic meaning of objects, persons, and episodes can evoke thoughts and feelings among recipients who are familiar with the relevant iconography and mythology (Gombrich, 1960). Viewers or readers can also empathize with characters and their dilemmas, thereby gaining a personal understanding of the relationship between emotions and life events (Crozier & Greenhalgh, 1992; Oatley, 1992). Styles such as dadaism, surrealism, and postmodernism, which juxtapose complex sets of semantic referents, promote interpretation and an appreciation of irony.

Subject matter can overwhelm style when pitted against it, particularly for naive viewers who tend to search for literal meaning in art (Cupchik & Gebotys, 1988). The results of a recent study (Cupchik, Winston, & Herz, 1992) showed that it was easier to discriminate "difference" than "similarity," and subject matter (e.g., still-life versus portrait) compared with style (e.g., impressionism, cubism, fauvism). "Different" subject matter interfered with the discerning of "similar" style, especially when subjects were instructed to approach the task in an objective manner. Art training must therefore overcome the automatic tendency to emphasize subject matter that generalizes from daily life and may interfere with an appreciation of style.

Differences between everyday and aesthetic processing are revealed in the attentional processes that explore and link physical, sensory, and semantic information. The Russian Formalists understood that everyday perceptual activity leads to automaticity and "habituation" (later analyzed by Sokolov, 1963). Aesthetic processing seeks to overcome the automatic cuing of semantic categories and reinvest attention in sensory experience. One way to "deautomatize" the process of perception is through "estrangement," locating familiar objects in novel contexts and thereby drawing attention to their stimulating value (see Van Peer, 1986). Shklovsky (1917/1988) expressed these ideas very clearly:

> The purpose of art is to impart the sensation of things as they are perceived and not as they are known. The technique of art is to make things "unfamiliar," to make forms difficult, to increase the difficulty and length of perception because the process of perception is an aesthetic end in itself and must be prolonged. *Art is a way of experiencing the artfulness of an object. The object is not important.* (p. 20)

The notion of reinvesting attention in perceptual processes is consistent with an emphasis on cognitive flexibility (Child, 1965) and attentional mobility (Lockhart, Craik, & Jacoby, 1976; Lockhart & Craik, 1990). Attention can be focused on physical–sensory as well as on semantic information, depending on the viewer's informational needs. This serves to increase the degree of "elaboration" at the target level of analysis, yielding a more distinctive and, hence, more memorable stimulus. Accordingly, in the Morellian method of "forensic" (i.e., objective) identification of artworks (Bryson, 1983), subtle physical features of an artwork define the artist's distinctive *signature* and provide proof of its authenticity.

If attention at the semantic level of analysis (in everyday processing) makes people conscious of objects, then investing attention at the physical–sensory level of analysis should make viewers *conscious of perceptual activity.* This point applies to artists in particular (Cupchik, 1992). The disciplined manipulation of a medium requires attention to physical–sensory cues and perceptual processes. One reason for this is that visual effects are founded on perceptual processes. Artists attend closely to the images that unfold through an application of these visual effects on a canvas or in clay. Thus, sensorimotor coordination in artistic production is predicated on an awareness of perceptual processes.

What kind of attentional process would facilitate the perception of second-order properties such as relative symmetry or the distribution of tonal variations across the surface of a canvas of natural scene? This is like asking about the processing skills required to discern distributional redundancy (Garner, 1962). Viewers must be able to abstract the quality in question from the visual field within which it is embedded and temporarily discount the remaining features. A holistic attitude would enable the viewer to attain a global appreciation of stimulus organization. This can be achieved through an abstracted gaze (Hochberg, 1979) founded on distributed attention (Triesman, 1985) and the use of short gazes (Nodine, Kundel, Toto, & Krupinski, 1992; Nodine et al., 1993). Peripheral attention takes on a more meaningful role within this process.

This analysis can contribute to our understanding of how connotative effects are achieved. Arnheim (1971) argued that "expression is an inherent characteristic of perceptual patterns" (p. 433) revealed in the patterning of "forces" that produce experiences of "expansion and contraction, conflict and concordance, rising and falling, approach and withdrawal" (p. 443). The traditional Gestalt field theory explanation for these effects has been discounted by modern researchers (Shaw & Turvey, 1981), but cognitive explanations involving "subliminal priming" (Marcel, 1983) have not solved the problem.

The problem arises because physical–sensory and semantic information are qualitatively different, and it is difficult to specify how they can interact. However, a multilevel account of aesthetic objects (Kreitler & Kreitler, 1972) and the principle of "unity in diversity" imply that there must be some way for the two domains to interact. Our explanation is based on the idea that both levels of information are activated during aesthetic processing. The by-products of second-order physical–sensory analyses, such as perceived asymmetry, *merge forward* with semantic information, producing an aesthetic "moment" involving, for example, the experience of "force." One can speculate that connotative effects are maximized when second-order properties are discerned but not consciously examined. This contrasts with an exclusive emphasis on specific first-order stimulus features (e.g., dark tonality) as the source of emotional connotation in art. The latter position is problematic because it presumes a limited set of stimulus features that have evocative properties.

In summary, artists and skilled viewers use attention flexibly, investing it in the structure of physical–sensory information. In doing so they overcome the automaticity and semantic bias that is evident in everyday processing. The density of physical–sensory processing of first- and second-order information makes the viewer conscious of perceptual experience. Attention to correlational redundancy is generally helpful for discriminating objects and yields a three-dimensional experience. Attention to distributional redundancy reveals the two-dimensional variations in isolated dimensions that are so important for producing mood effects when merged with semantic information.

B. Hierarchical Structure in Everyday and Aesthetic Stimuli

The problem of hierarchy is important both in everyday and aesthetic processing. Typically, the term *hierarchy* implies relative importance or generality among a set of features, dimensions, stimuli, or events. In everyday processing one can rank sources of information in terms of their importance and consult them accordingly, just as pilots observe gauges. However, in aesthetic processing the notion of hierarchy has to do with interrelations among multiple levels of organization in an artistic, literary, or other work and pertains to the unity in diversity problem. One would therefore assume that different criteria govern everyday and aesthetic treatment of hierarchical structures.

1. Everyday Structure

Psychologists have been very interested in the ontogenesis of percepts and concepts. Thus far we have seen that physical–sensory information is processed prior to semantic information in accordance with the epistemologial position of British empiricism. A related theoretical position holds that percepts develop from the *global* in the direction of the *particular* (Navon, 1977). The notion of global-precedence in stimulus decoding was first suggested in the 1920s by researchers of the *Aktualgenese* (perceptual microgenesis) school who believed that perception is intrinsically structured (see Flavell & Draguns, 1957). Neisser (1967) popularized the notion that a preattentive stage of global processing is followed by a focal attentive stage during which the details of stimuli are identified. A merging of these viewpoints implies that physical–sensory information is processed holistically, followed by specific (i.e., semantic) stimulus identification.

The problem of hierarchy in everyday processing (see Dodwell, 1993) is defined in terms of the level of generality of the initial and subsequent analyses. Stimuli are decomposed rather than built up (Navon, 1977) according to the global-precedence hypothesis. A first-pass cursory assessment of global properties or relationships can direct subsequent fine-grained local examination of a stimulus. Patterns possessing *good form* are processed faster when they are the target level of analysis (Sebrechts & Fragala, 1985). Although this process is useful because it takes advantage of low-resolution information, the nature of globality is not immediately apparent. Is it a product of Gestalt properties such as proximity and good configuration, or do global and local structures differ in terms of complexity, familiarity, or salience?

Kimchi's (1992) discussion of the role of stimulus properties stresses the importance of qualities such as relative size. Two levels of structure are distinguished; local elements and global configurations. The two levels are perceptually independent when the overall form of a pattern is associated with many small texture elements. In art this is exemplified by pointillism (i.e., the postimpressionist painting of Seurat), wherein innumerable colored dots are distinct from the overall configurations of objects. In patterns comprising a few large elements, the local elements are perceived as figural parts of the overall form (as in a figural sculptural composition by Henry Moore involving two or three interrelated bodily components).

An asymmetry in hierarchical patterns shows that local "elements can exist without a global configuration, but a global configuration cannot exist without local elements" (Kimchi, 1992, p. 33). Kimchi concluded that *globality* has to do with hierarchical structures and implies that properties of "higher" (i.e., larger) levels are more "global" than those at lower levels. The global advantage that favors a higher (i.e., larger) level is revealed at earlier stages of perceptual processing because of its relative salience (i.e., size).

The approach to hierarchies adopted by mainstream theorists to account for everyday perception is highly associationist. Nodes and arcs at the top of a hierarchy are more global than those at a lower level (Palmer, 1975), and comparisons are made in terms of these nodes (see Navon, 1977). Although size and relative salience of configuration are significant from the viewpoint of pure perception, this physicalist analysis is restricted to isolated images. These are also properties imposed on artificial stimuli by "detached" experimenters for whom relations among the elements are operationally defined (Giorgi, 1970). Comparative relations founded on contextualized meaning (and changing meanings) are more important for a theoretical examination of aesthetic process.

2. Aesthetic Structure

A recent debate between Hochberg (1986) and Arnheim (1986b), addressing relations between parts and wholes in visual illusions, offers contrasting views of hierarchical relations. Hochberg applies an everyday pragmatic attitude emphasizing the role of learning and takes issue with the seemingly autochthonous impact of the Gestalt "laws of organization" on object perception. His argument extends an earlier analysis (Hochberg, 1981) that focused on the role of local events in determining visual effects associated with *parts* of a figure.

Arnheim (1986b) countered Hochberg's analysis, arguing that Gestalt laws deal with visual shapes possessing structural unity rather than those that are piecemeal or atomistic: "Every Gestalt is generated by a two-way process operating downward from the comprehensive structure of the whole and at the same time upward from the structures of the constituent subwholes" (p. 283). Rather than talking about isolated parts, Arnheim emphasized "subwholes" or "isolable sections of contexts" which, "while clearly influenced by the context, retain considerable independence and by their conspicuous presence enrich the structural interaction of which the perceiver becomes aware" (p. 283). This implies that a meaningful hierarchy should be viewed in terms of "stepwise dependencies" among segments. In sum, a "part is a Gestalt embedded in a larger context. A whole, more often than not, is also a part of a larger context, which, however, is being ignored, with or without justification" (p. 284).

It is our general position that the empirical and nativist viewpoints are complementary in a comprehensive account of aesthetic processing. Hochberg emphasized the application of acquired knowledge for the purpose of decoding and exploring meaningful individual or isolated cues or events in an environment. Arnheim, on the other hand, stressed interrelations among different levels of organization in a hierarchy. The viewer's challenge is to discern the subwholes and appreciate their relations to other parts of the structure.

The problem of hierarchical relations, and wholes versus subwholes (or wholes versus parts), may appear confusing at first glance, but is in fact crucial for an appreciation of differences between everyday and aesthetic stimuli. In the simplest Gestalt examples of configurations, dots or lines are figurally "grouped" against blank backgrounds. These backgrounds are not bounded or framed in any way and have no "meaning." Thus, the parts are individual elements, tiny circles or small letters, and the whole groupings are physically larger configurations such as lines or larger letters. These elements or targets could hardly be considered "subwholes" in the sense implied by Arnheim (1986b).

One can search for a target against an irrelevant background in everyday processing. However, there is no such thing as a meaningless background in aesthetics. Negative spaces are significant subwholes and must be appreciated in relation to all others, both positive and negative. Decisions surrounding negative spaces that possess color, tone, texture, and so on, have a determining influence on the overall "beauty" of a piece, perhaps through the mediation of second-order properties like balance, harmony, contrast, and so on. This is where attention to "distributional redundancy" plays an important role in aesthetic processing. Both artists and skilled viewers must attend to figure and ground, to particular objects and relations among elementary qualities.

Relations between parts and wholes achieve a different kind of complexity in literary text processing where information is presented in an unfolding sequence. This contrasts with the purely spatial and simultaneous presentation of information in artworks. Reception theorists, such as Iser (1978), have explored the act of reading as a real-time process. Iser described the reader as synthesizing a text into an expanding network of connections integrating denoted references and contexts. The reader is constantly retrieving things from memory and modifying them while anticipating future events in the text.

Texts always challenge the Gestalt principle of *good continuation* because the fragmenting of text segments runs counter to the process of consistency building. Breaks in *good continuation* embedded in a text by its author through "fragmented, counterfactual, contrastive or telescoped sequence" (Iser, 1978, p. 186) mobilize interpretive activity. Thus, an "impeded text" promotes the production of diversified interpretive images.

Meaning emerges from the grouping of these interpretations with coherence serving as a criterion for interrelating "the polyphonic harmony of the layered structure" (Iser, 1978, p. 175) of the text. Relations between part and whole or "theme" and "field" are defined in terms of *"relevance"* (Gurwitsch, 1964). Themes can be juxtaposed against different contexts or "fields" in accordance with a sender's or receiver's goals, needs, and so on. The notion of *thematic relevance* is particularly important to the problem of multilayered meaning because a change of context can lead to the reconceptualization of meaning.

Iser maintains that the act of constituting meaning by selecting appropriate interpretive contexts heightens a reader's self-awareness. This is similar to an argument made by Pomerantz and Kubovy (1981) regarding awareness and stimulus processing. If

> changes are not occurring on the printed page but are generated internally, observers are *forced* into the metaperceptual mode. In this fashion they are made aware of the workings of their perceptual processes, of which they are usually unaware; processes that are normally transparent. (p. 425)

The further statement by Pomerantz and Kubovy to the effect, "We believe that this is the essence of the Gestalt phenomenological method" (p. 425) reveals a surprising convergence between mainstream perception and literary reception theory.

In summary, the problem of hierarchical relations is closely related to "unity in diversity." For everyday stimuli, these relations can be defined physically, in terms of properties such as relative size (Kimchi, 1992), or pragmatically, in terms of relative importance (Hochberg, 1978, 1979). The emphasis in mainstream accounts would appear to focus on diversity, individual elements, and local effects. Aesthetic processes challenge the theorist to account for various kinds of unity. In visual aesthetics, unity must be achieved between qualitatively different physical–sensory and semantic information (Berlyne, 1971; Moles, 1968). Unifying processes must also integrate figure and ground; there are no meaningless spaces in artworks. Literary aesthetics has provided a framework for considering the effects of shifting frames of reference (Iser, 1978) and competing dynamic structures (Sternberg, 1978) on the understanding of unfolding events. Relations between themes and fields are constantly changing in accordance with an author's plans and the reader's needs.

IV. CONCLUSIONS

Philosophical analyses of aesthetics from Shaftesbury, Hutcheson, Kant, and Schopenhauer had a major influence on theory and research in empirical aesthetics by directing scholarly attention to the issues of unity in diversity and aesthetic distance. These philosophical traditions have their modern expression in Beardsley's (1979) analysis, which identified five fundamental characteristics of aesthetic experience: fixed attention, release from concerns over past and future, detached affect, active discovery, and a sense of wholeness.

In our analysis, we have preserved some features of this philosophical tradition, while integrating concepts from mainstream psychology with aesthetically oriented processes. Psychologists have generally agreed that aesthetic activity is intrinsically motivated and free of practical concerns. However, less agreement has been in evidence regarding relations between everyday and aesthetic processes and the structure of the aesthetic object.

In everyday perception, combinations of physical–sensory qualities automatically signal (*cue forward*) the presence of objects. A similar process is evident in representational art where the focus is on emerging configurations of objects, people, and so on. The aesthetic process is most interesting when distinctive patterns of physical–sensory information do not refer to objects. This circumstance necessitates a dishabituation of the tendency toward object recognition combined with a reinvestment of attention in physical–sensory information. Artists attend to the process of perception itself, readily noticing patterns of physical–sensory information and manipulating media in order to produce the experience of visual effects.

Different kinds of attentional processes are implicated in aesthetic activity. A concern for local effects can be associated with focal attention to the structures linking physical–sensory features in the emergence of identifiable configurations. The appreciation of global meaning and hierarchical relations is more complex. At the physical level, broader attention to the distribution of qualities in a visual space can provide a sense for second-order properties such as balance, symmetry, and so on. This sense for structure can *merge forward* with semantic information to produce connotative effects.

Overall meaning is a function of the interpretive context that is brought to bear on it. It is here that the structure of an aesthetic object and the goals of the recipient meet. Aesthetic meaning is contingent on the interpretive context within which it is examined or understood (Cupchik, Shereck, & Spiegel, 1994). The interpretive context may be intentionally chosen by an author or artist within the work itself; may be a function of the setting within which the work is observed (e.g., a church for reverential viewers of a religious painting); or of the viewer or reader's purposive adopting of a frame of reference.

In sum, scholars understand that aesthetic episodes require *aesthetic distance,* a shift away from the pragmatics of everyday life. The problem of *unity in diversity* poses the greatest challenge. How do viewers learn to overcome the everyday tendency toward object recognition? What mechanisms are involved in reinvesting attention in the first- and second-order qualities of physical–sensory and semantic information? Having segmented the field into dynamically relevant properties, how do viewers and readers construct a unified aesthetic object that is coherent and personally meaningful? The shift between local concerns and an appreciation of global structure is particularly demanding. This interaction between viewers or readers and aesthetic works must be explicated by future generations of researchers in scientific aesthetics.

References

Anthony, Earl of Shaftesbury (1964). *Characteristics of men, manners, opinions, and times (Vol. 1).* Indianapolis, IN: Bobbs-Merrill. (Original work published in 1711)

Arnheim, R. (1971). *Art and visual perception.* Berkeley, CA: University of California Press.

Arnheim, R. (1985). The other Gustav Theodor Fechner. In S. Koch & D. Leary (Eds.), *A century of psychology as science* (pp. 856–865). New York: McGraw-Hill.

Arnheim, R. (1986a). A reply to Hochberg and Perkins. *New Ideas in Psychology, 4,* 301–302.

Arnheim, R. (1986b). The trouble with wholes and parts. *New Ideas in Psychology, 4,* 281–284.

Barasch, M. (1990). *Modern theories of art: Vol. 1. From Winckelmann to Baudelaire.* New York: New York University Press.

Barker, S. (1988). Kant on experiencing beauty. In J. Fisher (Ed.), *Essays on aesthetics: Perspectives on the work of Monroe C. Beardsley* (pp. 69–85). Philadelphia, PA: Temple University Press.

Baumgarten, A. G. (1750). *Aesthetica, Vol. 1. Teil.* Frankfort/Oder: Kunze.

Beardsley, M. C. (1958). *Aesthetics: Problems in the philosophy of criticism.* New York: Harcourt.

Beardsley, M. C. (1979). In defense of aesthetic value. *Proceedings and addresses of the American Philosophical Association, 52,* 723–749.

Beebe-Center, J. G. (1932). *The psychology of pleasantness and unpleasantness.* New York: Van Nostrand.

Bell, C. (1913). *Art.* London: Chatto & Windus.

Benjafield, J. (1985). A review of recent research on the golden section. *Empirical Studies of the Arts, 3,* 117–134.

Berlyne, D. E. (1960). *Conflict, arousal and curiosity.* New York: McGraw-Hill.

Berlyne, D. E. (1971). *Aesthetics and psychobiology.* New York: Appleton-Century-Crofts.

Berlyne, D. E. (1974). *The new experimental aesthetics: Steps toward an objective psychology of aesthetic appreciation.* Washington, DC: Hemisphere.

Berlyne, D. E., & Boudewijns, W. J. (1971). Hedonic effects of uniformity in variety. *Canadian Journal of Psychology, 25,* 195–206.

Berry, D. S., & MacArthur, L. Z. (1986). Perceiving character in faces: the impact of age-related craniofacial changes on social perception. *Psychological Bulletin, 100,* 3–18.

Beyerlein, S. T., Beyerlein, M. M., & Markley, R. (1991). Measurement of cognitive structure in the domain of art history. *Empirical Studies of the Arts, 91,* 35–50.

Biederman, I. (1987). Recognition-by-components: A theory of human image understanding. *Psychological Review, ,94,* 115–147.

Birkhoff, G. D. (1933). *Aesthetic measure.* Cambridge, MA: Harvard University Press.

Blunt, A. (1962). *Artistic theory in Italy.* Oxford: Oxford University Press.

Boselie, F. (1991). Against prototypicality as a central concept in aesthetics. *Empirical Studies of the Arts, 9,* 65–74.

Boselie, F. (1992). The golden section has no special aesthetic attractivity! *Empirical Studies of the Arts, 10,* 1–18.

Bryson, N. (1983). *Vision and painting: The logic of the gaze.* New Haven, CT: Yale University Press.

Bullough, E. (1912). 'Psychical distance' as a factor in art and as an aesthetic principle. *British Journal of Psychology, 5,* 87–98.

Child, I. L. (1965). Personality correlates of aesthetic judgments in college students. *Journal of Personality, 33,* 476–511.

Child, I. L. (1978). Aesthetic theories. In E. C. Carterette & M. P. Friedman (Eds.), *Handbook of perception: Vol. X. Perceptual ecology* (pp. 111–132). New York: Academic Press.

Cooper, L. A. (1980). Recent themes in visual information processing: A selected overview. In R. S. Nickerson (Ed.), *Attention and performance VIII* (pp. 319–345). Hillsdale, NJ: Lawrence Erlbaum.

Copplestone, F. (1946). *A history of philosophy: Vol. 1. Greece and Rome.* Westminster, MA: Newman Press.

Corter, J. E., & Gluck, M. A. (1992). Explaining basic categories: Feature predictability and information. *Psychological Bulletin, 111,* 291–303.

Cowan, P. S. (1992). Visual memory, verbal schemes, and film comprehension. *Empirical Studies of the Arts, 10,* 33–55.

Craik, F. I. M., & Lockhart, R. S. (1972). Levels of processing: A framework for memory research. *Journal of Verbal Learning and Verbal Behavior, 2,* 671–684.

Crozier, W. R., & Greenhalgh, P. (1992). Beyond formalism and relativism: The empathy principle. *Leonardo, 25,* 83–87.

Cupchik, G. C. (1992). From perception to production: A multilevel analysis of the aesthetic process. In G. C. Cupchik & J. Laszlo (Eds.), *Emerging visions of the aesthetic process* (pp. 83–99). New York: Cambridge University Press.

Cupchik, G. C., & Gebotys, R. J. (1988). The search for meaning in art: Interpretive styles and judgments of quality. *Visual Arts Research, 14,* 38–50.

Cupchik, G. C., & Shereck, L., & Spiegel, S. (1994). The effects of textual information on artistic communication. *Visual Arts Research, 20,* 62–78.

Cupchik, G. C., Winston, A. S., & Herz, R. S. (1992). Judgments of similarity and difference between paintings. *Visual Arts Research, 18,* 36–49.

Czikszentmihalyi, M., & Robinson, R. (1990). *The art of seeing: An interpretation of the aesthetic encounter.* Malibu, CA: J. Paul Getty Trust.

Danto, A. C. (1981). *The transfiguration of the commonplace: A philosophy of art.* Cambridge, MA: Harvard University Press.

Dickie, G. (1988). *Evaluating art.* Philadelphia: Temple University Press.

Dodwell, P. C. (1993). From the top down. *Canadian Psychology, 34,* 137–151.

Dufrenne, M. (1973). *The phenomenology of aesthetic perception.* Evanston, IL: Northwestern University Press.

Fechner, G. (1978). *Die Vorschüle der Aesthetik [The introduction to aesthetics]* (2 vols.) Hildesheim: Georg Holms. (Original work published in 1876)

Flavell, J., & Draguns, J. (1957). A microgenetic approach to perception and thought. *Psychological Bulletin, 54,* 197–217.
Garner, W. R. (1962). *Uncertainty and structure as psychological concepts.* New York: Wiley.
Garner, W. R. (1978). Aspects of a stimulus: Features, dimensions, and configurations. In E. Rosch & B. B. Lloyd (Eds.), *Cognition and categorization* (pp. 99–133). Hillsdale, NJ: Erlbaum.
Garner, W. R. (1981). The analysis of unanalyzed perceptions. In M. Kubovy & J. R. Pomerantz (Eds.), *Perceptual organization* (pp. 119–139). Hillsdale, NJ: Erlbaum.
Gibson, J. J. (1971). The information available in pictures. *Leonardo, 4,* 27–35.
Giorgi, A. (1970). *Psychology as a human science.* New York: Harper & Row.
Gombrich, E. H. (1960). *Art and illusion: A study in the psychology of pictorial representation.* Princeton, NJ: Princeton University Press.
Goodman, N. (1968). *Languages of art.* New York: Bobbs-Merrill.
Graham, N. (1981). Psychophysics of spatial-frequency channels. In M. Kubovy & J. R. Pomerantz (Eds.), *Perceptual organization* (pp. 1–25). Hillsdale, NJ: Erlbaum.
Gurwitsch, A. (1964). *The field of consciousness.* Pittsburgh, PA: Duquesne University.
Hekkert, P., & van Wieringen, P. C. W. (1990). Complexity and prototypicality as determinants of the appraisal of cubist paintings. *British Journal of Psychology, 81,* 483–495.
Herz, R. S., & Cupchik, G. C. (1992). An experimental characterization of odor-evoked memories in humans. *Chemical Senses, 17,* 519–528.
Hochberg, J. (1978). Art and perception. In E. C. Carterette & M. P. Friedman (Eds.), *Handbook of perception, 10* (pp. 225–258). New York: Academic Press.
Hochberg, J. (1979). Some of the things that paintings are. In C. F. Nodine & D. F. Fisher (Eds.), *Perception and pictorial representation* (pp. 17–41). New York: Praeger.
Hochberg, J. (1981). Levels of perceptual organization. In M. Kubovy & J. R. Pomerantz (Eds.), *Perceptual organization* (pp. 255–278). Hillsdale, NJ: Erlbaum.
Hochberg, J. (1986). Parts and wholes: A response to Arnheim. *New Ideas in Psychology, 4,* 285–293.
Hubel, D. H., & Weisel, T. N. (1962). Receptive fields, binocular interaction and functional architecture in the cat's visual cortex. *Journal of Physiology, 160,* 106–154.
Hutcheson, F. (1725). *An inquiry into the original of our ideas of beauty and virtue.* London: D. Midwinter and others.
Iser, W. (1978). *The act of reading.* Baltimore: Johns Hopkins University.
Julesz, B. (1981). Figure and ground perception in briefly presented isodipole textures. In M. Kubovy & J. R. Pomerantz (Eds.), *Perceptual organization* (pp. 27–54). Hillsdale, NJ: Erlbaum.
Kant, I. (1914). *Critique of judgment* (J. H. Bernard, Trans.). London: MacMillan. (Original work published 1790)
Kimchi, R. (1992). Primacy of wholistic processing and global/local paradigm: A critical review. *Psychological Bulletin, 112,* 24–38.
Kirk, G. S. (1954). *Heraclitus: The cosmic fragments.* Cambridge: Cambridge University Press.
Kreitler, H., & Kreitler, S. (1972). *The psychology of the arts.* Durham, NC: Duke University Press.
Lakoff, G. (1987). *Women, fire, and dangerous things.* Chicago: University of Chicago Press.
Lindauer, M. S. (1990). Reactions to cheap art. *Empirical Studies of the Arts, 8,* 95–110.
Locher, P., & Nodine, C. (1989). The perceptual value of symmetry. *Computers and Mathematics with Applications, 17,* 475–484.
Locher, P., & Wagemans, J. (1993). Effects of element type and spatial grouping on symmetry detection. *Perception, 22,* 565–587.
Lockhart, R. S., & Craik, F. I. M. (1990). Levels of processing: A retrospective commentary on a framework for memory research. *Canadian Journal of Psychology, 44,* 87–112.
Lockhart, R. S., Craik, F. I. M., & Jacoby, L. L. (1976). Depth of processing in recognition and recall: Some aspects of a general memory system. In J. Brown (Ed.), *Recognition and recall* (pp. 75–102). London: Wiley.
Lorenz, K. (1943). Die angeborenen Formen möglicher Erfahrung. *Zeitschrift für Tierpsychologie* [The innate form of possible experience], *5,* 233–409.

Marcel, A. J. (1983). Conscious and unconscious perception: An approach to the relations between phenomenal experience and perceptual processes. *Cognitive Psychology, 15,* 238–300.

Martindale, C. (1988). Aesthetics, psychobiology, and cognition. In F. Farley & R. Neperud (Eds.), *The foundations of aesthetics, art, and art education* (7–41). New York: Praeger.

Martindale. C. (1990). *The clockwork muse: The predictability of artistic change.* New York: Basic Books.

Martindale, C., & Moore, K. (1988). Priming, prototypicality, and perference. *Journal of Experimental Psychology: Human Perception and Performance, 14,* 661–670.

Martindale, C., & Moore, K. (1990). Intensity, dissonance, and preference for pure tones. *Empirical Studies of the Arts, 8,* 125–134.

Martindale, C., Moore, K., & Borkum, J. (1990). Aesthetic preference: Anomalous findings for Berlyne's psychobiological theory. *American Journal of Psychology, 103,* 53–80.

Martindale, C., Moore, K., & West, A. (1988). Relationship of preference judgments to typicality, novelty, and mere exposure. *Empirical Studies of the Arts, 6,* 79-96.

McWhinnie, H. (1987). A review of selected research on the golden section hypothesis. *Visual Arts Research, 13,* 73–84.

Moles, A. (1968). *Information theory and esthetic perception* (J. E. Cohen, Trans.). Urbana, IL: University of Illinois Press. (Original work published in 1958)

Moore, K. (1986). *Perceptual determinants of aesthetic unity.* Unpublished dissertation, University of Maine, Orono, Maine.

Navon, D. (1977). Forest before trees: The precedence of global features in visual perception. *Cognitive Psychology, 9,* 353–383.

Neisser, U. (1967). *Cognitive psychology.* New York: Appleton-Century-Crofts.

Neperud, R. W. (1988). A propositional view of aesthetic experiencing for research and teaching in art education. In F. Farley & R. W. Neperud (Eds.), *The foundations of aesthetics, art and art education* (pp. 273–319). New York: Praeger.

Nodine, C., Kundel, H. L., Toto, L. C., & Krupinski, E. A. (1992). Recording and analyzing eye-position data using a microcomputer workstation. *Behaviour Research Methods, Instruments, & Computers, 24,* 475–485.

Nodine, C., Locher, P., & Krupinski, E. A. (1993). The role of formal art training on perception and aesthetic judgments of art compositions. *Leonardo, 26,* 219–227.

Oatley, K. (1992). *Best laid schemes: The psychology of emotions.* New York: Cambridge University Press.

Osborne, H. (1986). Symmetry as an aesthetic factor. *Computers and Mathematics with Applications, 12B,* 77–82.

Palmer, S. E. (1975). Visual perception and world knowledge: Notes on a model of sensory-cognitive interaction. In D. A. Norman & D. E. Rumelhart (Eds.), *Exploration in cognition* (pp. 279–307). San Francisco: W. H. Freeman and Company.

Plotinus (1969). *The Enneads* (S. MacKenna, Trans.). New York: Farber & Farber.

Pomerantz, J. R. (1981). Perceptual organization in information processing. In M. Kubovy & J. R. Pomerantz (Eds.), *Perceptual organization* (pp. 141–179). Hillsdale, NJ: Erlbaum.

Pomerantz, J. R., & Kubovy, M. (1981). Perceptual organization: An overview. In M. Kubovy & J. R. Pomerantz (Eds.), *Perceptual organization* (pp. 423–456). Hillsdale, NJ: Erlbaum.

Schopenhauer, A. (1969). *The world as will and representation, Vol. 1.* (E. J. Payne, Trans.). New York: Dover. (Original work published in 1818)

Sebrechts, M. M., & Fragala, J. J. (1985). Variation on parts and wholes: Information precedence versus global precedence. *Proceedings of the Seventh Annual Conference of the Cognitive Science Society,* 11–18.

Shaw, R., & Turvey, M. T. (1981). In M. Kubovy & J. R. Pomerantz (Eds.), *Perceptual organization* (pp. 343–415). Hillsdale, NJ: Erlbaum.

Shklovsky, V. (1988). Art as technique. In D. Lodge (Ed.), *Modern criticism and theory* (pp. 16–30). New York: Longman. (Original work published in 1917)

Sokolov, E. N. (1963). A probabilistic model of perception. *Soviet Psychology and Psychiatry, 1,* 28–36.

Sternberg, M. (1978). *Expositional modes and temporal ordering in fiction.* Baltimore, MD: The Johns Hopkins University Press.

Stiny, G. (1978). Generating and measuring aesthetic forms. In E. C. Carterette & M. P. Friedman (Eds.), *Handbook of perception: Vol. X. Perceptual ecology* (pp. 133–152). New York: Academic Press.

Toulmin, S., & Leary, D. (1985). The cult of empiricism in psychology, and beyond. In S. Toulmin & D. Leary (Eds.), *A century of psychology as a science* (pp. 594–617). New York: McGraw-Hill.

Triesman, A. (1964). Monitoring and storage of irrelevant messages in selective attention. *Journal of Verbal Learning and Verbal Behavior, 3,* 449–459.

Triesman, A. (1985). Preattentive processing in vision. *Computer Vision, Graphics, and Image Processing, 31,* 156–177.

Tulving, E., & Madigan, S. A. (1970). Memory and verbal learning. *Annual Review of Psychology, 21,* 437–484.

Van Peer, W. (1986). *Stylistics and psychology: Investigations of foregrounding.* London: Croom Helm.

Wild, T. C., & Kuiken, D. (1992). Aesthetic attitude and variations in reported experience of a painting. *Empirical Studies of the Arts, 10,* 57–78.

Winston, A. S. (1992). Sweetness and light: Psychological aesthetics and sentimental art. In G. Cupchik & J. Laszlo, (Eds.), *Emerging visions of the aesthetic process* (pp. 188–136). Cambridge: Cambridge University Press.

Winston, A. S., & Cupchik, G. C. (1992). The evaluation of high art and popular art by naive and experienced viewers. *Visual Arts Research, 18,* 1–14.

Wolterstorff, N. (1980). *Works and worlds of art.* New York: Oxford University Press.

Woodworth, R. S. (1938). *Experimental psychology.* New York: Henry Holt.

CHAPTER 4

Music Perception and Cognition

Roger A. Kendall and Edward C. Carterette

I. INTRODUCTION

An explosion of research in the musical sciences is now well in evidence, brought about by technological advances, primarily in computing. The field of music perception and cognition has shared in this growth. Music technologies such as MIDI (Music Instrument Digital Interface), sampling keyboards, synthesizers of all kinds, CD and CD-ROM,[1] and the availability of high-powered computation for signal analysis and synthesis, connectionist modeling, and the design and execution of experiments, have dramatically increased the ecological validity of music perception research (Kendall & Carterette, 1992). Questions involving complex interactions of musical variables that would have been unapproachable only a few years ago have become routine.

Synthesized instrumental and orchestral music are now ubiquitous and widely

[1] CD refers to "compact disk," an optical system for storing data in digital form. CD-ROM simply refers to compact disk storage that can only be read (*R*ead *O*nly *M*emory) but not written onto. A Sony innovation, developed jointly with Philips (Netherlands), the "compact disc digital audio system" is considered by some analysts to be the most successful electronic product every introduced. One CD of 12-cm diameter, which holds about .75 gigabytes of data (275K pages of text or 74 minutes of very high-fidelity sound), can be duplicated cheaply and exactly in very large quantities. Based on optical principles so that the data are virtually lossless and distortionless, the CD is an ideal archival medium. Rossing (1990) gives a good description of CD recording and playback technology.

Cognitive Ecology

accepted;[2] often their ersatz nature is unsuspected. Multichannel systems of astonishing high fidelity control auditory images and motion in virtual and nonvirtual realities. The French composer Pierre Boulez in collaboration with the musical acoustics research group (IRCAM) at Centre Georges Pompidou in Paris has devised an expert-system "musician" which, after sensing and understanding a live orchestra's performance, then improvises original accompaniment in real time (Gerszo & Boulez, 1988). Computer models of music abound; Rowe's (1993) interactive music system for machine listening and composing is based explicitly on perceptual and cognitive principles (Rowe, 1993).

Also in evidence is a widespread interaction among scholarly disciplines. Psychologists have long found music an interesting and important stimulant for research on the operation of the mind. Musicologists increasingly are interested in understanding music as a product of cognitive operations, encouraged by the fruitful example of collaboration between such areas as linguistics and psycholinguistics. International congresses on music perception and cognition, themselves indicators of the growth of the field, routinely accept contributions from musicologists and psychologists of all types: neurophysiologists, cognitive scientists, educators, acousticians, aestheticians, and social scientists, among others. The membership of the Society for Music Perception and Cognition (U.S.), the Japanese Society for Music Perception and Cognition, and the European Society for the Cognitive Sciences of Music reflects the interdisciplinary orientation and suggests the international interest the field enjoys.

The interdisciplinary nature of the field of music perception demands a special type of researcher, one expert in multiple domains, as comfortable with an orchestral score as a statistical readout. Such a person is rare, and is a challenge to produce in a higher education system that turns out researchers who are knowledgeable in many areas but expert in none. One solution in current practice is to assemble a team of domain experts all of whom have substantial knowledge of each other's areas of expertise. There is increasing interest in university curricula such as systematic musicology, which combines musical expertise with acoustical, sociological, aesthetical, philosophical, computational, and psychological knowledge.

A. A Definition of Music

This chapter explores the multifaceted field of music perception and cognition from an ecological perspective as the interrelationship between composer, performer, and listener. Before proceeding, it is useful to have a working definition of the term *music* itself:

> Music is temporally organized sound and silence which is areferentially communicative within a context.

[2] Risset (1978), in the first edition of *The Handbook of Perception,* surveys the history and development of synthetic and electronic music from the mid-seventeenth century to the late 1970s.

This is not a complete definition, and other possibilities are not to be excluded; this is simply offered as a useful starting point. There are several important implications of this definition, however. First, an important criterion for music is its *temporality.* Suzanne Langer (1951) referred to music as based on "virtual time," to contrast it with other arts that appear to manipulate space. Second, the units of music—sound and silence—are organized; there is intent. Third, music is *areferential.* Perhaps a better word would be *self-referential.* Although music can be indexical and point to events and objects outside of itself, that is not its main criterion. In fact, the temporal syntactic of music qua music suggests that, principally, the meaning of music is embodied within itself (Meyer, 1956) (although this does not deny other levels of meaning). Fourth, music is a *process of communication* involving a chain of relationships among the composer, performer, and listener (see below). Fifth, the meaning of music is *context dependent.* Here it is useful to distinguish between *synchronic context,* for example the immediate environment of a particular concert at a particular time, and the *diachronic context,* involving enculturated knowledge acquired over the lifetime of the individual. Both types of context variables help establish mental set.

B. Music as Communication

Music is a process of communication among different behavioral systems; this process should be the central focus of an ecological psychology of music, involving the relations among composer, performer, and listener. The process of musical communication begins with an intended musical message that is recoded from ideation to a notational signal by the composer, then recoded from the notational signal to an acoustical signal by a performer (likely with an additional message of expressive intent), and finally recoded from an acoustical signal to ideation by the listener (Fig. 1). The recoding of musical messages from one frame of reference to another is at the heart of the elusive nature of music. The "structure" of music changes (more or less) as it is recoded from frame to frame. Any veridicality of communication implies a cognitive linkage, "a shared cultural contract" (Campbell & Heller, 1981) or an "alliance" (Lerdahl, 1988, pp. 255–256), worthy of research exploration in an ecological psychology of music.

The essential territory of music perception and cognition research is the empirical confirmation and disconfirmation of hypothesized relationships among the frames of reference involved in musical communication. It involves model building; theorization about the relationships between mental structures per se and manifestations of the operation of mental structures, such as the musical notation. Although some of these mental operations involve explicit knowledge, which, like the rules for assembling a child's toy, are overt and verbalizable, much musical activity emphasizes implicit, covert knowledge, which cannot be made explicit through introspection. Musicians "sense" tonal center, balance their part among others in an orchestra, make micromuscular adjustments in performance to create

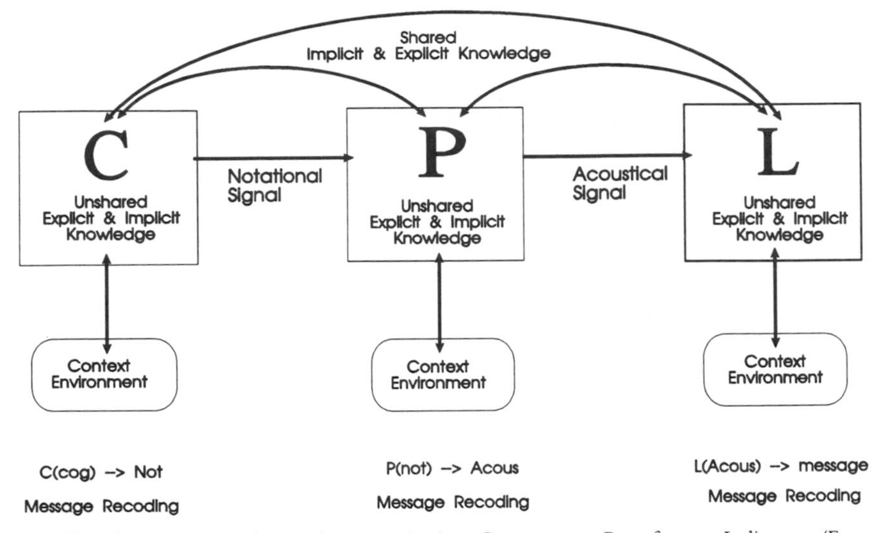

FIGURE 1 A model of musical communication. C, composer; P, performer; L, listener. (From Kendall & Carterette, 1991.)

expressive gestures, and create new musical forms in a manner which is immediate and automatic, like walking or talking (see Kendall & Carterette, 1990). For beginning performers, for example, each notational symbol is discrete, connected as stimulus to a motor response; the activity is one involving explicit knowledge. A sight-reading professional, however, has a different response to the musical notation, one that is holistic, pattern-based. An ecological music psychology is, therefore, a comparative music psychology.

Like all real-world domains for scientific research, the interaction of all the variables in the music communication process suggests a daunting task for the researcher in designing experiments. Removal of variables to simplify the situation may produce experiments that become ecologically invalid. Leaving all variables intact often produces results that cannot be attributed to anything specific. Like all research, there is a struggle between reliability and validity, and what we learn is dependent on what we ask and how we ask it (Kendall & Carterette, 1991).

In the sections that follow, we will be sensitive to these issues as we survey the field. The focus will be, where possible, on the ecological nature of music as communication, and therefore we include sections on musical communication per se, musical performance, orchestration, and music and other modalities that are not commonly found in such reviews. When dealing with the elements of music, such as pitch, we shall move from the simple to the complex, again always concerned with their function in a musical context. Therefore, issues of hearing and sensation, which are covered elsewhere in the *Handbook,* are not covered in detail. Two further caveats: First, the music psychology field is immense; an outline alone for the topics found in this chapter could easily exceed the space allotted in this volume. Therefore, the approach is highly selective and broad rather than in-depth. Second, music has its own frame of discourse, largely drawn from music theory. It is impossible to approach the field of music perception and cognition without dealing with some musically technical matters.

II. MUSICAL PITCH: SCALES AND TUNING

A. Basic Concepts

Perhaps no other area of music perception and cognition research has received more attention than the general area of pitch. This may be because of the phenomenological apprehension of its importance in Western music. It should be noted that in both Western and non-Western cultures musics exist (e.g., percussion ensembles) that do not rely on structural organizations of discrete frequencies. And in fact, much twentieth-century music has explored structural organizations of timbres and "sound structures," and are not "melodic" in the commonly accepted meaning of the term. As we shall see, there is much more to music than pitch alone.

Musical pitch is organized, that is to say, nonrandom. One of the first questions that arises is why there are not many more melodically functional pitches than there are, considering the thousands of pitches implied by a nominal just noticeable difference (JND) for frequency of pure tones of constant intensity of about 1% (ca. 1 cent) at frequencies around the tuning standard for Western music (ca. 440 Hz) (Shower & Biddulph, 1931). Burns and Ward (1982) suggested that frequency ratio JNDs, which imply event chains of order 1 (i.e., intervals between successive tones), might be more applicable to melodic information; such JNDs are wider than a single-frequency JND. More musicians can judge musical intervals (relative pitch) than can recall specific pitches (absolute pitch—see below). Houtsma (1968), using a two-temporal-interval, two-alternative forced-choice discrimination paradigm, found that the average JND for three subjects at the physical octave was far smaller (16 cents) than a tempered semitone. Clearly, mere discriminability does not lead to musical functionality.

Scholars have observed the choices of musicians in an effort to induce general principles governing music's structural organization. Burns and Ward (1982) and Dowling and Harwood (1986) marshaled converging lines of evidence to induce the basic principles of pitch organization. Musical cultures that use temporally linear pitch structures overwhelmingly utilize discrete pitch relationships to impart a musical idea (exceptions include timbral manipulation that is the primary carrier of information in Mongolian and some forms of Tibetan singing (Malm, 1977), or Australian aborigine music using the *didjeridu*). Helmholtz (1863/1954) suggested that a set of discrete pitches provides a cognitive standard through which temporally organized pitches could be parsed. The implication of this notion is that discrete pitches provide perceptual contrast points, separating the pitch continuum into parcels, thus facilitating judgment of internote pitch distances. Instruments that produce continuous pitch glides usually do so relative to these discrete pitch loci, moving up to and away from them; these pitch contour patterns (changes of pitch direction) serve to make discrete the pitch continuum. (In fact, a theremin or slide whistle relies on this to communicate a melodic idea. For more on pitch contour, see Section IV.) A cognitive structure (schema) for parsing the incoming musical message via discrete pitch categories would greatly reduce cognitive load, and form the basis for perceptual strategies of prediction and hence expectation.

Frequency ratios of 2:1, an *octave,* have a strong perceptual identity. Octave similarity provides redundancy, constraining the set of possibilities. In general, pitch sets are organized within the octave. Few world cultures fail to make use of octave similarity (Burns & Ward, 1982; Dowling & Harwood, 1986).

An additional constraint, apparently cognitive, involves Miller's (1956) concept of the magic number 7 ± 2. Miller (1956) tested observer categorization ability in several sensory modalities, concluding that, for unidimensional stimuli, information channel capacities of observers were on the order of 2.8 bits in all modalities. Most world musics employ systems that adhere to the 7 ± 2 rule. The musical systems of India, thought traditionally to have 22 possible intervals per octave, in reality use microtonal intervals (*shrutis*) in a nonfunctional manner (Jairazbhoy & Stone, 1963); these systems are similar to the Western 12-interval scales. As Burns and Ward (1982) pointed out, Western absolute pitch possessors (who are able to name pitches without referents) can process about 75 categories (6.2 bits), and relative pitch possessors (able to name pitches given a standard) can identify 32 (5 bits); however, octave similarity might be used to reduce the overall number of categories. They concluded that "these discoveries suggest that categorical perception and magic number 7 ± 2 results are not dichotomous phenomena but different manifestations of the processing of high information signals" (Burns & Ward, 1982, p. 253).

There is a general tendency in both the musicological and psychological literature to abstract from musical practice the total set of discrete frequencies and to arrange these ordinally from low to high. The terminology for sets and subsets of pitch collections varies from source to source. The term *scale* is a ubiquitous catchall. Dowling (1982) taxonomized pitch systems into four classes. The most abstract is a *psychophysical scale,* which maps frequency differences into pitch differences. The *tonal material* consists of all intervals used by a culture. The *tuning system* selects intervals used in melodic construction. A *mode* is a multidimensional concept, and includes the hierarchical functions of pitches within the system. Mode embraces the musical constraints of a given culture, including orchestral combinations, arrangements, performances, differing distributions of chromas and other imposed formulaic and hierarchical structures, which elsewhere we called *chromals* (Carterette, Kendall, & DeVale, 1993).

In Western music, a system of equal temperament is most often employed, based on the division of the octave into 12 equally spaced intervals based on the frequen-

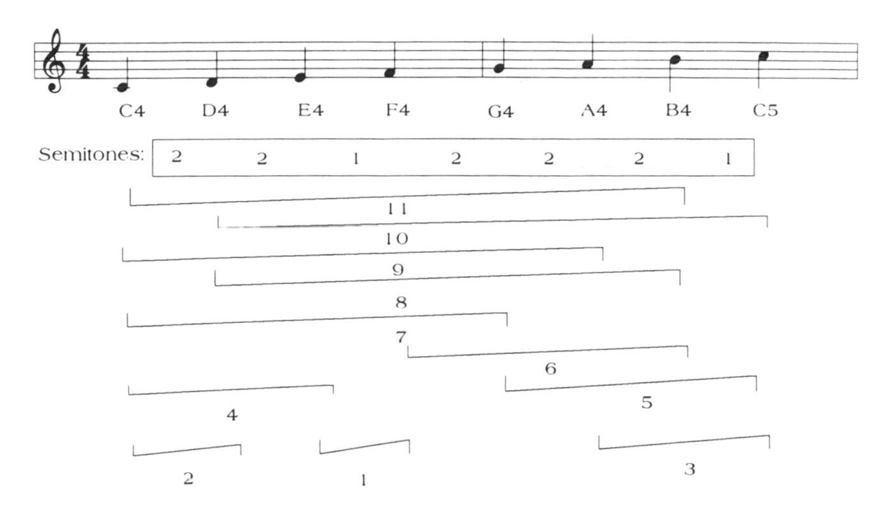

FIGURE 2 Notation of the diatonic major scale from C_4–C_5 showing intervals in semitones.

cy ratio $\sqrt[12]{2}$, with a tuning standard of A = 440 Hz. The *cent* is .01 of this *semitone* unit. The entire set is the *chromatic* scale. Individual pitch loci are alphabetically labeled, and the American Standards Association specifies a numerical designation of octaves such that A4 = 440 Hz (octaves change on C, therefore the C above A_4 is C_5, see Fig. 2). Subsets of the 12, in groups of 7, form *diatonic scales.* An important attribute of Western tuning, and the Chinese, is the use of a modular unit (semitone) in the creation of subsets; for diatonic scales the interval between adjacent tones is either one or two semitones (Fig. 2). Balzano (1980) demonstrated mathematically that the smallest number of pitches that provide maximum intervallic variety (termed *completeness*) is seven. He also noted another desirable attribute of scale systems, termed *coherence,* in which intervals of N scale steps are larger than any N - 1 step(s).

The entire area of tuning in Western music has provided endless fascination for musicologists and acousticians, who have argued for centuries about the 'best' tuning system. *Natural* tunings are so named because they, unlike equal temperament, theoretically evolved based on one or more of the following: (1) The harmonic series with its integer ratios of frequencies (Terhardt, 1974); (2) the lack of beating for simultaneously sounding intervals (consonance, see below) of complex tones of small-integer ratios such as the octave (2:1) or fifth (3:2), a position forwarded by Helmholtz (1863/1954); (3) a neural detector that prefers combinations of frequencies with a common periodicity (Lipps, 1883; Meyer, 1898). The Pythagorean tuning is based on beatless perfect fifths (actually combinations of fifths, fourths, and octaves), the result of which is a system that, when it cycles through 12 fifths to produce all intervals, fails to return to the octave; it is sharp by 24 cents (the Pythagorean *comma*). The system of just intonation uses the smallest whole-number frequency ratios: 2:1 (octave); 3:2 (fifth); 4:3 (fourth); 5:4 (major third); 6:5 (minor third); 9:8 or 10:9 (major second); 16:15 (minor second). This produces intervals of the diatonic major scale of seven tones, but problems arise for the remaining five. Table 1 presents interval and pitch notations, ratios, and cents for Pythagorean, just, and equal-temperament tunings.

B. Psychological Correlates

We turn now to perceptual correlates of intervals, scales, and tunings. This section is organized loosely on the basis of increasingly hierarchical models of pitch structure.

1. Absolute Pitch

The verbal identification of pitch chroma, in the absence of any other cue, is called absolute pitch (AP). Summaries of work in this area can be found in Burns and Ward (1982) and Takeuchi and Hulse (1993, note in particular Table 1, p. 347). AP possessors vary in their ability to identify pitches; their memory for pitches is mediated by verbal labels, and they do not have a superior recall memory for pitches (Takeuchi & Hulse, 1993). Apparently, some aspects of AP are highly correlated with timbral familiarity (one of the present authors [RAK] found that a well-known musicologist could identify key and pitch for natural instrument performances, but could not do so for synthesized; (see also Miyazaki, 1989). A typical experiment on AP is Miyazaki (1988). Listeners had to assign names to synthesized sine tones (removing timbral cues). Data indicated that listeners with AP had a high accuracy and short latency in naming the pitch. Further, microdiatonic deviations of frequency (steps of 20 cents) were identified at the nearest chroma, showing pitch categorization (see also Benguerel & Westdal, 1991; Ward, 1953). Subjects made octave errors, a common finding (Miyazaki, 1988; Miyazaki, 1989; Stumpf, 1883; Takeushi & Hulse, 1993; Ward & Burns, 1982).

Interactions of AP with other domains have been found. Miyazaki (1990) and Takeuchi and Hulse (1991) found that response latency and accuracy was better for white-key notes than black-key notes, with C being most quickly and accurately identified. This would suggest a hierarchical structuring (possibly due to familiarity) of pitch chroma for AP possessors. Crummer, Walton, Wayman, and Hantz (1994) found that AP musicians had shorter response latencies and peaks in neurophysiological measurements (P3) for timbre discriminations than nonmusicians and musicians without AP, again suggesting a connection between timbre and absolute pitch processing. Some authors suggest that AP reflects a unique, language-like representation of nonlexical musical notes in memory, citing neurophysiological measurements of P3 amplitudes that suggest different "underlying brain activity" for AP versus relative pitch subjects. In one study, downward shift of perceived pitch was found under a psychoactive drug and was reversible (Chaloupka, Mitchell, & Muirhead, 1994).

TABLE 1 Interval Comparison in Different Mathematical Tuning Systems[a]

Interval name	Solfeggio	Letter notation	Pythagorean tuning (PT)			Just intonation (JI)			Equal temperament (ET)	
			Numerical origin	Frequency ratio	Cents	Numerical origin	Frequency ratio	Cents	Frequency ratio	Cents
Unison	DO	C	1 : 1	1.000	0.0	1 : 1	1.000	0.0	1.000	0
Minor second		D♭	$2^8 : 3^5$	1.053	90.2	16 : 15	1.067		1.059	100
		C♯	$3^7 : 2^{11}$	1.068	113.7	16 : 15	1.067	111.7	1.059	100
Major second	RE	D	$3^2 : 2^3$	1.125	203.9	10 : 9[b]	1.111	182.4	1.122	200
						9 : 8[c]	1.125	203.9		
Minor third		F♭	$2^5 : 3^3$	1.186	294.1	6 : 5	1.200	315.6	1.189	300
		D♯	$3^9 : 2^{14}$	1.201	317.6	6 : 5	1.200	315.6	1.189	300
Major third	MI	E	$3^4 : 2^6$	1.265	407.8	5 : 4	1.250	386.3	1.260	400
Fourth	FA	F	$2^2 : 3$	1.333	498.1	4 : 3	1.333	498.1	1.335	500
Tritone		G♭	$2^{10} : 3^6$	1.407	588.3	45 : 32	1.406	590.2	1.414	600
		F♯	$3^6 : 2^9$	1.424	611.7	64 : 45	1.422	609.8	1.414	600
Fifth	SO	G	3 : 2	1.500	702.0	3 : 2	1.500	702.0	1.498	700
Minor sixth		A♭	$2^7 : 3^4$	1.580	792.2	8 : 5	1.600	813.7	1.587	800
		G♯	$3^8 : 2^{12}$	1.602	815.6	8 : 5	1.600	813.7	1.587	800
Major sixth	LA	A	$3^3 : 2^4$	1.688	905.0	5 : 3	1.667	884.4	1.682	900
Minor seventh						7 : 4[d]	1.750	968.8		
		B♭	$2^4 : 3^2$	1.788	966.1	16 : 9[e]	1.777	996.1	1.782	1000
		A♯	$3^{10} : 2^{15}$	1.802	1019.1	9 : 5	1.800	1017.6	1.782	1000
Major seventh	TI	B	$3^5 : 2^7$	1.900	1109.8	15 : 8	1.875	1088.3	1.888	1180
Octave	DO	C	2 : 1	2.000	1200.0	2 : 1	2.000	1200.0	2.000	1200

[a] Adapted from Martin, 1962

[b] Lesser

[c] Greater

[d] Harmonic

[e] Grave

Evidence for absolute frequency processing has been found among mallard ducklings (Evans, 1993). Such has also been found, to varying degrees, with starlings (Hulse, Page, & Bratten, 1990); however, Cynx (1993) found "no evidence for octave generalization, which is a hallmark of human absolute pitch perception" (p. 140). He continued that "avian absolute pitch perception must not be interpreted as identical with that in humans" (p. 140).

2. Scales

We will first address the collative aspects of scales and tunings before turning our attention to attributes of abstracted intervals. Relatively little literature has focused on the perception of different tunings. Ohgushi (1994) obtained subjective evaluations of the C-major diatonic scales (C_4 to C_5). He used seven temperaments: Pythagorean, just, meantone, Werkmeister, Kirnberger, Young, and equal. Subjects rated acceptability on a five-point scale ($N = 49$) and in paired comparisons ($N = 12$). He found that subjects preferred Pythagorean tuning, and that this corresponded to a multidimensional scaling (MDS) configuration derived from acoustic interval sizes. Ohgushi (1994, p. 290) suggested that a larger average interval from the whole tone (and a correspondingly shorter interval for the semitone) may make a more acceptable temperament. Contrived temperaments designed to test hypothesis appeared to confirm this (however, no statistically significant inferential analysis is offered for any of the data).

Seashore (1938) suggested that measurements of intonation in violin performance tended, if anything, toward Pythagorean tuning, providing marginal support for a preference theory such as Ogushi's (1994). String performers have long expressed a preference for beatless fifths in the tuning of their open strings. Ward (1970), however, statistically analyzed measurements from a number of sources (including Mason, 1960; Nickerson, 1948; Schackford, 1961, 1962). The mean interval tunings showed no consistent tendency to conform to either just intonation or Pythagorean tuning in either melodic or harmonic contexts. Musicians, however, would argue for context effects. With musical instruments that can vary intonation, performers generally contend that pitches will be played somewhat sharp for leading-tone melodic motion, and that an ascending augmented interval (e.g., C–Gb) will be played wider than an ascending diminished interval (e.g., C–F#). Burns and Ward (1982) suggested that the general tendency is to contract the semitone and to expand slightly all other intervals relative to equal temperament, which mirrors results of perceptual experiments (see below), which show a tendency to compress the scale for small intervals and stretch the scale for large intervals in both ascending and descending contexts. One must agree with Butler (1992) that "there is still some disagreement, however, on how much (and how consistently) performers bend pitches to fit the musical context" (p. 55).

What is clear is that there is a general tendency to stretch the octave interval (Ward, 1954; Dowling & Harwood, 1986) to a ratio approaching 2.009:1 at middle frequencies; some musical cultures stretch their entire scale precisely according to a power function (Carterette & Kendall, 1994) and with even greater stretch.

Some data are available on the perception of non-Western tuning systems. Carterette et al. (1993) investigated the *sléndro* tuning system for the Javanese gamelan. In theory, the system has five equal steps of 240 cents within the octave (equi-pentatonic). Small deviations in tuning away from 240 cents are in a patterned relationship; pairs of instruments share the same patterning (see measurements in Carterette et al., 1993, p. 391). Native musicians were asked to judge whether the second interval of a randomly drawn pair of natural gamelan tones was wider, narrower, or equal on a 100-point continuous scale. In addition, they used a protomusical context of four tones covering three adjacent scale degrees[3] (e.g., 1223). Data showed that only one professional musician of four could judge the small-cents deviations from the equi-pentatonic tuning.

These data seem to be consistent with studies of other cultures. North Indian musicians whose classical scale has (theoretically) 22 degrees per octave (*shrutis*) could not consistently identify these microtonal intervals, generally could not tell which of two intervals was the larger, and could do no better than identify an interval as one of the 12-note chromatic scale (Burns, 1974). Sampat (1978) compared Western and Indian musicians on three tasks: interval identification, interval discrimination, and categorical perception of pitch. The main results were that Western and Indian musicians were not different from each other on these tasks. In particular, it had been expected that Indian musicians, "having more categories in their musical pitch scales would exhibit finer discrimination" (p. 33) in categorical perception of pitch. But the outcome was like that of Burns (1974); Indian musicians could do no better than identify an interval as one of the notes of the 12-note chromatic scale.

In a comparative study, Western musician and nonmusician children (10–13 years old) were tested in detection of mistunings in a melody that was based on Western major, Western minor, or Javanese *pélog* scales (Lynch & Eilers, 1991). Child musicians and nonmusicians performed at chance level and no differently from each other, in the Javanese context. In the Western contexts child musicians detected mistunings better than in the Javanese context, and better than the child nonmusicians, who did better than chance. Lynch and Eilers interpreted their data as suggesting that informal musical acculturation leads to better perception of native than nonnative scales by 10–13 years of age, but that formal musical experience can facilitate the acculturation process.

The Western major, Western minor, and Javanese *pélog* scales figured again in a study by Lynch, Eilers, Oller, Urbano, and Wilson (1991). Adults who differed in musical sophistication listened to a melody that was based on interval patterns from

[3] Throughout this section we use Arabic numerals to indicate scale degrees (see Fig. 2), and roman numerals to indicate triadic function.

Western and Javanese musical scales. Threshold judgments were obtained by an adaptive two-alternative forced-choice method. Judgments of mistunings by the less sophisticated listeners were better for Western than for Javanese patterns, whereas the judgments of musicians did not differ between Western and Javanese patterns. The authors suggested that differences in judgments across scales are "accountable to acculturation" by listening exposure and musical sophistication gained from formal experience.

Further evidence for an implicit, templatelike tonal schema for the perception of tuning systems comes from several studies. Jordan and Shepard (1987) presented listeners with varying musical backgrounds an ascending or descending normal major diatonic scale or a distortion of that scale with stretched intervals (obtained by dividing a range of 13 consecutive semitones into 12 equal log-frequency intervals and taking from the resulting set of 12 stretched semitones the 7 corresponding to those of the normal diatonic major scale). A "probe-tone" with quarter-tone resolution followed 1.5 s after the sequence, and listeners rated the degree of fit on a seven-point scale (this is called the *probe-tone* method, developed by Krumhansl & Shepard, 1979). Results indicated that listeners, and particularly trained musicians, translated the pitch by the major part (⅔) of a semitone onto the normal major scale. They conducted an additional experiment with scales that were based on a $\sqrt[7]{2}$ frequency interval; an equi-heptatonic scale. Again, listener rating profiles, particularly in musicians, for the equi-heptatonic context correlated moderately with the normal scale context using both quarter-tone and equalized semiscale probes (.365–.805 across musicians and nonmusicians, probe resolution, and ascending vs. descending contexts, p. 501). This result compares well with that obtained by Kessler, Hansen, and Shepard (1984) using more complex (and methodologically problematic) diatonic musical contexts[4] and involving both Balinese and Western subjects. We return to examine the probe-tone method in the section on tonality.

3. Intervals, Consonance, and Dissonance

We have thus far examined research that involves linear presentations of tuning system elements. We now turn our attention to simultaneous, as well as successive intervals, that is, event chains of the first order. This area of study has been of interest because it involves a minimal and easily employed stimulus (however atomistic), the musical interval. Isolated musical intervals are often used as practice materials in the ear-training courses of university music departments; aural identification of intervals is considered a fundamental, prerequisite skill leading to score reading and more complex harmonic dictation.

[4] We believe there are problems with the context stimulus materials in these studies. For example, the frequency histogram of chromas for the major context in the study of Kessler, Hansen, and Shepard (1984, p. 141) would suggest A-aeolian rather than C-major and Eb-major instead of C-minor (aeolian). See Vaughn, 1993, p. 32, for a critical note regarding the stimulus materials in Castellano, Bharucha, and Krumhansl, 1984.

We noted in Section II.A. that consonance has been associated with the "simplicity" of integer-frequency ratios. Helmholtz (1862/1954) postulated that, for simultaneous vibrations of close but not identical frequency, the sensation of roughness or beating leads to dissonance; the lack of roughness, or smoothness, leads to consonance. For complex tones, interference among partials of close frequency difference yields beating, and noninteger ratios produce more beating partials than do integer ratios.

This approach to consonance and dissonance has been termed *sensory* or *tonal* or *acoustic* consonance. Stumpf's *tonal fusion* theory, based on the degree to which simultaneous tones loose individual identity (see also Section IV on Musical Timbre and Orchestration for a similar discussion related to timbral blend), also has proponents. Lundin (1953) and Cazden (1945, 1962) have noted that musical consonance, in contrast, is far more multidimensional, and is connected to cultural and music-stylistic factors related to the perceptual stability versus instability of temporally organized musical elements. Cazden (1962) provided a number of examples of how tonally consonant passages are musically unsettled (dissonant) in classical music (see also, Dowling, 1986, p. 88, who also includes listening examples), and the use of final chords with added major 7ths and 6ths (and even 9ths and 11ths), which should be extremely unsettled and dissonant, is commonplace in certain styles of popular and jazz music.

Recent studies of sensory consonance have expanded on Helmholtz's notion of beating and roughness to include interference among partials within critical bands (at midrange on the piano, ca. C_5, the critical bandwidth is approximately ⅓ octave). Plomp and Levelt (1965) proposed a model for sine tones in which frequency differences within a critical bandwidth are judged as dissonant, and those beyond the critical bandwidth as consonant. Expansions of this simple model are those of Kameoka and Kuriyagawa (1969), which (partly) are based on six harmonics of equal amplitude and a fixed reference frequency of A_4. Hutchinson and Knopoff (1978), in contrast, base their rankings on spectra of ten components with amplitudes $A_n = 1/n$. A comparative ordinal ranking (based on Krumhansl, 1990, p. 57) is useful here, arranged from most consonant to most dissonant (Table 2).

It is clear from Table 2 that simple integer ratio intervals (octave, 2:1; perfect fourth, 4:3; perfect fifth, 3:2) are more consonant than complex integer ratios (e.g., minor second, 16:15; major seventh, 15:8). Also, there appears to be a good deal of correspondence between approaches based on different signals (sines vs. complex), a fact borne out by Pearson correlations based on numerical values (Krumhansl, 1990, p. 60), which for the data represented by Table 2 produce a mean correlation of .84.

One would expect that AP possessors would perceive individual chroma categorically; highly trained musicians, drilled in aural interval recognition, would likely perceive musical intervals categorically, even though they might possess only relative pitch. Studies by Burns and Ward (1978) and Siegel and Siegel (1977a,b) support this hypothesis. In Burns and Ward (1978) intervals of a minor third (m3), major third (M3), and a perfect fourth (P4) (relative to C = 252 Hz) were

TABLE 2 Ordinal Rankings of Consonance for Isolated Intervals

Helmholtz equal-tempered	Helmholtz simple-ratio	Kameoka & Kuriyagawa (1969)	Hutchinson & Knopoff (1978)
Unison/octave	Unison/octave	Unison/octave	Unison/octave
Perfect fifth	Perfect fifth	Perfect fifth	Perfect fifth
Perfect fourth	Perfect fourth	Major sixth	Perfect fourth
Major third	Major sixth	Perfect fourth	Major sixth
Tritone	Major third	Major third/minor seventh	Major third
Minor sixth/major sixth	Minor third/ tritone/minor sixth	Minor third/major seventh	Minor sixth
Minor third/minor seventh	Minor seventh	Minor sixth	Tritone
Major second	Major second	Tritone	Minor seventh
Major seventh	Major seventh	Major second	Minor third
Minor second	Minor second	Minor second	Major seventh
			Major second
			Minor second

presented for an identification task. Results showed categorical boundaries between intervals. In a discrimination task, subjects were asked to judge which of two intervals was wider when differences between them were 25 cents (¼ semitone), 27.5 cents (⅜ semitone), and 50 cents (½ semitone) over a range of 250 to 550 cents. Results showed most acute discrimination at the boundaries between the musical intervals. Nonmusically trained subjects did not evidence categorical perception. Siegel and Siegel (1977a) also performed an identification task similar to that of Burns and Ward (1978). They found that musicians with relative pitch assigned interval names to nondiatonic intervals, always forcing amusical intervals onto the diatonic interval set. They also found no evidence for categorical perception among untrained listeners. This is hardly surprising, because interval identification obviously is a trained aural skill crossing sensory, cognitive, and semantic spaces (the listener must connect a sound with a verbal label). Siegel and Siegel (1977b) also had listeners describe intervals that were incrementally sharp or flat. Intervals were presented in isolation. Even musically trained listeners rated high and low stretching of frequency ratios as "in-tune," and were often unable to distinguish the direction, sharp or flat, of out-of-tune intervals. However, when a musical context was provided by Wapnick, Bourassa, and Sampson (1982), consisting of a 10-note passage, musicians were somewhat better able to judge following intervals as either sharp or flat (58% vs. 51%), supporting the contention of Kendall (1986) that all experimentation with isolated musical units must be viewed with caution.

4. Pitch as a Multidimensional Variable

Psychological studies of pitch phenomena largely have used artificially generated sines or complex tones in no or impoverished musical contexts. However, converging lines of evidence suggest that for many Western listeners there are one or more internalized implicit schemata for pitch structure.

Since the nineteenth century (Drobisch, 1852) it was suggested that pitch had multidimensional properties, specifically that there was *pitch height,* the perceived "highness" versus "lowness" of pitches relative to one another, versus *pitch chroma,* the quality of a pitch locus within the octave, that is to say, its perceptual uniqueness as C, or D, or Bb. Figure 3 illustrates the idea in a three-dimensional spiral. The top part of the figure shows the two-dimensional chroma circle, a geometry for relations among chroma within the octave. It is instructive to note that the tritone (e.g., C–F#), a distance of six semitones, is maximally distant and therefore appears on the radii of the circle (in music theory this interval is sometimes called the "devil in music" and was prescriptively avoided. Berlioz, in *La Damnation de Faust* [1846], uses the tritone symbolically at the entrance of Mephistopheles). Octave equivalence and height is shown by the upwards unwinding of the circle.

Much controversy has raged over the psychological reality of this pitch geometry (see Burns, 1981; Burns & Ward, 1982; Shepard, 1982). Shepard (1964) generated the now famous "auditory barber pole" illusion, where presentation of a cyclically repeating sequence of complex tones composed of 10 octave partials with a bell-shaped amplitude profile appears endlessly to rise or fall without octave jumps; this has been suggested as partial verification of the model. By moving the amplitude envelope up or down in log frequency and holding the frequencies of components fixed, a change in pitch height independent of chroma can be achieved. By moving the frequencies of components up or down in log frequency and holding the amplitude envelope fixed, a change in pitch chroma can be achieved independent of pitch height (Shepard, 1982, p. 371). Risset (1978) used the apparent orthogonality of the dimensions to create patterns that change height and chroma in opposite directions (e.g., raising pitch height while simultaneously lowering the chroma). The pattern appears to continuously ascend but, paradoxically, ends on a lower pitch than when the pattern started. However, Burns (1981) demonstrated in experiments with 21 college students that inharmonic complexes of stretched or compressed octaves could also give rise to the illusion, suggesting that octave equivalence was not a necessary condition.

An additional confirmation of part of the geometry comes from the presentation of isolated tritones using Shepard tones. You will recall that tritones bisect the chroma circle. Shepard (1964) noted that intervals of a tritone (C–F#) could be heard as either ascending or descending, but not both. He played pairs of tones and asked subjects to judge whether what they heard ascended or descended. When the tones were near each other in chroma, judgments were made by proximity (i.e., C–D was heard as descending). This result weakened as the distances along the circle

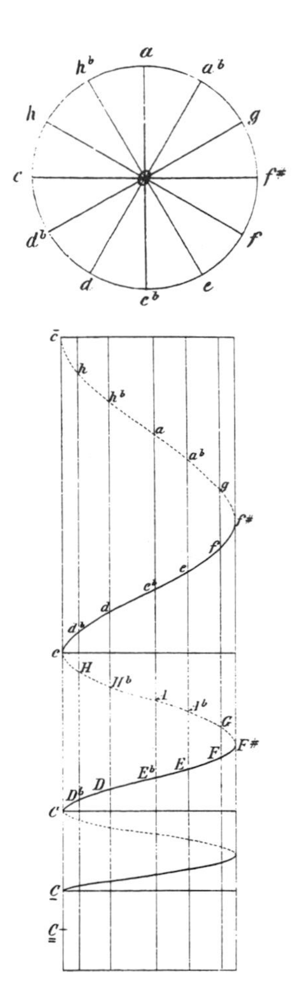

FIGURE 3 The pitch spiral. (From Drobisch, 1852).

became greater, until an equal number of ascending and descending judgments were made to the tritone (half-octave). Deutsch (1986; see also Deutsch, 1987; Deutsch, Moore, & Dolson, 1984, 1986) extended this work using six-component Shepard tones. She, however, varied the starting pitch chroma and analyzed data by subject. Deutsch (1986, 1987) found that judgments were dependent on *which* pitch was first, that is, its position along the chroma circle. Furthermore, the direction of pitch height related to chroma was subject-dependent (Deutsch, 1987). Deutsch (1987, p. 574) suggested that this reflects an internalization of pitch chroma. Although this does not lead to absolute pitch, subjects are sensitive about where intervals start on the chroma circle, a capability that parallels a musician's ability to judge the correct key of musical passages (Terhardt & Ward, 1982;

Terhardt & Seewann, 1983). Deutsch (1991, 1994) also found that Californian Americans tended to hear the pattern as descending, persons who grew up in southern England as ascending. Perception of the tritone paradox and the pitch range of the listener's spontaneous speaking voice were correlated, which "indicates strongly that the same culturally acquired representation of pitch classes influences both speech production and perception of this musical pattern" (p. 335). Deutsch (1984) showed that the strength of the relationship between pitch class and perceived height depended on the overall heights of the spectral envelopes under which the tones were generated, which bolstered her view that the tritone paradox is related to the processing of speech sounds.

Intuitively, musicians find the helical pitch model perplexing: Chroma-forming intervals of a perfect fifth appear too far apart (the octave and fifth are used as focal points of tuning-system construction), as do members of the diatonic scale of a particular key relative to nonmembers. Krumhansl (1979) had musically sophisticated subjects rate similarity between sequentially presented pairs of notes that followed a minimal musical context—an ascending or descending diatonic major scale (Fig. 2) or a major triad (scale degrees 1, 3, 5 of Fig. 2). She found no significant differences between these contexts. The similarity judgments were subject to multidimensional scaling. A two-dimensional solution separated in order, on the second dimension, the notes of the triad (C-E-G-C′), the rest of the seven-tone diatonic collection (D-F-A-B), and the five pitches outside of the diatonic C-major scale (C#-D#-F#-G#-A#). The principle dimension appears strongly related to pitch height. Krumhansl (1979) imposed a conelike geometry on the scaling plot consisting of three concentric rings (triad, scale, outside [chromatic] tones). Clear evidence for octave similarity is evident. Furthermore, she found that nondiatonic to diatonic note pairs were judged more similar than diatonic to nondiatonic note pairs. This judgment asymmetry confirms a musician's intuitions: Nondiatonic tones are less "stable" or more "active" tending, in linear motion, to move toward more stable, diatonic tones. The motion away from stability is, for the same frequency ratio, perceptually greater than the identical interval in the opposite direction.

Krumhansl and Shepard (1979) played a major scale minus the last tone as an antecedent context. The consequent tone was drawn from the set of 12 chromatic pitches, and the subject rated the degree to which the consequent fit with the antecedent. Results were analyzed extensively by Shepard (1982). One reasonable representation accounts for the relatedness of fifths via the well-known circle of fifths, which forms the base of the double-helix geometry (Fig. 4). The six tones (per octave) of each strand are separated by two semitones; planes drawn horizontally through the figure would connect adjacent semitones. The diatonic series is outlined (using white circles) at the right of the figure; the remaining tones (black circles) are nondiatonic. For both musically inexperienced and experienced subjects, octave equivalence was in evidence; this was the only tonal feature utilized by inexperienced subjects.

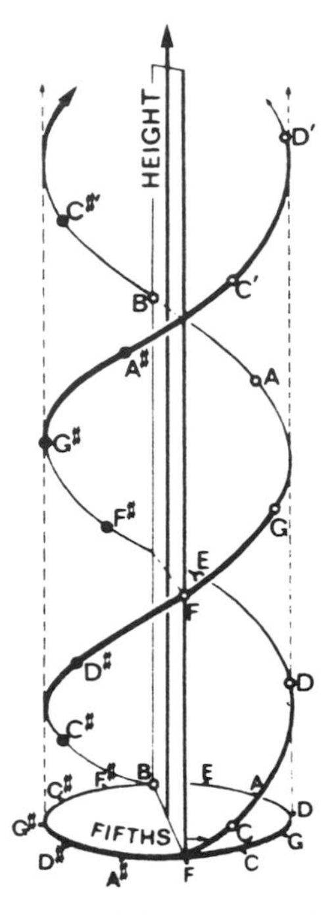

FIGURE 4 The pitch helix, relating pitch height, pitch chroma, and the circle of fifths. (From Shepard, 1982.)

Vaughn (1993) conducted experiments in which entire modes were rated in pairs for similarity. She used 10 Indian modes that correspond largely with traditional Western church modes. The dimensional configuration of the scales when presented without the traditional background (a tambura drone) is circumplicial; modes having tones in common are closest to one another, corresponding with the Indian music theory, a Circle of *Thats* (modes). However, with the drone context, the circular configuration disintegrated into a set of clusters based on successive gaps in interval between scales. Vaughn (1993, p. 30–31) suggested caution in interpreting results of scaling studies that are ecologically impoverished.

These geometries suggest hierarchical relations among pitches including the idea of a center, foundation pitch around which all others are arranged, that is, the *tonic* pitch within a tonal framework. We turn now to the issue of tonality.

5. Tonality

Tonality is one of those musical concepts that, because we have a single noun word to identify our experience, is easily oversimplified. The word is used in a multiplicity of contexts and with a large number of diverse implications. Lewis Rowell (cited in Thomson, 1991) noted with humor that metaphors have included "*focus,* as in optics; *homing,* as in pigeons; *attraction,* as in magnets; *vectoring,* as in airport approaches; and *vanishing point,* as in perspective" (Rowell, 1983, p. 235). Thomson (1991, p. 4) then suggested that all of these factors relate to "what we shall call *pitch focus* (or better, *pitch-class focus*). . . . Tonality is a set of pitches operating as a resolutional hierarchy. It is the pitch nucleus, the drawing together of the members of a pitch collection as aural vectors toward a tonic" (p. 5).

Some pitches within the diatonic set are theorized to be more active, tending to move successively to less active, more stable tones. The tonic, dominant (scale degree 5), and mediant (scale degree 3) are often proposed to be more stable; sounded simultaneously they produce (1-3-5) a tonic triad in a given key. Krumhansl and Shepard's (1979) approach, outlined above, is called the *probe-tone technique* and has been applied in an extensive series of studies (see Krumhansl, 1990, for a review) designed to ferret out the relationships among diatonic pitches in a tonal context. Krumhansl and Kessler (1982), in a seminal study, extended Krumhansl and Shepard's (1979) work to include various contexts which, according to notational theory, ought to establish a *key,* or tonal center. These contexts included diatonic scales (this time included the tonic at both starting and ending positions), single triads (1-3-5 scale degrees), and three cadential chord progressions (in music notation, IV-V-I; VI-V-I; II-V-I)[5] in both major and minor keys. Subjects rated the degree to which Shepard tonelike probes fit into the preceding context. The result is a profile of fit that suggests the kinds of hierarchical relations among diatonic pitches (Fig. 5). One can clearly discern the prominence given by listeners to the tonic, third, and fifth scale degrees. This profile of results correlates well with consonance and dissonance rankings, particularly for major modes (Krumhansl, 1990, p. 60), with the strongest correlation (.839) being that between the profile and the study of Kameoka and Kuriyagawa (1969). Krumhansl and Kessler (1982) also interpolated their results to different musical keys, cross-correlated these, and found that the results were a reasonably close match to music theoretic notions of key relationships. You will recall that the circle of fifths is an important component of the multidimensionality of pitch (see Fig. 4). A musical key refers to the fact that a given intervallic pattern, say major (Fig. 2), minor, dorian, or whatever, may begin on any given pitch chroma. Keys that are closest in relation to one another on the circle of fifths share the most common tones, for example, the major keys of C and G share all tones but one (F#), and are a single step apart on the circle of fifths. It is therefore possible to speak of a "key distance" in terms of shared chroma and relative distance along the circle of fifths (this concept becomes important in research on the perception of melody, see below). Krumhansl and Kessler's data recovered, via MDS, the circle of fifths in a two-dimensional solution, including relationships between parallel major and minor modes (e.g., C-major and A-minor).

[5] Chord symbols are rendered in all capital letters, and should not be read as indicating chord quality (major/minor/diminished/augmented) as is traditional in music theory.

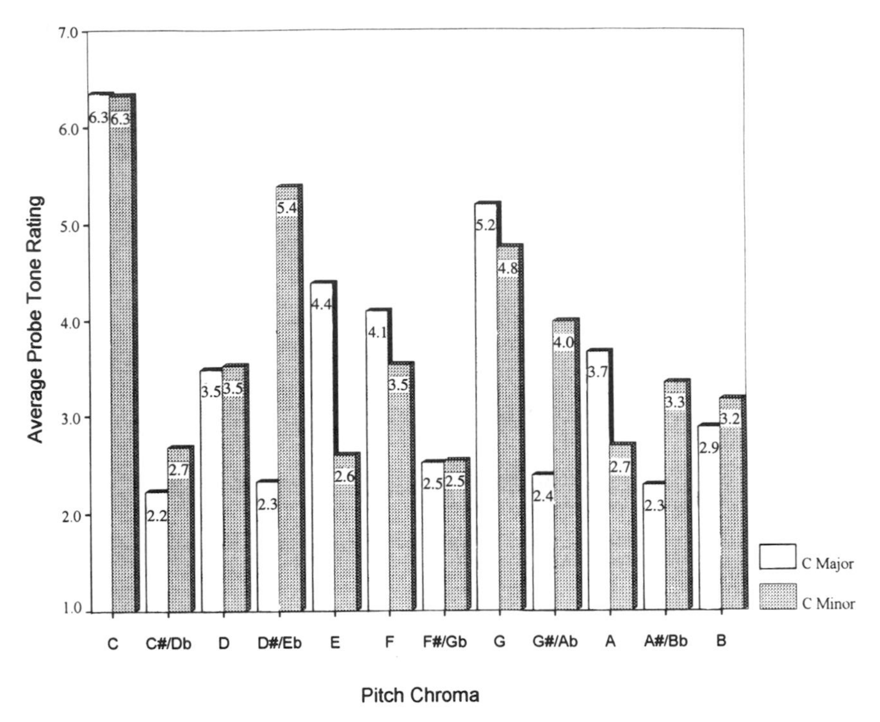

FIGURE 5 Histogram of the tonal hierarchy. (Created from data in Krumansal & Kessler, 1982.)

Krumhansl, Bharucha, and Kessler (1982) extended this work to chords. They played an ascending scale (C-major, G-major, and A-minor, all closely related) followed by two chords. The subject's task was to rate how well the second chord followed the first (i.e., how "related" they were). These relatedness ratings were treated as similarities in MDS analysis. Results supported the music theoretical relations among chords, with the tonic (I: 1-3-5)[2] and dominant (V: 5-7-2) being most closely related, followed by the subdominant (IV: 4-6-1), followed by more peripheral (VI, II, III, VII) chords. Bharucha and Stoecking (1986) suggested that a given chord implies, through its hierarchical relation with other chords, what will follow, an idea they call *priming.* Of course, that is the implication of any nonrandom system: redundancy is the catalyst for prediction.

Additional work with probe tones has been applied to a bitonal (C–F#) passage from Stravinsky's *Petrouska* (Krumhansl & Schmuckler, 1986). Although evidence of the two keys was found in the profile, additional experimentation failed to demonstrate that the listeners could selectively attend to either one of the two keys. Unfortunately, the single passage used was homophonic, and much bitonal music relies on registral, timbral, and temporal contrasts for polyphonic emphasis of different keys (for example, Milhaud's *Le Beouf sur le Toit*), so the jury is out as to whether more than one key can be selectively attended to under more complex musical conditions.

Kessler, Hansen, and Shepard (1984) applied the probe-tone technique to relatively extended passages drawn from the music of Bali. Their subjects included 27 native villagers unfamiliar with Western diatonicism and Western musicians unfamiliar with the equi-pentatonic *sléndro* scale (240 cents/interval) or nonequi-heptatonic *pélog* scale. Subject probe-tone ratings sometimes produced were style appropriate even when they were unfamiliar with the style, suggesting that the frequency distribution of pitch chroma in the extended passage was sufficient information to invoke the "tonal hierarchy." A similar result was found by Castellano, Bharucha, and Krumhansl (1984) for music of Northern India.

There has been some controversy surrounding the operationalizing of tonality in terms of the probe-tone technique. Oram's (1989) finding that changes in frequency distribution of chroma alone establishes the structure of a "tonal hierarchy" would suggest that there is a kind of "tonal buffer" that accumulates the frequency distribution. Krumhansl has been careful to point out that her work is directed toward a *tonal* hierarchy as distinguished from an *event* hierarchy (Bharucha, 1984), and that it is but one facet of tonality. What is missing here is an operationalization of the temporally dynamical nature of tonality as a hierarchy of resolution patterns (see also Deutsch, 1984; Dowling, 1984). Butler (1983, 1989a,b) showed that the temporal ordering of tones is as powerful an influence on tonality as the frequency distribution itself. He introduced the idea of *intervallic rivalry,* stating that "any one tone will suffice as a perceptual anchor—tonal center—until it is supplanted by a better rival" (1989b, p. 427). An additional component of his approach is based on Browne's (1981) observation that rare intervals within the diatonic pitch set, particularly the minor second that occurs in only two cases (scale degrees 3–4 and 7–1 in the major mode), might somewhat unambiguously define a tonal center. H. Brown and Butler (1981) tested the hypothesis with various temporal orderings using short (i.e., three-tone) patterns with different combinations of common and rare intervals. They found that rare intervals did indeed lead to identification of tonal center even with such short patterns, and that temporal ordering of the rare intervals was important; patterns of scale degree 4–7 were more accurately identified as to tonal center than 7–4, for example (see also Butler, 1992, pp. 118–122; Krumhansl, 1990, also noted this asymmetricality in her own studies using the probe-tone technique). Other scholars have suggested that multifaceted approaches to tonality—including probe tones, rare interval studies, and MDS, many of which are presented above—converge on the nature of tonality. All show the relationships among pitch chromas in fairly isolated musical contexts and suggest a tonal center (tonic) and the significance of tonic (1-3-5), subdominant (4-6-1), and dominant (5-7-2) tonal elements (see Cook, 1994; Kendall, 1994; Kendall & Carterette, 1992, for further discussion).

III. MUSICAL GESTALTS

We have seen an emerging realization, even among researchers in intervals, scales, tuning, and tonality, of the importance of temporal factors in musical perception. After all, the patterns that form the substance of music are organized in time. We turn now to research on the fundamental grouping mechanisms through which such temporally organized events can be mentally apprehended and coded. The literature on streaming and grouping of auditory sequences is vast (see Jones & Yee, 1993; Warren, 1993; Bregman, 1990; Deutsch, 1982) and we will touch on only a few areas of particular musical relevance.

A. Fission

Anyone familiar with Baroque music has experienced aural fission and streaming, as a single instrument often produces several perceptually distinct melodic lines. Figure 6 illustrates the idea in musical notation. Gestalt psychologists first called attention to principles underlying figural organization, including proximity, similarity, and common direction, and these appear to operate in musical contexts as well. Frequency proximity is one such example. It is no accident that, over all cultures, pitch interval sizes of five semitones or less account for about 90% of all observed interval sizes (Dowling & Harwood, 1986, Fig. 6.2), and for the difficulty in apprehending a coherent gestalt in the pointillistic music of Anton Webern.

It is well known to musicians that rapid alternations between sequential pitches (a trill) can split apart into separate perceptual streams if the musical interval is larger than approximately three semitones (a minor third—perhaps accidentally about equal to the critical bandwidth in middle frequency ranges). Miller and

FIGURE 6 An example of melodic fission in a Concerto Grooso of G. F. Handel. Even though the notes are played singly, one after another (original), many people begin to hear two streams (top stream and bottom stream).

Heise (1950) determined this trill threshold experimentally, but van Noorden (1975)[6] greatly extended this work.

Figure 7 provides a schematic of the general idea of fission. With long time intervals between tones (IOI—interonset interval), both relatively small and larger frequency differences are perceptually tied together in succession. When the IOI is decreased (the speed or tempo of the pattern is increased), the larger frequency differences split into two perceptual streams, high and low frequencies grouped together, respectively. Van Noorden (1975) manipulated subject attentional set via subject instructions. Some subjects were told to follow the pattern of highs and lows (integrative attending), others were told to focus selectively on the low tones alone (selective attending). The physical independent variables were the IOI in milliseconds and the frequency difference between tones. The point at which the trill percept broke into two separate streams under integrative attending was called the temporal coherence boundary. The temporal coherence boundary was about three semitones (minor third) for IOIs of about 125 ms or less (sixteenth notes at MM = 120, i.e., eight notes/s); fast patterns broke into separate streams when frequency differences were relatively small. As the IOI becomes longer, the slower patterns require much larger frequency separations in order to break into streams. For example, with an IOI of about 200 ms, the temporal coherence boundary is exceeded only by patterns that differ by more than an octave. However, with selective attention focusing on the lower tones alone, the fission boundary remains around two semitones for relatively fast IOIs of 400 ms or less, and then becomes gradually larger in frequency difference up to about 5 semitones at 800 ms. Therefore, when subjects are asked to perceptually hold the tones together, there is a strong interaction of frequency and speed. When subjects are asked to perceptually separate a sequence, they are able to do so over a wide range of speeds as long as there is a frequency separation of a few semitones.

These results suggest that at an intermediate IOI value of 300 ms or so, a listener could hear a four-semitone alternation as either a coherent trill pattern when using selective attention or as separate streams, when using integrative attention. Van Noorden (1975) suggested that, with regard to the temporal coherence boundary, frequency and time coalesce to form a single higher order relation; they are interlocked. Space does not permit restatement of models of attention adequately outlined elsewhere in this handbook, but it is worth noting that van Noorden's conception differs from channel theories where independent capacity limits are postulated for frequency and time (see Jones & Yee, 1993, 78ff). Bregman (1993, p. 24) interpreted these results in a compelling manner:

> When listeners are trying to integrate the sequence, the segregation is involuntary, acting in *opposition* to their intentions. . . . On the other hand, the segregation that occurs when listeners are trying to segregate the sounds, i.e. when the segregation is

[6] See Jones and Yee (1993) for an extensive exposition on van Noorden's (1975) work.

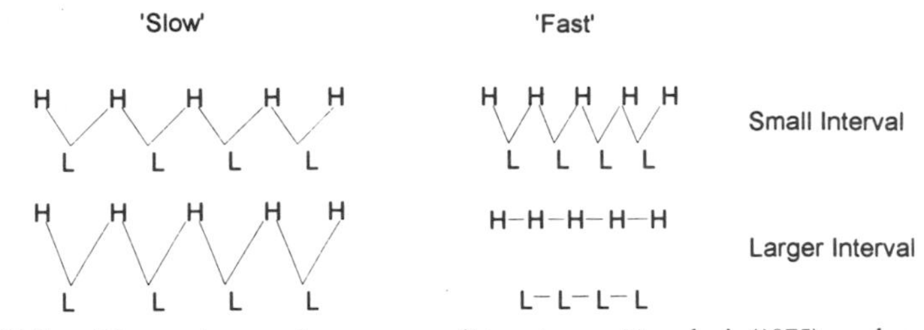

FIGURE 7 Diagram interpreting some conditions in van Noorden's (1975) work on temporal coherence and fission.

> *consistent* with their intentions, is a product of a selection process carried out by attention. (Bregman, 1990, Ch. 4)

Further experiments by van Noorden (1975) investigated the JND in time for the lower tones of the alternating sequence with a base IOI of T ms. The Weber fraction $\Delta T/T$ was small (good discrimination) when subjects could integrate high- and low-frequency tones into a trill. Furthermore, the Weber fraction was not constant with change of T; time discrimination did not conform to Weber's law and was influenced by large changes in frequency. Dowling (1986, p. 157) provided further musical examples of these phenomena based on van Noorden's (1975) work.

One might suspect that if the contrast between frequencies in such alternating sequences could be lessened through means other than interval contraction, fission might be reduced or disappear. Bregman and Dannenbring (1973) studied patterns in which frequency glides were placed between the alternating notes, and demonstrated that fission is thus reduced or disappears.

B. Similarity and Contrast

Similarity implies contrast, that is, sonic events that are similar may be grouped together relative to other, contrasting or dissimilar events. Timbre provides a means of grouping by similarity. Orchestration practice is highly dependent on the fact that some instruments, such as oboe and muted trumpet, sound more similar to each than do others, such as oboe and flute. Melodic passages that extend themselves outside the playing range of one instrument can be continued successfully by the application of another instrument of similar timbre. Substitutions in school band scores readily exploit timbral similarity through the use of cross-cues, which instruct soprano saxophone to play in the absence of oboe or alto saxophone in the absence of French horn. Contrasting instrumental color, in addition to frequency range contrasts, can provide the means to create figure-ground effects. In the music of India, for example, a tambura provides a time-variant, continuous spectral wash as the background for the melodic instrument (see Carterette, Vaughn, & Jairazbhoy, 1989). A similar Western musical concept is that of the *pedal tone,* either a higher or lower frequency sustained tone over which melodic and harmonic material forms the figure.

Bregman and Pinker (1978) showed how fission involving two timbres, a pure tone and a two-component tone, could stream based on timbral similarity, even with relatively small frequency intervals. McAdams and Bregman (1979) demonstrated that perceptual segregation of two pure tones was aided when the third harmonic was added to one of the tones. Warren, Obusek, Farmer, and Warren (1969) conducted an experiment with repeating sequences consisting of four different timbres and frequencies of 200-ms duration. Subjects were unable to name the ordering of the sounds. However, when the sound durations were increased to 500 ms or more, ordering was possible. One can hypothesize that competing aspects of frequency proximity and timbral dissimilarity made rapid sequence orderings difficult to discern. Furthermore, subjects were not familiar with the timbres, and therefore recognition for patterns required more time than that found when repeating verbal sequences must be ordered (Warren & Warren, 1970). In real music, subjects are generally familiar with their culture's timbral and melodic formulae, and cyclical patterns varying in multiple octaves and dissimilar timbres are uncommon (in fact, much of the literature on streaming explores what subjects can be *made* to do, rather than what they naturally do in ecologically valid musical settings). We turn now to additional studies involving multiple cues for grouping.

C. Multiple Cues

It seems clear that when multiple cues for grouping exist, there is some selectivity among perceptual strategies for decoding the musical message. Although one may argue about the mental architecture for this accomplishment, whether it is a series of hierarchical stages involving primordial, Gestalt-like mechanisms that feed higher order processing or some other form, numerous examples from the perceptual literature suggest that there is a feedback relationship between knowledge structures and the strategies used for decoding the musical message (Kendall & Carterette, 1990). Dannenbring (1976) reported experiments with sine-tone glides that alternatively ascend and descend. Gaps of silence are placed at the points of frequency direction change and filled with a block of white noise. Most subjects hear the pitch motion pass through the block of white noise, an example of good continuation. Our experience with the tapes provided by Bregman (and produced at McGill University) for classroom use is that many students can selectively turn on or off their ability to perceive the continuation. Others are influenced by whether one plays the gapped example minus the white noise, where the gaps are clearly present, prior to playing the example with the noise. We report elsewhere (Carterette & Kendall, in press) that the Shepard auditory barber pole illusion, described earlier, may or may not be heard as continuously rising, depending on

FIGURE 8 Musical notation of "Frère Jacques" (filled notes) interleaved with those of "Twinkle, Twinkle" (open notes) in the same pitch range (A) and in different pitch ranges (B). (Dowling, 1986, p. 125.)

the cultural background of the student. You will also recall that in the fission experiments by van Noorden (1975), where streams are based on frequency proximity, subjects could direct their attention to either stream. Some stimulus situations are ambiguous, and higher order cognition can employ many different strategies for resolving the ambiguity. Erickson (1982, p. 528–29) described his composition *LOOPS,* in which cyclically repeating isochronous pitch structures are pointillistically spread between five different instrumental timbres. The relationship of pitch to timbre over time is in conflict, creating an unclear and shifting sense of structure as the listener tracks different features. Deutsch (1982) remarked: "It therefore appears that although there are strong involuntary components in the formation of groupings, ambiguous stimulus situations may be set up were voluntary attention can be the determining factor" (p. 128).

Wessel (1979) provided an example of the interaction of timbre and pitch in the perception of tone sequences. He presented a repeating sequence of three ascending tones, C-E-G (C=major triad) with alternate tones differing in timbre. For small timbre differences, the ascending pattern remained intact. For large timbre differences, however, a *descending* sequence based on the alternate tones (G-E-C) was heard.

An excellent example of voluntary attention guided by knowledge structures or schemata (Bartlett, 1932; Neisser, 1976) is that of Dowling (1973, 1986, p. 125ff). Two common tunes, such as "Frère Jacques" and "Twinkle Twinkle," are played one note after another with the original rhythm altered so that there is an isochronous series at 7.5 tones/s (Figure 8). The tunes are interleaved so that the first note of one tune is followed by the first note of the other and so on. The two tunes are generated either in the same pitch range or different pitch ranges (ca. an octave apart). When in different pitch ranges, a listener can hear one melody or the other, but not both at once. Frequency proximity allows for grouping. When in the same pitch range, differences in loudness, timbre, spatial separation, or any other contrasting feature allow for perceptual grouping in the absence of frequency proximity. Finally, if the listener is told what the tunes are ahead of time, the listener can use this knowledge to search for critical contour cues in the interleaved melodies and through this resolve them, even when there is no other contrasting physical

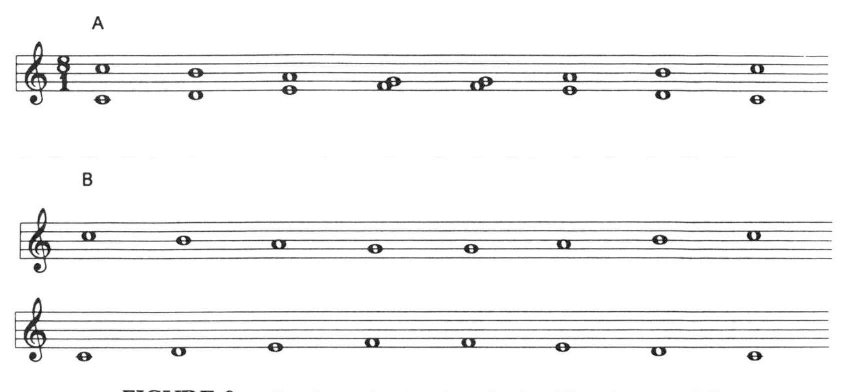

FIGURE 9 Crossing scales showing stimulus (A) and percept (B).

factor. Selective attention is facilitated by prior knowledge of the "scrambled" melodies.

D. Polyphony

We noted the prevalence of fission in the solo string works of Baroque composers, among others. Polyphonic music, composed of two or more simultaneous melodic lines or tunes, is another example of musical practice that depends on selective attention facilitated by elements of perceptual contrast. Often in such music, melodic lines cross, as in Figure 9A. The empirical question is whether one hears two separate melodic lines that pass through one another, guided by the principle of continuity, or whether, near the crossing point, frequency proximity directs attention back in the direction of the original melodic motion. In the latter case, the listener would hear two half-scales (Fig. 9B). Gregory (1994) conducted an extensive study of this figure. When the scale pattern was presented with little or no timbre difference between the scales (timbre differences were quantified via MDS; see Section VI), listeners perceived the upper and lower half-scales. As the timbre difference increased, the crossing-scale pattern emerged. The crossing-scale percept was also aided by temporally offsetting the tones of one line relative to the other (similar to Dowling's interleaving mentioned previously) and playing one scale in a key relatively far along the circle of fifths. These results extend those of Tougas and Bregman (1985), who found that pure tones contrasted with four-tone harmonic complexes led to crossing streams.

Deutsch (1975) conducted a famous experiment using crossing scales[7] played

[7] Many examples in psychological research are analogs, rather than copies, of musical practice. The crossing scale, often used in streaming research, is one example. In real music, polyphonic lines often are distinguished by pitch/time interactions, harmonic/melodic function, as well as orchestration/registration. As was mentioned earlier regarding the perception of polytonality, examples that remove temporal variation greatly limit the generalization of findings to an ecologically valid musical context.

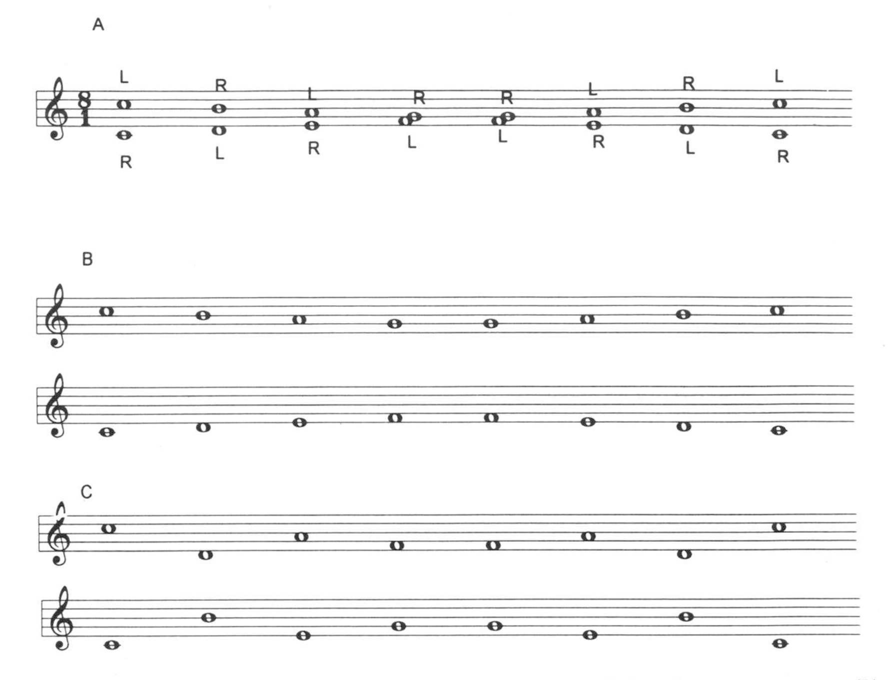

FIGURE 10 A version of Deutch's (1975) scale illusion (A) including the common percept (B) and the less common, more-veridical percept (C).

such that successive notes went to different ears. Figure 10A presents musical notation showing the binaural switching of notes to each ear. This scale illusion or paradox, as it has become known, demonstrates the primacy of grouping by frequency proximity over spatial location for most listeners. Generally, listeners hear a pattern similar to Figure 10B, with the half-octave localized in each ear, rather than a more veridical pattern such as Figure 10C, which is, except for the scale center, organized according to the pitch sequence presented to a given ear. Deutsch (1982, p. 114) found that right-handed listeners tended to perceive the upper half-scale pattern at the right ear, and the lower half-scale pattern at the left ear; left-handed listeners did not exhibit this tendency. Butler (1979a,b) extended this work with other patterns using loudspeakers in a free sound field. He found grouping by frequency proximity over spatial location. Butler (1979a) also switched timbres between ears. Listeners reported that a kind of new, composite quality appeared from both speakers. Kendall (1988) found that, using headphones and loudspeakers, timbres tracked along with pitch, forming the half-scale sequence (Fig. 10B). An oboe sound presented to the left ear and a clarinet sound to the right remained in that ear, even though frequency location was altered to conform to frequency proximity. An exception was when the timbres became very dissimilar because of the addition of time-variant characteristics such as vibrato. Then musicians and nonmusicians alike heard the more veridical pattern of Figure 10C. Davidson, Power, and Michie (1987), using a methodologically controversial priming procedure, found that musical background altered the perception of the ambiguous Deutsch scale pattern. Contemporary composers of art music had a very high rate of perceiving the more veridical pattern (Fig. 10C), as did nonmusicians. This was in contrast to listeners and composers of more traditional classical music. These results suggest that, once again, the perceived figure is the result of a perceptual synthesis based on incoming elements. Perceptual strategies solve the problem provided by the stimulus situation. Although there are a few short examples of composed music that involve competition between spatial location and frequency (e.g., Tchaikovsky's Sixth Symphony), it is not common, and these results mostly have been useful as general explanations and for the composition of electronic music.

IV. MELODY AS PITCH PATTERN

Psychologists have often studied musical melody in terms of isochronous pitch patterns. This has utility for design of experiments, because temporal patterning can be eliminated as a variable. However, it must be admitted that very few real melodies are isochronous strings of pitches.

It also is clear that cognitively a melody is not simply a serial sum of pitches, such that "Frère Jacques" is $C_4+D_4+E_4+C_4+C_4+D_4+E_4+C_4$, and so on. Davies (1969) designed a musical experiment analogous to one conducted by Miller and Selfridge (1953) that used written sentences. Musically trained subjects were asked to write the next note given an antecedent context of one or more notes. It was hypothesized that larger chains of preceding notes would provide "contextual restraint," increasing the organizational cohesion of the resulting pitch patterns. The melodies created from this procedure were given to listening subjects for a recognition task; the more organized melodies based on higher order chains were more easily recognized.

A. Melodic Contour

This type of experiment does not tell us about the fundamental schemas or knowledge structures involved in the perception of melody. You will recall the salient importance of the octave, described previously. What happens if pitch-pattern melodies are created with scrambled octaves? Deutsch (1972) examined this in detail using the first half of the tune "Yankee Doodle." stimuli were unaltered versions, versions with successive pitches randomly varied among three different octaves, and clicks in the rhythm of the tune without pitch. Subjects easily identified the unaltered version, but the scrambled octave version produced poor performance, no better than the clicks-only version. Dowling (1978) found similar

FIGURE 11 Two melodies have the same contour if the sequence of up and down intervals are the same even if the interval sizes differ. (A) and (B) have the same contour, but (C) has a different contour from either (A) or (B).

results, and expanded the work to include the pattern of ups and downs between notes in a melody, known as melodic contour. Figure 11 illustrates the idea of pattern contour. Dowling and Hollombe (1977) found that if listeners were informed of the presence of contour information it helped somewhat, yielding 65% accuracy. If pitch chroma is altered in addition to scrambling of octaves, performance falls to 10% (Kallman & Massaro, 1979). It would appear that both the label and contour help to retrieve a particular pitch-pattern schema, and that pitch chroma is an important element in long-term memory. Play Figure 12 on the piano and determine what tune it is.

Melodic contour dominates the immediate recognition of atonal pitch patterns that are not drawn from a well-established scale. Dowling and Fujitani (1971) had listeners respond to pairs of five-note atonal melodies. The comparison melody was an exact transposition, a same contour imitation, or a different contour comparison (Fig. 13). Listener's could easily recognize transpositions and imitations from the pitch patterns with altered contours, but could not tell them from each other. This demonstrates that in short-term memory recognition tasks listeners use contour similarity. In additional research, Dowling (1978) used tonal, rather than atonal, pitch patterns and found similar results except that tonal transpositions and atonal imitations could be told apart, especially by experienced musicians. Therefore, tonality itself was used as a cue by experienced listeners. Dowling and Fujitani (1971) found that contour was an important feature in the recognition of familiar

FIGURE 12 An octave scrambled melody that is particularly easy to recognize, perhaps because of its redundancy.

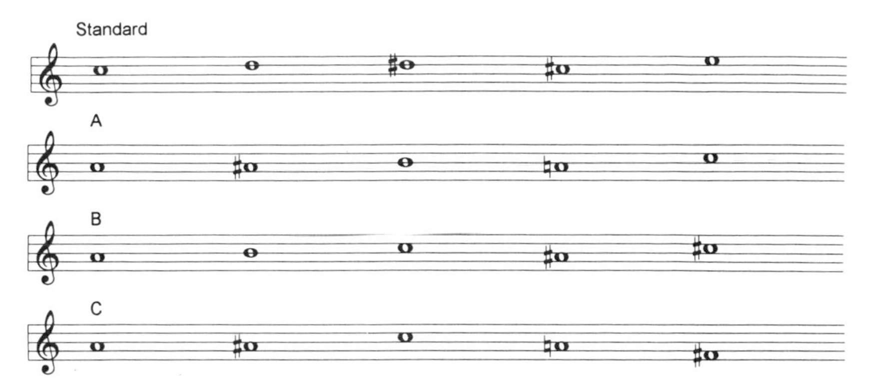

FIGURE 13 "Atonal" melodies similar to those used by Dowling and Fujitani (1971). (A) same contour imitation; (B) exact transposition; (C) different-contour comparison.

pitch patterns as well. Undistorted versions of familiar tunes were recognized almost perfectly, whereas comparisons with distorted contours were recognized at about the chance level of 30%. However, when the distortion preserved contour, recognition performance increased to about 60%. Recognition performance also increased if larger interval sizes in the standard were relatively large in the comparison as well.

B. Key Distance

Another influence on recognition memory for novel melodies is key distance (Bartlett & Dowling, 1980; Cuddy, 1982; Cuddy & Cohen, 1976; Cuddy & Lyons, 1981; Cuddy, Cohen, & Mewhort, 1981). You will recall from the discussion of tonality, above, that one key is related to another mainly through the number of pitches each shares. Two keys that are close to each other on the circle of fifths (Fig. 4, floor) share more pitches than those that are far apart. Therefore, the keys of C- and G-major are relatively close to each other when compared to C and F#. Bartlett and Dowling (1980) investigated key distance in the recognition of pitch patterns. A standard melody was heard followed by a comparison that was either identical to the standard, transposed, or a tonal imitation in either a near or a far key. Results showed that as the key distance of transformation went from same to near to far, it was easier to reject the imitations. This key-distance effect was mainly due to listeners' better rejection of far-key imitations, rather than to better recognition of far-key transpositions. If keys of a standard and comparison are close, both contour and chroma are shared in imitations, so that they are difficult to recognize as different. If the keys of a pair are distant, chroma is not shared; hence the imitation is not confusing. Bartlett and Dowling (1980) suggested that listeners solve the recognition task in part by using schematic scale information.

C. Memory

Although contour dominates the listener's immediate memory for novel melodies, it does not appear to be as important in long-term (semantic) memory, which requires retrieval from among the large number of items stored there. If you try to pick out a well-known melody on the keyboard and make a mistake, you instantly sense a distortion of interval size, even with contour preserved. Bartlett and Dowling (1980) studied immediate recognition of melodic phrases drawn from familiar and unfamiliar folk songs, finding that listeners scored very much better with familiar phrases in telling same-contour imitations and transpositions apart. Dowling and Bartlett (1981) further studied the relationship between long- and short-term memory and contour versus interval information using excerpts from Beethoven's string quartets. Subjects heard sequences made up of themes, repetitions and imitations, and excerpts unrelated to the themes as standards, followed by a 5-min pause. Comparisons were exact replications of the themes, their imitations, and completely different unrelated excerpts. Listeners were told to try to recognize the imitations. Results showed that listeners had 75% accuracy in recognizing exact repetitions, but performed at chance level (50%) at recognizing contour imitations and completely different items. This was surprising because imitation and theme had the same contour, were in close proximity within the piece, and therefore shared similar tempo, texture, and loudness. It appeared that long-term memory was used in recognizing chromas or interval-size patterning but could not be used in recognizing contour. Additional experiments confirmed that interval or chroma information was stored the first few times the melodies were encoded in long-term memory. Interval information was found to be difficult to encode, but was apparently retained with high efficiency in long-term memory. Dowling (1986, p. 142) concluded that listeners probably store in long-term memory a sequence of relative pitch chromas rather than a set of intervals between tones, because when the intervals are distorted but the chromas are left intact, recognition is still possible (Kallman & Massaro, 1979).

Therefore, in subject tasks that require immediate judgment of novel pitch patterns, contour is the dominant cue. But in long-term memory, many melodies share the same contour, and relative pitch chroma information becomes an important feature in retrieval.

As with all research on complex musical topics, there have been additional studies using different methods that suggest caution in accepting all of the results just previously discussed. Croonen and Kop (1989) noted that Bartlett and Dowling (1980) and Dowling and Bartlett (1981) used pitch patterns with many different contour patterns. Boltz and Jones (1986, discussed below) found that as the number of directional contour changes increases, it becomes more difficult for subjects to recognize a sequence; this forms an important element of their work on melodic construction using rule-based operations. In addition, Croonen and Kop (1989) argued about the issue of tonality, offering a concept of tonal clarity: pitch sequences that outline triads of tonally defining progression (e.g., I-IV-V-I) have "clear tonal structure" (see also Cuddy, 1982; Cuddy, Cohen, & Mewhort, 1981, for a similar approach). Croonen and Kop (1989) used five different retention times (1, 5, 8, 15, and 30 s), clear tonal schemes, and two different contour patterns of two or four directional changes. Subjects in their experiments were able to discriminate tonal imitations and transpositions at all retention times, regardless of contour pattern. They concluded that interval information was extracted quickly from sequences with strong tonal cues, and that "one of the major dimensions of tone sequences is not tonality or nontonality, but rather the degree of tonal clarity" (p. 49). Croonen (1994) later expanded on this line of research using a series of 7 or 10 tones, strong or weak tonal structure, and few or many contour reversals (but not including retention time). He found that 7-tone sequences were better recognized than 10-tone series, tonally strong structures were better recognized than weak ones, and that contour complexity did not influence the responses (see Edworthy, 1985, for additional experiments on pitch pattern using many of these variables).

V. RHYTHM

So far we have come in contact only with temporal variables in the discussion of fission. Melody as previously discussed is better termed *pitch pattern* because temporal patterning is absent. Yet, a hundred years ago, William James (1890) noted that a tune could be recognized from its rhythmic pattern alone. The idea that rhythm is tonality's poor relation (Davies, 1978) was prevalent until relatively recently. It is becoming increasingly clear that the fundamental, underlying mechanisms of music cognition are very much concerned with temporal variables. For example, time patterning (rhythm) was found to be a more salient dimension than pitch when subjects rated the similarities of proto-melodies (Carterette, Monahan, Holman, Bell, & Fiske, 1982).

A. Basic Concepts

Notationally, time in music is represented on an ordinal scale from left to right on the musical staff. Specific note durations are represented by different symbols; spacing is not generated according to an interval scale of note length. Events of equal length are spaced evenly on the staff, and patterns of longer and shorter notes are located so the shorter note is further from the longer note than would be the case with isochronous series. This convention is presumably to aid visual parsing of the musical notation. Temporal values can be expressed in terms of IOI and the filled interval of an event. In musical notation, the filled interval is indicated by the note time value; silence is indicated by a rest symbol.

Much music presumes an underlying, periodic pulse train, often called the *beat*. A meter signature, often at the start of a section or composition, indicates beat

FIGURE 14 Musical notation showing temporal patterning in $\frac{4}{4}$ time. Each successive symbol, from whole, to half, to quarter, to eighth, to sixteenth, is in the time ratio of 2:1.

groupings. For example, a meter signature of $\frac{4}{4}$ indicates that four quarter notes are grouped together in a measure demarked by a bar line (numerator) and that a quarter note gets one beat (denominator). Meter, like mode in pitch systems, implies a hierarchical relationship among the beats; some are more important than others. Perceptual salience of musical elements, including beats, is often called an *accent*. In a waltz, for example, the beat pattern is ONE, two, three, ONE, two three, with emphasis (salience/contrast/accent) on beat one. Accents can emerge from any number or combination of musical variables (musicians most commonly think of a dynamic accent, which is a loudness contrast).

Rhythmic patterns are indicated in notation by using symbols for subdivisions as well as compounds of the beat. Figure 14 shows musical notation for a temporal pattern with successive lengths that decrease in the ratio of 2:1 (recall that such a frequency ratio is the octave). Monahan (1984) suggested a parallel between temporal duration ratios and frequency ratios in pitch intervals. In fact, Stevens (1967) had subjects produce temporal durations one-half as long as a standard. He found that the exponent n in the function $\psi = \alpha\phi^n$ was about one, thus that psychological judgments of duration were a linear function of the physical duration values. This conclusion was confirmed by Fraisse (1978) and Bovet (1968), who used successive bisections in order to obtain an interval scale of duration.

B. Tempo

The *tempo* of the music is the rate at which the beat onsets occur relative to Maelzel's metronome, hence a tempo of M.M. 120 for the quarter note indicates 120 beats (quarter notes) per minute; each beat would last 500 ms. Spontaneous or personal tempo refers to a natural speed of a psychomotor response, such as tapping. Fraisse (1982) noted the great interindividual variability of this tempo, which ranges from 200 to 1400 ms; he asserts that 600 ms is the most representative value. This tempo of ca. 1.7 events/s is somewhat slower than march tempo (2 events/s = M.M. 120); however, it fits well with musical intuition about what constitutes a moderate tempo. This value of 600 ms also corresponds with preferred tempi of periodic patterns of lights and sounds (Dowling & Harwood, 1986; Fraisse, 1982). It is interesting to note that, at 600 ms, a listener can detect a change of about 15 ms in spacing, or about 2.5%; you will recall that the JND for pitch was about 1%.

C. Temporal Ratios and Accent

In Western music, temporal ratios of 2:1 constitute close to 90% of counted values of pairs of tones, and the longer of the pair was generally between 300 and 900 ms in length, spanning the preferred and spontaneous tempo of ca. 600 ms (Fraisse, 1982, p. 172). Povel (1981) had subjects tap to repetitions of a repeated rhythmic pattern with intervals that varied over ratios of 4:1 to 5:4. Subjects demonstrated a strong tendency to tap in a 2:1 ratio. As the model patterns approached 5:4, subjects produced incorrect tapping split between the ratios of 2:1 and 1:1. Complex patterns with ratios of 4:4:1 tended to have responses that assimilated the second and third event to the ratio 2:1. Povel argued that the listener encodes a rhythmic pattern in a dual schema, a hierarchy of beat schema with rhythmic subschemas. This "clock" theory assumes a periodic rate within rate.

Vos (cited in Fraisse, 1982) investigated tapping of subjects familiar with classical art music to commercial recordings of Bach preludes. Subjects were asked to tap at the beginning of each perceived rhythmic pattern, presumably indicated in notation by the bar line. In 2/4 meter, 40% of the subjects tapped with the first beat of the measure, 45% with the second beat, and 10% tapped on each beat. For 3/4 meter, 80% tapped every two beats, a result that trained musicians can hardly believe, and 20% tapped on every beat. Subjects tended to tap at intervals shorter than the notated bar length. Fraisse (1982, p. 173) explained these results in terms of limited short-term memory capacity.

The question arises as to what variables influence accent structure for varying rhythmic sequences. Povel and Okkerman (1981) generated isochronous equi-interval sequences and gradually increased the length of every third interval. At first, listeners reported hearing triples of intervals with an accent on the tone ending the lengthened interval (ssl-Ssl-Ssl, where s = short, l = long, capital letters denote accent). As the ratio of the longer to shorter onset approached 2:1, the pattern ssL-ssL emerged, with some weaker accenting still on the tone ending the longer interval. In general, the strongest natural temporal accent tends to appear on tones that begin relatively longer intervals (Povel & Essens, 1985; Povel & Okkerman, 1981). Vos (1977) generated sequences of two and three events where the longer to shorter IOI was at the ratio 4:1. Subjects judged whether the two-duration sequences were iambus (ua; u = unaccented, a = accented) or trochee (au), and for the three-duration sequences whether they were dactyl (auu), anapest (uua), or amphibrach (uau). Vos (1977) concluded that tones that are separated by shorter IOIs are grouped together, that the first tone of a perceptual group is that which follows a long IOI, and that long tones are perceived as accented and shorter ones as unaccented. Thus, amphibrach accenting (uau) was produced by the short-long-short triple of IOIs. This latter point flirts with Gestalt principles and with the work of Garner (1974), who termed a similar finding *the gap principle.* Vos (1977) also found he could create perceptual accents for strings of isochronous IOIs by varying the length of the sound-filled interval. In musical notation, this is similar to a melody played with all notes legato (long-filled) contrasted with the same melody played with some notes legato and others staccato (short-filled). In Vos's case, the long-filled tone sounded accented, with the shorter tone, followed by the longer silent interval, ending the group. Of course, performers have the ability to vary the length of the sound-filled interval in the course of expressive performance, and may thus create accents and grouping cues by varying the articulation, and hence length, of the sound-filled interval.

Povel and Essens (1985) hypothesized a set of rules based on previous research and connected to the clock principle mentioned above. They assumed that the listeners attempt to generate an internal clock on the basis of naturally accented events in equation sequences. The pattern of IOIs determines whether an internal clock can be generated and which internal clock is best, or most likely to be induced. They suggested three important rules for determining the best clock: (1) The best clock must divide the total period of the pattern integrally; (2) the underlying metric (pulse) should avoid coincidence with rests, that is, time intervals that have no event onsets; (3) the metric should also avoid coincidence with intervals that have naturally unaccented events. Therefore *metrical* patterns have a higher order fixed time span, the clock beat period, and *non metrical* patterns fail in this regard. Essens and Povel (1985) tested the clock model using the subject production task of tapping. Metrical and nonmetrical patterns were generated from 50-ms bursts of square-wave signals (830 Hz) with IOIs of ratios 2:1 and 3:1. A metrical pattern, for example, might consist of the time pattern 2-2-3-1-2-1-1-4, where 1 is a single time unit. The overlying metrical pulse is four time units, which corresponds with the rules above—there is no violation of the perceptual hypothesis of a four-time-unit clock. The pattern 3-2-2-1-3-2-3 is nonmetrical; there is no "best clock." Essens and Povel (1985) found that metrical structure facilitated tapping accuracy with temporal groupings based on 3:1 time ratios, as the clock model would predict.

D. Similarity Scaling

Gabrielsson (1973) conducted a series of studies using various monophonic and polyphonic rhythms generated by drums, on piano, and electronically. He uncovered three important groups of rhythmic dimensions: (1) the cognitive-structural dimension, including meter, accent pattern and clarity, and degree of complexity; (2) the movement-motion dimension, which included tempo and rapidity (number of events per unit time), forward motion (longer durations followed by shorter durations yielding a feeling of acceleration), and movement (less clearly defined); (3) the emotional dimension, which distinguished rhythms on affective variables, such as excited, solemn, rigid, and calm. MDS analysis of six rhythmic patterns presented for similarity estimates in pairs produced a solution with Dimension 1 related to the forward motion concept, separating patterns that started with long tones (lsslss; lssll, etc.) versus those that began with short tones (sslll; ssll, etc.). Dimension 2 represented meter, separating duple and triple patterns. Dimension 3 was tempo, and separated slower from faster patterns.

E. Polyrhythms

More than one pattern of accents may be sounded simultaneously, as in many Western orchestral scores (hemiola, or accent patterns of two against three, was a favorite device of Brahms, for example) or the indigenous drum music of West Africa. Handel and Lawson (1983) investigated simultaneous patterns having two or three tones at different frequencies and which repeated at different rates. These patterns are called *polyrhythms.* For example, a high-frequency pattern might repeat twice per metrical group along with a lower frequency pattern that repeated three times in the same time span. This produces a 2 × 3 polyrhythm; note onsets jibe at the start of the metrical group or measure. The subjects were asked to tap along with the pattern in any manner they wished. For slower tempos (ca. 3.0 s) many listeners followed the individual note onsets of the superimposed lines, tapping at every note onset in either line, particularly for 2 × 3, 3 × 4, and 3 × 5 polyrhythms. As the tempo increased to 1.6–1.2 s/measure, tapping shifted to follow one or another of the individual streams. Individual variations of strategy were noted. Some listeners tapped with the lower frequency stream, others chose which pattern to follow based on tempo. Subjects rarely followed patterns with accents more than 800 ms apart. Therefore, a 2 × 3 pattern at 1.6 s per pattern (measure) caused most listeners to follow the shorter interval triple-accent line, rather than the longer interval duple-accent line. Tapping generally tended to regress towards the "preferred" tempo of 600 ms, discussed previously. Therefore, at slow tempos, individual IOIs are tapped to, and at faster tempos a longer clock is imposed on subdivisions, much as Povel (1981) would suggest.

Handel and Lawson (1983) added a third stream to the patterns, creating 2 × 5 × 7 polyrhythms, for example. At tempos too fast to resolve the seven-accent pattern, listeners tended to group the five- and seven-accent streams together and tap to the two-accent stream. However, when the pattern was 2 × 3 × 7, listeners tended to follow the 2 × 3 composite set of accents, in contrast to the seven-accent stream. Results showed the contextual effect of temporal patterning that yielded complex interactions between levels of accent structures. Different subjects displayed different strategies for resolving ambiguous or difficult perceptual tasks, although general trends could be identified.

F. Melody as the Interaction of Pitch with Time

Just as layers of frequency-separated accent structures can be studied as polyrhythms, it is possible to conceive of musical melody as the superposition of pitch-level-accenting based on contour inflection and interval size and temporal accent-

FIGURE 15 Notation of "America." Below that, the rhythmic pattern of a tango. Below that, the combination of the pitch pattern of "America" with the tango rhythm.

ing based on the principles outlined above. A melody in this sense is conceived of as consisting of two superimposed layers of accents—an interaction between pitch and time. This stratification is derived from ideas embodied in Yeston (1976).

Consider the tune "America" (Fig. 15; after Monahan & Carterette, 1985). It is notated in 3/4 meter, and has a pitch contour patterning or accent structure that corresponds to the triple-metric "best clock." The staff underneath this first line, notated in 2/4 meter, is the rhythmic patterning of a tango. Underneath that appears the superposition of the tango rhythmic pattern on the "America" pitch pattern. In essence, there is a 3 (pitch level) × 2 (temporal level) polyrhythm. If you play or sing the resulting composite, it sounds very different from the original (and in fact, in university classes where we have played this almost no one, on first hearing the composite, identifies it with "America").

Now consider "The Star-Spangled Banner," which is notated on the first staff of Figure 16. When most nonmusicians are asked to clap to this as they sing, they do so in patterns of two, as suggested by the second staff of Figure 16, an observation that startles musicians. "The Star-Spangled Banner" has pitch contour inflections and long tone accent points (based on the research outlined above) that do *not* violate strongly a listeners' hypothesis of a duple meter or clock. It is a metrically ambiguous relationship of pitch to time, and this may be one reason why it is so difficult to learn.

Monahan (1984) and Monahan and Carterette (1985) studied pitch and time interaction through a MDS study. Subjects rated the similarity of pairs of brief melodies on a nine-point scale. Four melodies with an increasing number of contour inflections and their inversions were played in each of four rhythmic patterns (ssl, anapest; lss, dactyl; sl, iamb; ls, trochee, see Fig. 17), creating 1024 pattern pairs. MDS and cluster analysis indicated that at least five dimensions were needed for a good accounting of the perceptual space of these melodies. Figure 17 shows the solution for the first two principal dimensions, which are correlated with

FIGURE 16 Two possible interpretations of the metrical structure of the "Star-Spangled Banner."

rhythmic variables. Dimension I separates two-element patterns (iamb and trochee) from three-element patterns (anapest and dactyl). Dimension II separated patterns with initial long tones (trochee and dactyl) from those with initial short tones, and thus, final long tones (iamb and anapest). This result parallels that of Gabrielsson (1973), although Monahan (1984) assumed that the long tones are sources of melodic accent rather than yielding of "forward motion." Dimension III (not shown) separated pitch patterns on the basis of pitch direction, rising versus falling. Dimension IV separated pitch patterns on the basis of the number of contour inflections. Rhythm was more salient than pitch in the similarity of the types of melodies used in this study.

Monahan, Kendall, and Carterette (1987) used a short-term recognition memory task to explore further the interactions of pitch and rhythmic variables. They noted that most theorists and researchers have "concentrated on the organization of

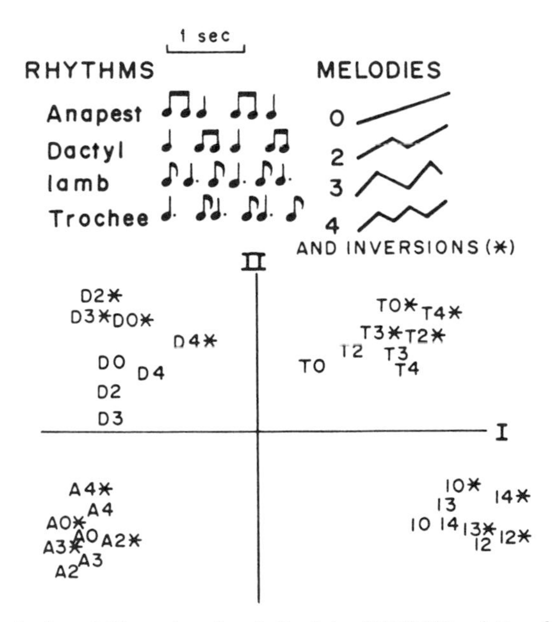

FIGURE 17 A plot of Dimensions I and II of the INDSCAL solution from Monahan and Carterette, 1985. See text for explanation.

pitches or pitch intervals by rule, and not on pitch-level accent as a basis for determining melodic grouping or simplicity" (p. 580). The rule-based approach is based on the observation that melodic patterns are often highly redundant. For example, "Three Blind Mice" consists of a descending pitch pattern with the rhythm pattern of s-s-l. This is repeated at a higher pitch level, with a small rhythmic variation. These four measures that form the first phrase are very similar to one another. Dowling and Harwood (1986, p. 179) show how a tree can model this pitch structure. Deutsch and Feroe (1981) produced an involved symbolic system that includes alphabets of the pitch material and operations on those alphabets. Presumably, the less complicated (shorter alphabet, fewer operations) a sequence is, the more easily it is remembered. Jones (1981) and Boltz and Jones (1986) produced a rule system with group theoretic properties that are applied to pitch alphabets—rules such as identity, complement, reflection, and transposition. However, these approaches are largely driven a priori by an aesthetic of symmetrical, elegant, formal design, not from the experimental literature. Dowling and Harwood (1986) pointed out that melodies are rarely as symmetrical as the systems of rules or homogeneous tree structures would suggest. Trees in particular can appear visually balanced and symmetrical, when they are in fact the relationships they subsume are decidedly not so (see other arguments in Kendall & Carterette, 1990, p. 134). Monahan et al. (1987) chose to combine ideas regarding sources of pitch pattern and rhythmic accent outlined above. Essentially, strata of pitch-level accent are superimposed on temporal accents. The resulting relationships between strata suggest the ease or difficulty of melodic perception and coding.

Four nine-tone pitch patterns varying in contour pattern and redundancy were combined with three-interval temporal patterns for the first and last three tones (dactyl, anapest, amphibrach, termed *rhythmic frames*) and these patterns plus four silent-interval rest versions for the center three tones (termed *center rhythms*) were used in all combinations of rhythmic frames, center rhythms, and pitch patterns. In essence, pitch and temporal variables could be either periodic, aligned and nested (consonant), periodic at the same interval yet never aligned (out of phase), or polyrhythmic (dissonant). Figure 18 shows a schematic of these three relationships.

Subjects heard a standard melody and an octave-raised comparison that was either identical or had one pitch changed at a given position in the nine-tone series. Subjects indicated whether the comparison was the same as the standard. Areas under the memory-operating characteristic were subject to statistical analysis.

Results showed that significantly more accurate responses were made to the consonant conditions than either the out-of-phase or dissonant conditions. Pitch inflection (contour direction change) was more salient than unidirectional pitch skips in signaling accent. Some evidence was obtained suggesting that musicians use regular patterning early in consonant patterns to predict pitch differences at points of contour inflection. Musicians and nonmusicians shared only 21% of the total variance, suggesting different perceptual strategies. Monahan et al. (1987) found some evidence that there were interactions between fonal function and the other variables; this indicates the enormous multidimensional complexity of even simple nine-tone patterns such as they employed. Recently, Lipscomb and Kendall (1995) applied the idea of strata of consonant, out-of-phase, and dissonant accent structures to pitch patterns and their alignment with visual animations. They found that subjects could reliably tap to visual accent strata. Changing direction, size, color, and shape all served as reliable periodic accent points. Melodic accents were similarly investigated and then combined. Pilot studies show that consonant relations are easier to code and respond to than dissonant or out-of-phase relationships. This work forms the basis for studying the important but neglected syntactical relationships between visual accent and musical accent in film.

VI. MUSICAL TIMBRE AND ORCHESTRATION

Orchestration is "the employment of . . . various sonorous elements, and their application" according to Berlioz (1856, p. 4). Only relatively recently have experimental studies begun to address musically valid questions. Although there is a significant body of work on the psychoacoustics of timbre, which manipulates the components of the Fourier equation, relatively little work has been done using natural instrument timbres with their time-variant spectra; even less study has been directed toward how natural instrument timbres function in musical contexts. However, the ready availability of computing technologies promises to be a catalyst to increasing the ecological validity of timbre studies.

A. Signal Partitions

Most experiments in orchestral timbre have used single, isolated musical tones. As Handel (1989) noted "The results from studies on the accuracy of instrument identification are inconsistent and reinforce the notation that single-tone presentation may be a poor context" (p. 240). Important studies in this area include Berger (1964), Clark, Luce, Abrams, Schlossberg, and Rome (1963), Grey (1975), Saldanha and Corso (1964), and Wedin and Goude (1972). In absolute identification of instruments using verbal labels, correct answers range across the four studies from 33 to 85%, and the ordinal rankings of different instrument identification accuracies are not even the same. However, the isolated tones were of different lengths across these studies, and Berger (1964) used pitch chroma F_4, Wedin and Goude (1972) used A_4, and Grey (1975) used Eb_4. Saldanha and Corso (1964) included variables of frequency, vibrato, presence of transients, and length of steady state and found a significant main effect across these variables. The percentage of correct identifications at F_4 (43%) was significantly higher than those at either C_4 (37%) or A_4 (38%).

Many of these studies included a condition that severed the attack and/or decay portions (transients) of the acoustical signal. Removal of decay transients did not significantly affect identification. However, absence of the attack transient universally led to poorer identification. In Grey's (1975) MDS analysis of ca. 300 ms, Eb_4 signals from 16 orchestral instruments, the "cut attack" condition is separated from all others in the perceptual space. Hence, the idea that severing attacks from instrumental signals degrades identifiability became a universally accepted notion.

However, the definition of *attack* in these studies most often was a constant time value, which ignores the differences in attack characteristics among woodwind, brass, and bowed string signals. Furthermore, the lack of musical context is a concern. Kendall (1986) addressed these issues, and systematically studied the contribution of acoustical signal partitions in the identification of clarinet, violin, and trumpet performances. Musical phrases of three folk songs were employed. Two phrases were digitally edited, creating six conditions: unaltered signal, steady state with gaps (preserving overall duration), steady state with elision, transients alone, static steady state (replications of a representative single period) with transients, and static steady state without transients. These same conditions were created for isolated tones drawn from the first note of the phrases, affording comparisons to previous work. These conditions permitted convergence on the necessary and sufficient conditions for instrument categorization. Transients were operationally defined in terms of a three-part hierarchical set of criteria, and thus varied in length from instrument to instrument and note to note. A matching paradigm using the third folk song was employed. Data from single tones produced listener response accuracies similar to those found in previous studies; response accuracy in single-tone presentations was worse for the conditions without attack transients. For musical phrases, mean identification of the unaltered and natural steady state alone conditions were statistically equivalent; these means were statistically higher than the transients alone and the static steady-state means. Therefore, Kendall (1986, p. 208) concluded that "transients are not necessary for the categorization . . . [they are] in addition, not sufficient for the categorization of the three instruments." The mean responses to whole phrases was significantly higher than to single notes, suggesting that important information was provided by the musical context. How-

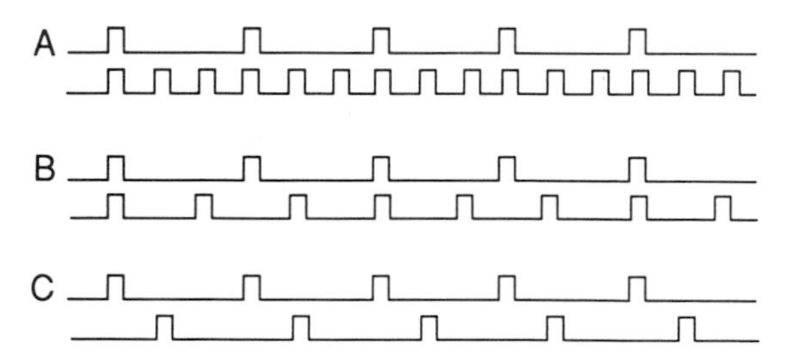

FIGURE 18 Stratification of two accent structures: (A) consonant; (B) dissonant; (C) out-of-phase.

ever, musical context provided no additional information for listener categorizations with transients or static spectra alone.

There is some evidence that acoustic signal components have varying importance by instrument. Strong and Clark (1967), using synthetic tones, interchanged spectral and temporal envelopes among instruments. They concluded (p. 285) that in tones for which the spectral envelope was unique (e.g., oboe, clarinet) the spectral envelope was more important than the temporal envelope. For tones in which the spectral envelope was not unique (e.g., flute, trombone), the temporal envelope was of equal or greater importance. Hajda, Kendall, & Carterette (1995) found that temporal envelope was very important for the perception of impulse-envelope tones, and noted that scaling studies (see below) that incorporate both impulse and continuant tones have solutions that are biased in favor of temporal attributes.

B. Multidimensional Scaling

Wedin and Goude (1972) provided an early example of the application of MDS techniques to instrumental timbres. All possible pairs of natural tones (A_4) for oboe, trumpet, violin, cello, clarinet, flute, bassoon, French horn, and trombone, with and without transients, were presented to subjects who rated the degree of similarity. It should be noted that, as in many studies of this type, the pitch A_4 is high for the cello, bassoon, and trombone; this is not their most characteristic tessitura. Figure 19 is a plot of the two-dimensional solution from their data ($N = 36$; 92.5% of the variance accounted for). Kendall and Carterette (1993a, p. 496) noted that this scaling corresponds well to their interpretation of dimensional axes in their broad range of studies (see below). The first dimension is, left to right, nasal (e.g., oboe) versus not nasal or rich (e.g., French horn); the second dimension is, top to bottom, reedy (e.g., clarinet) versus brilliant (e.g., trumpet). Relative strengths of spectral components mapped into the perceptual space most strongly, including successively decreasing intensity of the upper partials, and a low fundamental intensity with increasing intensity of the first few overtones (p. 239). This is, in essence, a spectral centroid, defined mathematically as:

$$\frac{\Sigma f_n A_n}{f_o \Sigma A_n},$$

where f_n is the frequency of the nth partial, A_n is the amplitude of the nth partial, and f_o is the frequency of the fundamental. High spectral centroids have more energy in upper partials relative to the fundamental. Low spectral centroids have more energy in lower partials, particularly the fundamental, relative to upper partials. As we shall see, evidence points to spectral centroid as one of the principal physical variables that maps onto the timbral dimension for steady-state, instrumental signals.

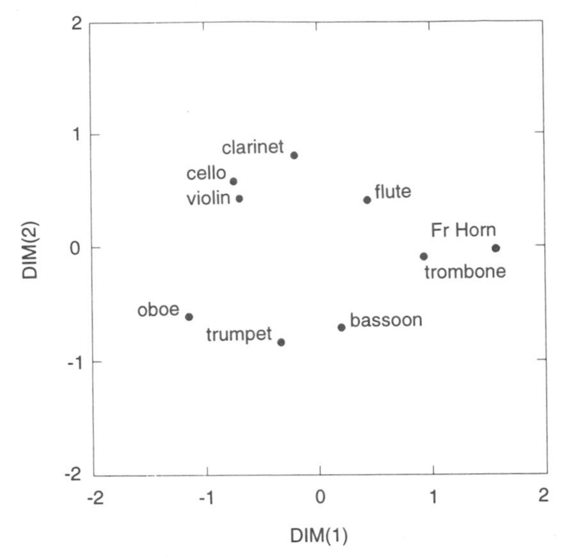

FIGURE 19 Two-dimensional timbral configuration obtained using classical MDS on the data of Wedin and Goude (1972).

Wessel (1973) also scaled the timbral similarities of 1.5-s stimuli from three brass (trumpet, French horn, trombone), woodwind (oboe, clarinet, bassoon), and string (violin, viola, cello) instruments (ca. Eb_4). In the two-dimensional solution, onset characteristics mapped into the first dimension and spectral centroid (called *brightness* by Wessel) onto the second. Trumpet, oboe, violin, and viola all had higher centroids than bassoon, clarinet, French horn, trombone, and cello. This largely confirms the results of Wedin and Goude (1972). Using synthesized instrument tones, Miller and Carterette (1975) varied attack and decay of amplitude-time envelope, number of partials, and the relative onset time of partials. MDS of similarity judgments of the stimuli revealed dimensions that mapped onto the number of partials, the overall attack envelopes (e.g., piano, brass, smooth), and relative onset times, again confirming the role of spectral content and time-variant envelope functions as primary determinants of timbre.

Grey (1975) conducted an extensive, multifaceted study of musical timbre, primarily directed toward finding timbral differences among natural, complex resynthesis based on many analysis frames of data, and data-reduced signals, based on 6–8-line segments for each partial. Sixteen instrumental sounds (Eb_4 ca. 311 Hz) were employed, which were equalized psychophysically in duration, pitch, and loudness. The physical durations were from 280–400 ms and were thus quite brief. It is instructive to note the limitations of these signals, because they have become, for worse rather than better, a canonical set of stimuli for modern timbre studies.[8]

[8] For a further discussion of problems with the Grey (1975) stimuli and study, see Kendall (1986) and Kendall and Carterette (1991).

Grey (1975) conducted a discrimination study of the natural versus complex resynthesis and line-segment data-reduced signals, and claimed that it was difficult to discriminate between complex resynthesis and line-segment resynthesis. However, the mean discrimination of 62.7% (Hajda, 1995) indicates the contrary (chance was 50%). The synthetic versions of some instruments, such as soprano saxophone, were discriminated from the natural condition nearly 90% of the time. Grey (1975) argued that the high level of discrimination may have been due to tape hiss signatures, but this never has been fully investigated. Instead, researchers have routinely used the line-segment data-reduced synthetic signals, equalized for duration, loudness, and pitch, as representations of natural instrument signals.

Grey (1975) had subjects rate the similarity of all pairs of the line-segment data-reduced stimuli, and subjected the data to individual differences scaling (INDSCAL). The resulting three-dimensional solution has been reproduced dozens of times (e.g., Butler, 1992, p. 132; Dowling and Harwood, 1986, p. 77; Handel, 1989, p. 236), thus we will not do so here. The first dimension separates out muted trombone and oboe from French horn and cello *sul tasto,* and is related to the spectral energy distribution (spectral centroid), as in previous research. The second dimension separated flute and cello (normally bowed) from the unidentified saxophones and clarinet, and is related to the synchronicity of the upper harmonics and general amount of spectral fluctuation. The third dimension separated trumpet and muted trombone and the clarinets and cello *sul tasto,* and appears to be related to the presence of high-frequency energy in the attack versus those without this characteristic. Hajda (1995, p. 27) noted that Grey's configuration does not appear to corroborate those found by Wessel (1973), because trumpet and bassoon are adjacent in Grey's (1975) configuration and maximally distant in Wessel's (1973). Kendall and Carterette (1993a) noted that their data corresponds more closely with Wedin and Goude (1972) than with Grey's (1975).

C. Timbral Combinations and Timbral Intervals

Recently, Kendall and Carterette (1989, 1991) extended the MDS technique to combinations of orchestral instruments, hoping to increase the ecological validity of such studies. Five wind instruments (alto saxophone, oboe, flute, trumpet, and clarinet) were played by professional instrumentalists in duet, creating ten dyads. Musical contexts were Bb_4 unisons, unison melodies (D_5-Eb_5-F_5-D_5), major thirds (Bb_4-D_5), and harmonized melodies. Each instrument was used as the soprano in the homophonic contexts, creating six musical contexts. Subjects rated timbral similarity of all possible pairs of loudness-equalized examples within each context. Then the resulting mean matrices for each context were entered as "subjects" into INDSCAL analysis, producing a composite representation of the timbral relations among the ten wind instrument dyads. Figure 20 shows the resultant two-dimensional configuration, which approaches a circumplex. In terms of verbal attributes, Dimension 1 was, from left to right, nasal versus not nasal or rich; Dimension 2, top to bottom, was reedy versus brilliant (see Kendall and Carterette,

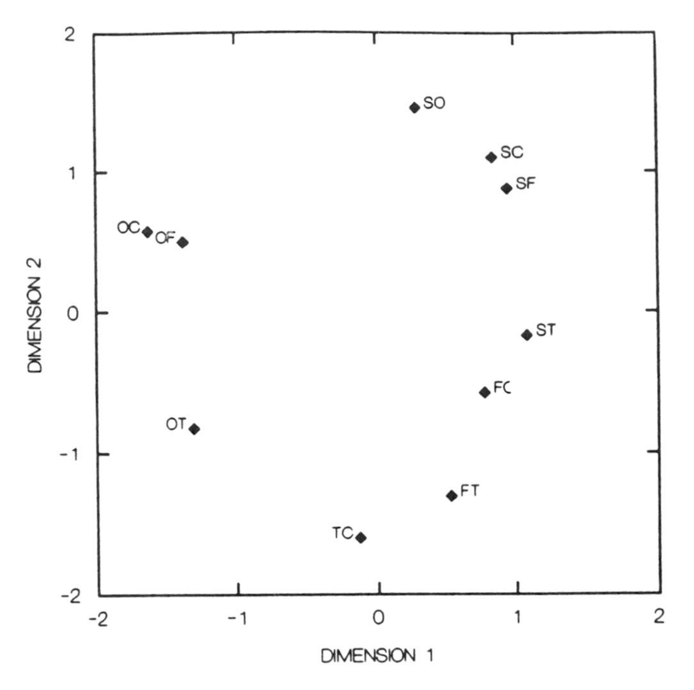

FIGURE 20 Timbral circumplex for wind instrument dyads. S, saxophone; O, oboe; F, flute; T, trumpet; C, clarinet. (From Kendall & Carterette, 1991, p. 389.)

1993a, for extensive verbal attribute experiments). For example, adding an oboe to another instrument increased a dyad's nasality; adding alto saxophone increased its richness and reediness. The ordinality of dyads on Dimension 1 was correlated with the spectral centroid; Dimension 2 mapped onto the mean coefficient of variation, a quantification of the degree of time variance in spectral trajectory. These results confirm, quantitatively rather than phenomenologically, aspects of Wedin and Goude (1972) and Wessel (1973) above, this time with regard to simultaneously sounding instruments.

Kendall and Carterette (1991) also analyzed the vector algebraic properties of their perceptual space (Fig. 20), treating the individual points in the space as point vectors and connecting points with line vectors. They decomposed the dyads into individual instruments, producing plots of single instruments, and created new spaces using theorized positions for instruments not included in the original study, such as English horn. They then had a highly trained music professor place points in a space labeled with the verbal attributes. They found a high correlation between the decomposed dyad space and the professor's placement of instruments using timbral imagery. Their other experiments (Kendall & Carterette, 1993a,b) provide additional support for the veracity of the timbral circumplex.

Ehresman and Wessel (1978) forwarded the idea of *timbral analogies,* where timbre intervals are constructed according to their positions in multidimensional space. For example, one can set up an analogy: timbre A is to timbre B, therefore timbre C is to timbre X, where X is a choice from a pair of possibilities. McAdams and Cunible (1992) extended the idea and used timbre vectors to determine likely X candidates using a parallelogram geometry. They found that, with their "naturalistic" synthesized tones, both musicians and nonmusicians were able, more or less, to respond to timbres that fit the analogy according to the parallelogram model. This provides additional evidence that timbre vectors in perceptual space are a good model for creating and manipulating timbre intervals.

The concept of blend in orchestration was studied by Kendall and Carterette (1993b) using the stimulus materials of their previous experiments, discussed above. Blend was operationalized as the subject judgment, on a continuous response scale, of the degree of oneness versus twoness of a stimulus. Music-major subjects' data across contexts ordered the dyads of Figure 20 largely according to their position on Dimension 1. "Nasal" combinations such as oboe–flute, saxophone–oboe, and oboe–clarinet were less blended than "brilliant" or "rich" dyads, such as flute–trumpet, flute–clarinet, and trumpet–clarinet. Additional experiments requiring verbal identification of the constituents showed a moderate, inverse relationship between identification and blend (−.731 for unisons), which indicated that as blend increased, identification became poorer. A relatively large spectral centroid of one instrument relative to another yields poor blend, a conclusion similar to that of Sandell (1989), who arithmetically combined Grey's (1975) stimuli.

The last fifteen years has seen the development of new commercial musical genera based on synthesized and emulated sound. Keyboards, integrated piano keyboards and various digital sound-producing components, have become ubiquitous and cheap. New Age music is well known for combinations of synthesized, emulated (produced from samples of real instruments or voices), and live performers. Music for television and radio programs has benefited from the low-cost performance ratio and high flexibility of individuals who compose, arrange, and perform in their home electronic music studios entire scores using emulated orchestras.

It is worth noting that a keyboard, no matter how ingeniously conceived, is not a clarinet or violin. The performer's degrees of freedom are not at all the same. A violinist can vary bow velocity, pressure, and position. A clarinetist can vary wind velocity and pressure and the manner in which his or her fingers move from note to note. These performer degrees of freedom are not identical to keyboard pressure and velocity sensitivity. The expressive dynamic range may be greater than, or at least different from, emulated instruments relative to natural instruments. Perhaps that is one reason why "unplugged" (acoustic) guitar performances in rock music have reemerged recently.

In addition, the electronic keyboard is connected to the acoustical environment through loudspeakers (or headphones). In contrast, natural instruments have coup-

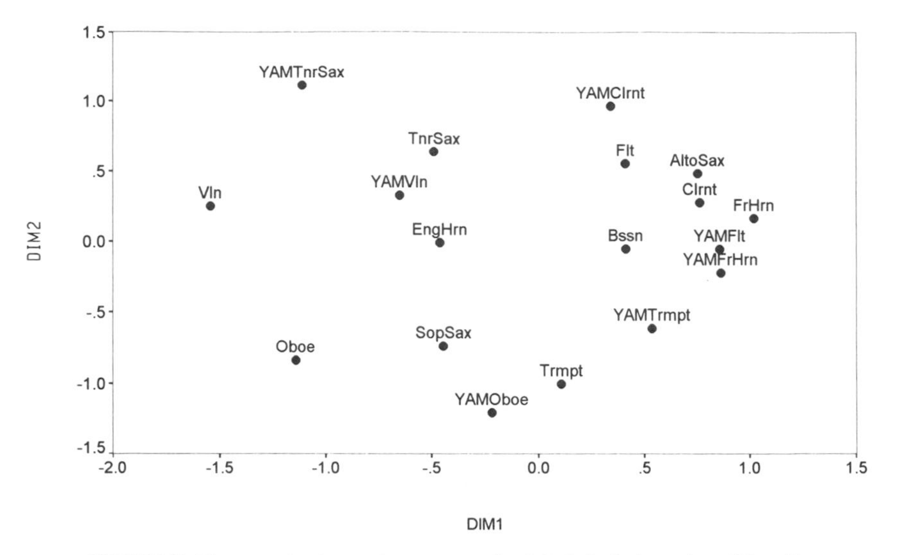

FIGURE 21 MDS of natural versus Yamaha DX-7 similarity ratings ($N = 8$).

lings between their air columns and the environment. It is not stretching to say that natural clarinet timbre is dependent on the performer–environment interaction, including the radiation pattern from the instrument that provides feedback for the performer.

Interesting studies have begun on the relationship between emulated or synthesized instrumental timbres and natural timbres. In one study (Kendall, Carterette, & Hajda, 1994), eleven natural instruments (flute, oboe, clarinet, soprano, alto, and tenor saxophones, French horn, English horn, violin, bassoon, and trumpet) were played (ca. Bb_4) by professionals in a concert hall. A professional synthesist emulated these instruments on a Yamaha DX-7 (FM synthesis), Roland D-50 (hybrid synthesis), and E-mu EMAXII/Proteus 2 (sampling) module. Listeners were asked to rate the degree of similarity of natural and counterpart tones (among other tasks). Figure 21 presents the results for natural versus Yamaha DX-7 timbres. Again, as in other work described earlier, Dimension 1 is nasal (−2) versus rich (+2), and Dimension 2 is reedy (+2) versus brilliant (−2). Therefore, you can see that the Yamaha DX-7 tenor saxophone is both too reedy and nasal to be in closer proximity to the natural instrument, and the Yamaha DX-7 oboe may not be reedy or nasal enough. In fact, mismatches along Dimension 1 (nasality) can be attributed to mismatches between synthetic instruments and their natural counterparts in terms of spectral centroid.

VII. MUSICAL COMMUNICATION

A. Affect and Meaning

If music communicates, what does it mean? We can suggest three types of musical meaning, adapted from Meyer (1956): (1) *Referentialism,* the association of music with external objects and events, such as the use of national anthems in the "1812 Overture"; (2) *Formalism,* the meaning of music that arises from explicit knowledge about musical concepts and ideas, such as the structure of sonata form; (3) *Expressionism,* the meaning of music that arises from the tensions and releases embodied within the music itself. Meyer (1956) coined the term *embodied meaning* to suggest the power of music to refer to relationships (patterns) within itself. His concept of emotion is connected to this idea: When a tendency to respond is blocked or inhibited, affect results. Composers, therefore, "play" with and manipulate the expectations of listeners. They set up situations, internal to the music itself, where tension is created based on these expectations. Unexpected or unusual resolution of these "tendencies to respond" lead to affect. Of course, this approach has a parallel in experimental aesthetics that employs information theory—much redundancy or extreme complexity would not yield a situation where a tendency to respond can be inhibited. Meyer (1956) illustrated his ideas using Gestalt principles applied to musical notation. Expectancy is an important component of Narmour's (1990) influential notational theory. Carlsen (1981) and Unyk and Carlsen (1987) conducted experiments where two-tone intervals were presented and had respondents extend these prototypical beginnings. They found general agreement on the "best" extensions; however, these varied from culture to culture. Unyk and Carlsen (1987) also conducted a melodic-dictation exercise and found that listeners made more errors in dictation when melodies violated their expectations. Schmuckler (1989) found that both listeners who rated continuations of melodic and harmonic patterns and performers who improvised continuations of patterns had consistent melodic and harmonic expectancies. The role of these expectations in affect or emotion is less clear.

Affective response to music has been studied using physiological measurements of heart rate, galvanic skin response, muscle tension, and breathing since the late nineteenth century (Radocy & Boyle, 1979). Dainow (1977) conducted a critical review of these studies and found little evidence to support an isomorphism between musical variables, such as tempo and heart rate, for example. The lack of prediction in these types of physiological studies may explain why few studies in the area have been conducted since the mid-1940s.

Mood responses to music are often operationalized by adjective checklists or the semantic differential. Checklist procedures have been used since Gatewood (1927) and recently have been rediscovered (Namba, Kuwano, Hatoh, & Kato, 1991). The seminal work in this area is by Hevner (1935), who developed a circular grouping of 67 adjectives into eight mood clusters (later factor analyzed and redesigned by

Farnsworth, 1954). From subjects' responses to recorded music, she concluded that the major mode was *happy, graceful,* and *playful* whereas the minor mode was *sad, dreamy,* and *sentimental.* Firm rhythms were *vigorous* and *dignified,* flowing rhythms were *happy, graceful, dreamy,* and *tender.* Consonant harmonies were *happy, graceful,* and *lyrical,* and dissonant harmonies were *exciting, agitating,* and *vigorous* although inclined toward *sadness* (Hevner, 1936).

The major–minor distinction was studied among children fairly recently (Kastner & Crowder, 1990). Thirty-eight children ages 3–12 listened to 12 musical passages in a counterbalanced design of unharmonized and harmonized major and minor modes. Children pointed to one of four faces that spanned happy to sad. Results showed that the conventional happy–sad association with major and minor mode was instilled in even the youngest children.

R. Brown (1981) conducted a study, part of which was directed toward subtle distinctions in mood. He called the study "Twelve Variations on Sadness." Musical pieces theorized to span a space of sadness, from *reflective* to *funereal,* were presented to 32 nonmusicians and 22 musicians. Subject response tasks included grouping the actual musical examples into like pairs and adjective-cluster checklists. Results indicated poorer than chance performance for both musicians and nonmusicians at the pairing task, and only musicians scored marginally above chance on selecting the right adjective cluster. One must conclude that mood associations are broadly gauged, relatively imprecise, and vary greatly from individual to individual.

B. Performance and Expression

Although the study of performance has accelerated recently as the result of available computer technology, particularly the MIDI, few studies of the communication process between composer, performer, and listener can be found. The scientific study of performance largely has focused on analysis of timing patterns of pianists (e.g., Clarke, 1988; Palmer, 1989; Shaffer, Clarke, and Todd, 1986; Sloboda, 1983). In general these studies show that deviations from clock time are suggested by structural features in music notation, such as phrase boundaries, and that multiple performances, even when separated by a long time, are remarkably consistent. Clynes (1983) suggested that there exist "composer's pulses," patterns of timing deviations that correspond to a given composer's style. Even if music structure dictates performer-generated patterns of expressive timing, one would suspect that the great invention and chronological development of a composer's style would interact greatly with a "composer's pulse," and Repp (1989) provided perceptual data that suggest that Clynes's (1983) approach is at least an oversimplification.

Seashore (1938, p. 247) had a pianist perform the first 25 measures of Chopin's Nocturne, op. 27, no. 2 in "artistic" and "attempted metronomic" time. Patterns of accumulated measure and phrase durations were similar between the two renditions. In general, attempts by pianists to play metronomically (without expression) reveal internal schema that react to musical features and are outside of volitional control, thereby generating decidedly nonmetronomic timing profiles (Kendall & Carterette, 1991; Palmer, 1989).

The study of the process of musical communication among composer, performer, and listener is perhaps the best manifestation of ecologically valid research in music perception and cognition. Unfortunately, the number of studies is limited, perhaps because of the difficulties involved with natural music contexts. Ecologically valid musical contexts involve an audience that is actively participating in the communication process; such an intrusion would no doubt reduce the ecological validity. Natural contexts also involve musically trained composers and performers, who must be persuaded (and often paid) for their participation. Uncontrolled and interactive variables in natural contexts are legion. In short, real musical contexts and their analogs are, at every level, an expensive resource in experimentation.

The study of performer contributions to musical perception and cognition is also limited. This is unfortunate, as the performer is a central figure in the process of musical communication. The performer does not merely translate musical notation into acoustical signal, but adds his or her own message directed towards the listener. Unfortunately for the research scientist, the schemata that guide the performer in musical interpretation is largely at the level of implicit, rather than procedural or explicit, knowledge. No amount of introspection allows the performer to bring to conscious awareness the vast mental resources that are used so effortlessly and naturally. Perhaps it is for this reason that a mystique has enveloped the skilled performer (as well as the composer). In fact, many performers eschew scholarly research into their domain, and often musical conservatories sublimate psychomotor over cognitive skills.

Ethnomusicologists have noted the prevalence of oral tradition for passing on knowledge in non-Western cultures (see Malm, 1977). They should be reminded that most instruction in Western performance takes place via oral tradition. It is not possible to write down in a book the procedures for becoming a great performer. Most instruction in Western music schools takes place one-on-one; modeling is extremely important; much of the instruction is nonverbal.

Any performer knows that there is an interaction between cognition of musical structures and psychomotor schema. For example, try singing "America" starting at the thirteenth note! Mental rehearsal using long-term memory must be employed; there is no way simply to "index" the perceptual store at the thirteenth element. Instrumentalists are aware of how easy it is to encode incorrect (relative to notation) psychomotor schemas for technically challenging passages. These schema are often accurate at salient points in the phrasal structure, but inaccurate in between. Only slow, conscious repetitive rehearsal can correct the performance errors.

Sloboda (1985) noted that "performance is the result of an interaction between a mental *plan* which specifies features of the intended output and a flexible programming system" (p. 89). This might explain why a performer can know how a passage should sound but be unable to play it correctly—the mental plan is fine but the motor program is inadequate. Sloboda (1985) provided additional anecdotal evidence about performance skills that suggest many areas for additional research.

Some research has been conducted on performer reading of the musical score. Weaver (1943) demonstrated that musical texture governs the sequence of eye fixations in pianists' score reading. The performer cannot read both staves of the piano score at once. In chordal (homophonic) music, the eye fixation pattern moves vertically, scanning from top to bottom the pitches of a chord, before moving to the top note of the next chord horizontally. In polyphonic music with two or more melodic lines, the eye fixation pattern moves first horizontally, capturing a perceptually significant phrase of melodic material, before returning and scanning the subsequent melodic line that is notated underneath the first. Recent extensions of this work (Goolsby, 1994a,b) using modern equipment show that eye movement is reduced when performing melodies with more concentrated visual information, and that skilled music readers look farther ahead in the notation than the unskilled. Unskilled readers scan on a note-by-note basis using long fixations, parsing individual characteristics of the notation such as note heads and stems. Skilled sight readers, on the other hand, direct attention to all areas of the notation, including blank regions of the visual field.

Nakamura (1987) investigated the communication of dynamics (patterns of loudness) between performer and listener. An example of Baroque music (Sonata in G Major by Willem de Fesch) was interpreted by performers on oboe, violin, and recorder. Each performer noted their score with dynamic markings. Listeners were given a score with various possible dynamic marking patterns. As they listened to the music, their task was to pick among the possible dynamic patterns, or reject any of the suggestions and generate their own response. Results indicated that crescendos (gradual loudness increases) were easier to play and recognize than decrescendos (loudness decreases). Perception of a crescendo was enhanced by rising pitch, and perception of a decrescendo was enhanced by falling pitch. Even though dynamic markings on a musical score mean relative, not fixed, intensity levels, a performer's intent to observe the markings was conveyed fairly well to the listeners.

Senju and Ohgushi (1987) investigated the ability of a performer to communicate a playing style to listeners. A violinist played Mendelssohn's Violin Concerto in ten different interpretations characterized by expressive terms: weak, powerful, bright, sad, sophisticated, beautiful, dreamy, fashionable, simple, and deep. Listeners used semantic differentials to rate each of the stylistic performances. An MDS analysis of the data revealed that, on the whole, the player's intention was communicated to the listener.

Campbell and Heller (1979) investigated the ability of a cellist and a violist to communicate detailed notational changes in dynamics, bowings, fingerings, and articulations. They found that musicians could identify 79, 65, and 50% of normal, amplitude compressed, and synthesized notational interpretations, respectively.

Recently, Kendall and Carterette (1990) conducted an integrated study of the

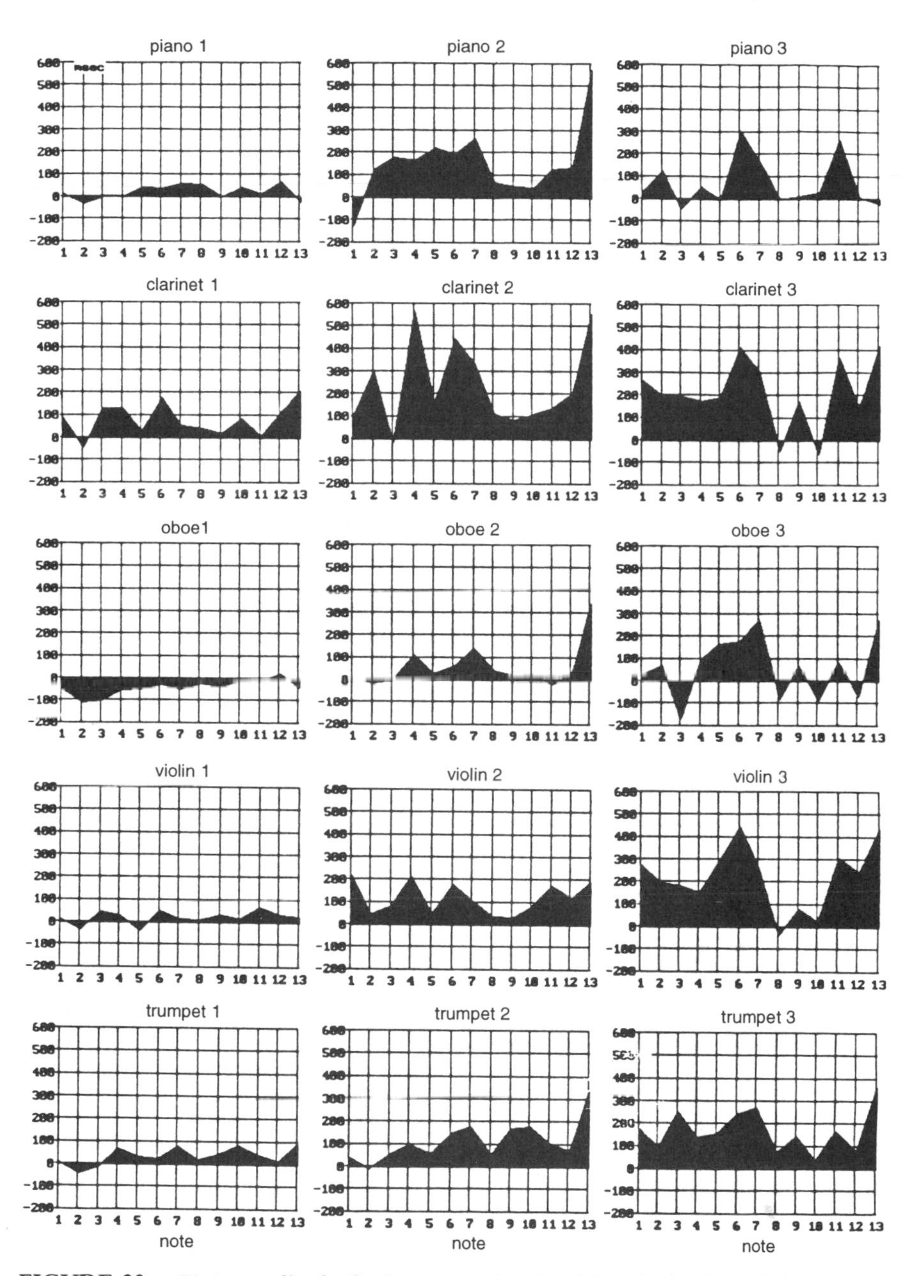

FIGURE 22 Timing profiles for five instruments (rows) and three levels of intended expression (columns): mechanical, appropriate, and exaggerated. A performance that adhered to clock time would produce a straight line at the 0 coordinate on the y axis. Values are in millisecond deviation from clock time.

communication of musical expression between performer and listener. In their study, an internationally recognized pianist performed a short musical phrase at three levels of expressiveness: without expression, with appropriate expression, and with exaggerated expression. Professional artists on clarinet, trumpet, violin, and oboe listened to these examples, notated their music, and created their own versions of the three expressive performances. Listeners were asked to categorize the performances according to perceived expressiveness. Both musicians and nonmusicians matched the performer-intended level of expression with good accuracy (mean = 65%, range 43–81%), a finding confirmed under other experimental conditions. Acoustical signals of the performances were analyzed and produced the internote timings of the different levels of expressiveness. Figure 22 shows a comparative plot of the timing patterns. In the two expressive levels note times are virtually always longer than ontological times, and the basis of expressivity appears to be patterned contrasts in the lengthening and shortening of notes. An important finding, which can be verified by visual inspection of Figure 22, was the near-zero correlation among the three timing patterns produced by the virtuoso pianist. On the other hand, moderately large correlations were found between timing profiles of the other professional, yet less skilled, performers. This suggests that a significant capability of the virtuoso is the ability to vary expressive timing patterns at will so as to produce highly unique expressive variations, yet remain cognizant of musical features. Of course, more study with a larger number of skilled pianists of international stature could confirm this hypothesis. Finally, synthesized performances based only on the timing profiles of Figure 22 were accurately categorized by listeners into the three expressive levels, which points to the importance of timing deviation patterns in communicating expressive intent.

VIII. CODA

The use of ecologically valid contexts stimuli has become commonplace as the technology for the manipulation of music has made extraordinary advances. Difficult areas of research such as musical performance, orchestration, and multimedia art are now approachable. Yet, several research biases, which we have discussed here, are prevalent, as is evident from the large number of studies devoted to musical constructs like pitch, scale, and tonality, and in contrast to the small number of studies devoted to such crucial issues as timbre, affect, aesthetics, and acoustical–physiological correlates of perception and cognition.

There is a pervasive belief in academia that the manipulation of highly developed notational symbol systems—musical, mathematical, connectionist, grammatical—illuminates fundamental principles and processes. A deep problem with most of these systems is that there is no linking of frames of reference—only the most superficial attention is given in these symbol systems to the mapping of acoustic, physiological, perceptual, and cognitive processes across domains. Also, when a concept is not well understood, the systematist assigns the problem to the domain of brain function.

Additional biases stem from a Western, Euro-centric music-theoretic tradition. Therefore, there are far more studies using Western scales, tonality, orchestration, and performance than those which take a comparative stance. There is great value in taking a comparative stance because important aspects of context—personal, cultural (political, social, religious, and the like)—that influence cognitive schemata can be analyzed. We have noted previously (Kendall & Carterette, 1990) that music conceived as structure is continuously deformable and varies in response to the social–cultural context. This domain is certainly amenable to empirical investigation and surely deserves more attention. (For further discussion of these issues from a comparative stance, see Carterette & Kendall, in press.)

Music itself arises from the fundamental properties of human cognition and is not arbitrary. Human information processing capacity is limited as codified in notions like The Magical Number Seven Plus or Minus Two (Miller, 1956)[9], as instantiated in the fact that world musics are generally restricted to using no more than seven functional pitch classes. Because biological cognitive systems must operate on low information-transmission rates, categorization in music of such concepts as timbres, instrument classes, melodic formulas and contours, tonal center, rhythms, and styles is fundamental, just as is the case in speech for concepts like vowels, phonemes, intonation, and source identity. In essence, perception samples data streams from the world at differing rates and according to many different criteria.

For music, abstraction is the rule; areferential and self-referential structures abound. At one end of the continuum is the content (semantic), at the other end is the form (syntactic). Different art forms mix the referential and the areferential in different ways. Music emphasizes pure, abstract areferential patterns—form rules over content. Speech emphasizes the referential or indexical, in short, the content. We believe that it is now time to return to the rigorous study of musical meanings—including affective, emotive, aesthetic, and cultural.

References

Balzano, G. (1980). The group-theoretic description of 12-fold and microtonal pitch systems. *Computer Music Journal, 4*(4), 66–84.

Bartlett, F. C. (1932). *Remembering.* Cambridge: Cambridge University Press.

Bartlett, J. C., & Dowling, W. J. (1980). The recognition of transposed melodies: A key-distance effect in developmental perspective. *Journal of Experimental Psychology: Human Perception & Performance, 6,* 501–515.

[9] Note, however, that Miller was very aware that his rule applied to single continua and that much larger amounts of information can be processed for combinations of several independent dimensions. Generalizing this principle to musical scales has problems because, first, musical scales are multidimensional; second, melodies in Western music are interactions of pitch and time alphabets; and third, most Western music has in theory both vertical (chords, harmony) and horizontal (melody, rhythm) components. The implied research problem is to dissect the interacting alphabets in order to determine how and why we end up with seven-plus-or-minus-two pitch classes.

Bengurel, A., & Westdal, C. (1991). Absolute pitch and the perception of sequential musical intervals. *Music Perception, 9*(1), 105–119.

Berger, K. W. (1964). Some factors in the recognition of timbre. *Journal of the Acoustical Society of America, 36*(10), 1888–1891.

Berlioz, H. (1856). *A treatise on instrumentation and orchestration.* Mary Cowden Clarke (Trans). London: Novello.

Bharucha, J. J. (1984). Event hierarchies, tonal hierarchies and assimilation: A reply to Deutsch and Dowling. *Journal of Experimental Psychology: General, 113,* 421–425.

Bharucha, J. J., & Stoecking, K. (1986). Reaction time and musical expectancy: Priming of chords. *Journal of Experimental Psychology: Human Perception and Performance, 12,* 403–410.

Boltz, M., & Jones, M. R. (1986). Does rule recursion make melodies easier to reproduce? If not, what does? *Cognitive Psychology, 18*(4), 389–431.

Bovet, P. (1968). Echelles subjective de durées obtenues par une méthode de bissection. *Anée Psychologique, 68,* 23–26.

Bregman, A. S. (1990). *Auditory scene analysis: The perceptual organization of sound.* Cambridge, MA: The MIT Press.

Bregman, A. S. (1993). Auditory scene analysis: Hearing in complex environments. In S. McAdams and E. Bigand (Eds.). *Thinking in sound* (pp. 10–36). Oxford: Clarendon Press.

Bregman, A. S., & Dannenbring, G. L. (1973). The effect of continuity on auditory stream segregation. *Perception & Psychophysics, 13,* 308–312.

Bregman, A. S., & Pinker, S. (1978). Auditory streaming and the building of timbre. *Canadian Journal of Psychology, 32,* 19–31.

Brown, H. (1988). The interplay of set content and temporal context in a functional theory of tonality perception. *Music Perception, 5,* 219–250.

Brown, H., & Butler, D. (1981). Diatonic trichords as minimal tonal cue-cells. *In Theory Only, 5* (6–7), 37–55.

Brown, R. W. (1981). Music and language. In *Documentary report of the Ann Arbor Symposium* (pp. 244–265). Reston, VA: Music Educator's National Conference.

Browne, R. (1981). Tonal implications of the diatonic set. *In Theory Only, 5*(6–7), 3–21.

Burns, E. M. (1974). In search of the shruti. *Journal of the Acoustical Society of America, 56* (S), 25–26.

Burns, E. M. (1981). Circularity in relative pitch judgments for inharmonic complex tones: The Shepard demonstration revisited, again. *Perception & Psychophysics, 30*(5), 467–472.

Burns, E. M., & Ward, W. D. (1978). Categorical perception-phenomenon or epiphenomenon: Evidence from experiments in the perception of melodic musical intervals. *Journal of the Acoustical Society of America, 63*(2), 456–468.

Burns, E. M., & Ward, W. D. (1982). Intervals, scales, and tuning. In D. Deutsch, (Ed.), *Psychology of music* (pp. 241–269). New York: Academic Press.

Butler, D. (1979a). A further study of melodic channeling. *Perception & Psychophysics, 24,* 264–268.

Butler, D. (1979b). Melodic channeling in a musical environment. *Research Symposium on the Psychology and Acoustics of Music,* University of Kansas.

Butler, D. (1983). The initial identification of tonal centers in music. In J. Sloboda & D. Rogers, (Eds.), *Acquisition of symbolic skills* (pp. 251–261). New York: Plenum.

Butler, D. (1989a). Describing the perception of tonality in music: A critique of the tonal hierarchy theory and a proposal for a theory of intervallic rivalry. *Music Perception, 6,* 219–242.

Butler, D. (1989b). Mapping the listener's recognition of tonality: A delineation of the theory of intervallic rivalry. In *Proceedings of the First International Conference on Music Perception and Cognition* (pp. 425–430). Kyoto: Japanese Society of music Perception and Cognition.

Butler, D. (1992). *The musician's guide to perception and cognition.* New York: Schirmer.

Campbell, W., & Heller, J. (1979, February). *Judgments of interpretations in string performance.* Paper presented at the Research Symposium on the Psychology and Acoustics of Music, University of Kansas, Lawrence, Kansas.

Campbell, W., & Heller, J. (1981). Psychomusicology and psycholinguistics: Parallel paths or separate ways? *Psychomusicology, 1*(2), 3–14.

Carlsen, J. (1981). Some factors which influence melodic expectancy. *Psychomusicology, 2,* 12–29.

Carterette, E. C., & Kendall, R. A. (1994). On the tuning and stretched octave of Javanese gamelans. *Leonardo Music Journal, 4,* 59–68.

Carterette, E. C., & Kendall, R. A. (in press). Comparative music perception and cognition. In Deutsch, D. (Ed.), *The psychology of music* (2nd Ed.). San Diego, CA: Academic Press.

Carterette, E. C., Kendall, R. A., & DeVale, S. C. (1993). Comparative acoustical and psychoacoustical analyses of gamelan instrument tones. *Journal of the Acoustical Society of Japan (E), 14*(6), 383–396.

Carterette, E. C., Monahan, C. B., Holman, E., Bell, T., & Fiske, R. A. (1982). Rhythmic and melodic structures in perceptual space. *Journal of the Acoustical Society of America, 72,* S11.

Carterette, E. C., Vaughn, K., & Jairazbhoy, N. A. (1989). Perceptual, acoustical and musical aspects of the Tambura drone. *Music Perception,* 7(2), 75–108.

Castellano, M. A., Bharucha, J. J., & Krumhansl, C. L. (1984). Tonal hierarchies in the music of North India. *Journal of Experimental Psychology: general, 113,* 394–412.

Cazden, N. (1945). Musical consonance and dissonance: A cultural criterion. *The Journal of Aesthetics and Art Criticism, 4,* 3–11.

Cazden, N. (1962). Sensory theories of musical consonance. *Journal of Aesthetics and Art Criticism, 20,* 301–319.

Chaloupka, V., Mitchell, S., & Muirhead, R. (1994). Observation of a reversible, medication-induced change in pitch perception. *Journal of the Acoustical Society of America, 96*(1), 145–149.

Clark, M., Luce, D., Abrams, R., Schlossberg, J., & Rome, J. (1963). Preliminary experiments on the aural significance of parts of tones of orchestral instruments. *Journal of the Audio Engineering Society, 11*(1), 45–54.

Clarke, E. C. (1988) Generative principles in music performance. In J. A. Sloboda (Ed.), *Generative processes in music* (pp. 1–26). Oxford: Clarendon Press.

Clynes, M. (1983). Expressive microstructure in music, linked to living qualities. In J. Sundberg (Ed.), *Studies of music performance* (pp. 76–181). Stockholm: Royal Swedish Academy of Music, #39.

Cook, N. (1994). Perception: A perspective from music theory. In Rita Aiello and John Sloboda (Eds.), *Musical perceptions* (pp. 64–95). Oxford: Oxford University Press.

Croonen, W. L. M. (1994). Effects of length, tonal structure and contour in the recognition of tone series. *Perception & Psychophysics, 55,* 623–632.

Croonen, W. L. M., & Kop, P. F. M. (1989). Tonality, tonal scheme and contour in delayed recognition of tone sequences. *Music Perception, 7,* 49–68.

Crummer, G., Walton, J., Wayman, J., & Hantz, E. (1994). Neural processing of musical timbre by musicians, nonmusicians, and musicians possessing absolute pitch. *Journal of the Acoustical Society of America, 95*(5), 2720–2727.

Cuddy, L. L. (1982). On hearing pattern in melody. *Psychology of Music, 10*(1), 3–10.

Cuddy, L. L., & Cohen, A. J. (1976). Recognition of transposed melodic sequences. *Quarterly Journal of Experimental Psychology, 33,* 148–157.

Cuddy, L. L., Cohen, A. J., & Mewhort, D. J. K. (1981). Perception of structure in short melodic sequences. *Journal of Experimental Psychology: Human Perception & Performance,* 7, 869–882.

Cuddy, L. L., & Lyons, H. I. (1981). Musical pattern recognition: A comparison of listening to and studying tonal structures and tonal ambiguities. *Psychomusicology, 1*(2), 15–33.

Cynx, J. (1993). Auditory frequency generalization and a failure to find octave generalization in a songbird, the European starling (*Sturnus vulgaris*). *Journal of Comparative Psychology, 107*(2), 140–146.

Dainow, E. (1977). Physical effects and motor responses to music. *Journal of Research in Music Education, 25,* 211–221.

Dannenbring, G. L. (1976). Perceived auditory continuity with alternately rising and falling frequency transitions. *Canadian Journal of Psychology, 30,* 99–114.

Davidson, B., Power, R. P., & Michie, P. T. (1987). The effects of familiarty and previous training on perception of an ambiguous musical figure. *Perception & Psychophysics, 41*(6), 601–608.

Davies, J. B. (1969). *An analysis of factors involved in musical ability, and the derivation of tests of musical aptitude.* Unpublished doctoral dissertation, University of Durham, Great Britain.

Davies, J. B. (1978). *The psychology of music.* Stanford: Stanford University Press.

Deutsch, D. (1972). Octave generalization and tune recognition. *Perception & Psychophysics, 11,* 411–412.

Deutsch, D. (1975). Two-channel listening to musical scales. *Journal of the Acoustical Society of America, 57,* 1156–1160.

Deutsch, D. (1982). *The psychology of music.* New York: Academic Press.

Deutsch, D. (1984). Two issues concerning tonal hierarchies: Comment on Castellano, Bharucha, and Krumhansl. *Journal of Experimental Psychology: General, 113,* 413–416.

Deutsch, D. (1986). A musical paradox. *Music Perception, 3,* 275–280.

Deutsch, D. (1987). The tritone paradox. *Perception & Psychophysics, 41*(6), 563–575.

Deutsch, D., & Feroe, J. (1981). The internal representation of pitch sequences in tonal music. *Psychological Review, 88,* 503–522.

Deutsch, D., Moore, F. R., & Dolson, M. (1984). Pitch classes differ with respect to height. *Music Perception, 2,* 265–271.

Deutsch, D., Moore, F. R., & Dolson, M. (1986). The perceived height of octave-related complexes. *Journal of the Acoustical Society of America, 80,* 1346–1353.

Dowling, W. J. (1973). The perception of interleaved melodies. *Cognitive Psychology, 5,* 322–337.

Dowling, W. J. (1978). Listeners' successful search for melodies scrambled into several octaves. *Journal of the Acoustical Society of America, 64,* S146.

Dowling, W. J. (1982). Musical scales and psychophysical scales: Their psychological reality. In T. Rice & R. Falck (Eds.), *Cross-cultural perspectives on music* (pp. 20–28). Toronto: University of Toronto Press.

Dowling, W. J. (1984). Assimilation and tonal structure: Comment on Castellano, Bharucha, and Krumhansl. *Journal of Experimental Psychology: General, 113,* 417–420.

Dowling, W. J. (1986). *Music cognition.* New York: Academic Press.

Dowling, W. J., & Bartlett, J. C. (1981). The importance of interval information in long-term memory for melodies. *Psychomusicology, 1,* 30–49.

Dowling, W. J., & Fujitani, D. S. (1971). Contour, interval, and pitch recognition in memory for melodies. *Journal of the Acoustical Society of America, 49,* 524–531.

Dowling, W. J., & Harwood, D. L. (1986). *Music cognition.* New York: Academic Press.

Dowling, W. J., & Hollombe, A. W. (1977). The perception of melodies distorted by splitting into several octaves: Effects of increasing proximity and melodic contour. *Perception & Psychophysics, 21,* 60–64.

Drobisch, M. (1852). Über musikalische Tonbestimmung und Temperatur. In Abhandlungen der Könglich sächsischen Gesellschaft der Wissenschaften zu Leipzig. Leipzig: S. Hirzel. [On the stability of musical pitch and temperature. In *Treatises of the Royal Saxony Academy of Sciences, Leipzig.*]

Edworthy, J. (1985). Melodic contour and musical structure. In P. Howell, I. Cross, & R. West (Eds.), *Musical structure and cognition* (pp. 169–188). London: Academic Press.

Ehresman, D., & Wessel, D. (1978). Perception of timbral analogies. *IRCAM Technical Report 13.* Paris, France: Centre Georges Pompidou.

Erickson, R. (1982). New music and psychology. In D. Deutsch (Ed.), *The psychology of music* (pp. 517–536). New York: Academic Press.

Essens, P. J., & Povel, J. D. (1985). Metrical and nonmetrical representations of temporal patterns. *Perception & Psychophysics, 37,* 1–7.

Evans, C. (1993). Recognition of contentment cell spectral characteristics by mallard ducklings: Evidence for a consistent perceptual process. *Animal Behaviour, 45*(6), 1071–1082.

Farnsworth, P. R. (1954). A study of the Hevner adjective list. *Journal of Aesthetics and Art Criticism, 13,* 97–103.

Fraisse, P. (1978). Time and rhythm perception. In E. C. Carterette & M. P. Friedman (Eds.), *Handbook of perception: Vol. 8. Perceptual coding* (pp. 203–254). New York: Academic Press.

Fraisse, P. (1982). Rhythm and tempo. In D. Deutsch (Ed.), *The psychology of music* (pp. 149–180). New York: Academic Press.

Gabrielsson, A. (1973). Similarity ratings and dimensional analysis of auditory rhythm patterns. I and II. *Scandinavian Journal of Psychology, 14*(2 and 3), 138–160, 160–176.

Garner, W. R. (1974). *The processing of information and structure.* Potomac, MD: Erlbaum.

Gatewood, E. L. (1927). An experimental study of the nature of musical enjoyment. In M. Schoen (Ed.), *The effects of music* (pp. 78–103). New York: Harcourt Brace.

Gerszo, A., & Boulez, P. (1988). Computers in music. *Scientific American, 258*(4), 44–50.

Goolsby, T. W. (1949a). Eye movement in music reading: Effects of reading ability, notational complexity, and encounters. *Music Perception 12*(1), 77–96.

Goolsby, T. W. (1949b). Profiles of processing: Eye movements during sight-reading. *Music Perception 12*(1), 97–123.

Gregory, A. H. (1994). Timbre and auditory streaming. *Music Perception, 12*(2), 161–174.

Grey, J. M. (1975). *An Exploration of Musical Timbre.* Report STAN-M-2, CCRMA, Department of Music, Stanford University.

Hajda, J. M. (1995). *The relationship between perceptual and acoustical analyses of natural and synthetic impulse signals.* Unpublished master's thesis, University of California, Los Angeles.

Hajda, J. M., Kendall, R. A., & Carterette, E. C. (1994). Perceptual and acoustical analyses of impulse tones. In I. Deliege (Ed.), *Proceedings of the 3rd International conference on Music Perception and Cognition* (pp. 315–316). Liege, Belgium: European Society for the Cognitive Sciences of Music.

Handel, S. (1989). *Listening.* Cambridge, MA: MIT Press.

Handel, S., & Lawson, G. R. (1983). The contextual nature of rhythmic interpretation. *Perception & Psychophysics, 34,* 103–120.

Helmholtz (1863/1954). *Die Lehre von den Tonempfindungen als physiologische Grudlage für die Theorie der Musik.* F. Vieweg & Sohn, Braunschweig. English translation by A. J. Ellis, On the sensations of tone as a physiological basis for the theory of music. Reprinted by Dover Publications, New York, 1954.

Hevner, K. (1935). Expression in music: A discussion of experimental studies and theories. *Psychological Review, 42,* 186–204.

Hevner, K. (1936). Experimental studies of the elements of expression in music. *American Journal of Psychology, 48,* 246–268.

Houtsma, A. J. M. (1968). Discrimination of frequency ratios. *Journal of the Acoustical Society of America, 44,* 383 (A).

Hulse, S., Page, S., & Bratten, R. (1990). Frequency range size and frequency range constraint in auditory perception by European starlings (*Sturnus vulgaris*). *Animal Learning & Behavior, 18*(3), 238–245.

Hutchinson, W., & Knopoff, L. (1978). The acoustic component of Western consonance. *Interface, 7,* 1–29.

James, W. (1890). *The principles of psychology* (Vol. 1). New York: Holt.

Jairazbhoy, N. A., & Stone, A. W. (1976). Intonation in present-day North Indian classical music. *Journal of the Indian Musicological Society, 7,* 22–35.

Jones, M. R. (1981). A tutorial on some issues and methods in serial pattern research. *Perception & Psychophysics, 30*(5), 492–504.

Jones, M. R., & Yee, W. (1993). Attending to auditory events: The role of temporal organization. In S. McAdams & E. Bigand (Eds.), *Thinking in sound, the cognitive psychology of human audition* (pp. 69–112). Oxford: Clarendon Press.

Jordan, D., & Shepard, R. (1987). Tonal schemas: Evidence obtained by probing distorted musical scales. *Perception & Psychophysics, 41*(6), 489–504.
Kallman, H. J., & Massaro, D. W. (1979). Tone chroma is functional in melody recognition. *Perception & Psychophysics, 26,* 32–36.
Kameoka, A., & Kuriyagawa, M. (1969). Consonance theory Part II: Consonance of complex tones and its calculation method. *Journal of the Acoustical Society of America, 45,* 1460–1469.
Kastner, M., & Crowder, R. (1990). Perception of the major/minor distinction: IV. Emotional connotations in young children. *Music Perception, 8*(2), 189–202.
Kendall, R. A. (1986). The role of acoustic signal partitions in listener categorization of musical phrases. *Music Perception, 4*(2), 185–214.
Kendall, R. A. (1988, February). *Perceptual strategies in music cognition.* Paper and audio tape presented at the Los Angeles Chapter of the Acoustical Society of America and Audio Engineering Society, Sherman Oaks, CA.
Kendall, R. A. (1994). Issues and problems for theory and research in musical science. In I. Deliege (Ed.), *Proceedings of the Third International Conference for Music Perception and Cognition* (pp. 25–27). Liege, Belgium: European Society for the Cognitive Sciences of Music.
Kendall, R. A., & Carterette, E. C. (1989). Perceptual, verbal, and acoustical attributes of wind instrument dyads. *Proceedings of the 1st International Conference on Music Perception and Cognition* (pp. 365–370). Kyoto, Japan: Japanese Society of Music Perception and Cognition.
Kendall, R. A., & Carterette, E. C. (1990). The communication of musical expression. *Music Perception, 8*(2), 129–164.
Kendall, R. A., & Carterette, E. C. (1991). Perceptual scaling of simultaneous wind instrument timbres. *Music Perception, 8,* 369–404.
Kendall, R. A., & Carterette, E. C. (1992). Convergent methods in psychomusical research based on integrated, interactive computer control. *Behavior Research Methods, Instruments, & Computers, 24*(2), 226–231.
Kendall, R. A., & Carterette, E. C. (1993a). Verbal attributes of simultaneous wind instrument timbres: I. von Bismarck's adjectives; II. Adjectives induced from Piston's orchestration. *Music Perception, 10*(4), 445–468; 469–502.
Kendall, R. A., & Carterette, E. C. (1993b). Identification and blend of timbres as a basis for orchestration. *Contemporary Music Review, 9*(1 & 2), 51–67.
Kendall, R. A., Carterette, E. C., & Hajda, J. M. (1994). Comparative perceptual and acoustical analyses of natural and synthesized continuant timbres. In I. Deliege (Ed.), *Proceedings of the Third International Conference for Music Perception and Cognition* (pp. 317–318). Liege, Belgium: European Society for the Cognitive Sciences of Music.
Kessler, E., Hansen, C., & Shepard, R. (1984). Tonal schemata in the perception of music in Bali and in the West. *Music Perception, 2*(2), 131–165.
Krumhansl, C. (1979). The psychological representation of musical pitch in a tonal context. *Cognitive Psychology, 11,* 346–374.
Krumhansl, C. (1990). *Cognitive foundations of musical pitch.* New York: Oxford University Press.
Krumhansl, C., Bharucha, J., & Kessler, E. (1982). Perceived harmonic structure of chords in three related musical keys. *Journal of Experimental Psychology: Human Perception and Performance, 32,* 96–108.
Krumhansl, C., & Kessler, E. (1982). Tracing the dynamic changes in perceived tonal organization in a spatial representation of musical keys. *Psychological Review, 89,* 334–368.
Krumhansl, C., & Schmuckler, M. (1986). The *Petroushka* chord. *Music Perception, 4,* 153–184.
Krumhansl, C., & Shepard, R. (1979). Quantification of the hierarchy of tonal functions within a diatonic context. *Journal of Experimental Psychology: Human Perception and Performance, 5,* 579–594.
Langer, S. K. (1951). *Philosophy in a new key* (2nd ed.). New York: New American Library.
Lerdahl, F. (1988). Cognitive constraints on composition systems. In J. A. Sloboda (Ed.), *Generative processes in music: The psychology of performance, improvisation, and composition* (pp. 231–259). New York: Oxford.

Lipps, T. (1883). *Psycholgische Studien*. Heidelberg: Weiss.

Lipscomb, S. D., & Kendall, R. A. (1995). Sources of accent in musical sound and visual motion. In I. Deliege (Ed.), *Proceedings of the Third International Conference for Music Perception and Cognition* (pp. 451–452). Liege, Belgium: European Society for the Cognitive Sciences of Music.

Lundin, R. (1953). *An objective psychology of music* (2nd ed.). New York: Wiley.

Lynch, M. P., & Eilers, R. E. (1991). Children's perception of native and nonnative musical scales. *Music Perception, 9*(1), 121–132.

Lynch, M. P., Eilers, R. E., Oller, K. D., Urbano, R. C., & Wilson, P. (1991). Influences of acculturation and musical sophistication on perception of musical interval patterns. *Journal of Experimental Psychology: Human Perception and Performance, 17*(4), 967–975.

Malm, W. P. (1977). *Music cultures of the Pacific, the Near East, and Asia* (2nd ed.). Englewood Cliffs, NJ: Prentice-Hall.

Martin, D. W. (1962). Musical scales since Pythagoras. *Sound, 1,* 22–24.

Mason, J. A. (1960). Comparison of solo and ensemble performances with reference to Pythagorean, just, and equi-tempered intonations. *Journal of Research in Music Education, 8,* 31–38.

McAdams, S., & Bregman, A. (1979). Hearing musical streams. *Computer Music Journal, 3*(4), 26ff.

McAdams, S., & Cunible, J. (1992). Perception of timbral analogies. *Philosophical Transactions of the Royal Society, London, Series B,* 383–389.

Meyer, L. B. (1956). *Emotion and meaning in music.* Chicago: University of Chicago Press.

Meyer, M. (1898). Zur Theorie der Differnztöne and der Gehorseempfindugen überhaupt. In C. Stumpf (Ed.), Volume 2, *Beiträge zur Akustik und und Musikwissenschaft* (pp. 25–65). Leipzig: J. A. Barth. [On the theory of the difference tone and the general hearing sense. In C. Stumpf (Ed.), Volume 2, *Contributions to acoustics and systematic musicology.*]

Miller, G. A. (1956). The magical number seven, plus or minus two: Some limits of our capacity for processing information. *Psychological Review, 63,* 81–97.

Miller, G. A., & Heise, G. (1950). The trill threshold. *Journal of the Acoustical Society of America, 22,* 637–638.

Miller, G. A., & Selfridge, J. A. (1953). Verbal context and the recall of meaningful material. *American Journal of Psychology, 63,* 176–185.

Miller, J. R., & Carterette, E. C. (1975). Perceptual space for musical structures. *Journal of the Acoustical Society of America, 58,* 711–710.

Miyazaki, K. (1988). Musical pitch identification by absolute pitch possessors. *Perception & Psychophysics, 44,* 501–512.

Miyazaki, K. (1989). Absolute pitch identification: Effects of timbre and pitch region. *Music Perception, 7,* 1–14.

Miyazaki, K. (1990). The speed of musical pitch identification by absolute-pitch possessors. *Music Perception, 8*(2), 177–188.

Monahan, C. B. (1984). Parallels between pitch and time: The determinants of musical space. *Dissertation Abstracts International, 45,* 1942B. (University Microfilms No. 84–20, 214).

Monahan, C. B., & Carterette, E. C. (1985). Pitch and duration as determinants of musical space. *Music Perception, 3,* 1–32.

Monahan, C. B., Kendall, R. A., & Carterette, E. C. (1987). The effect of melodic and temporal contour on recognition memory for pitch change. *Perception & Psychophysics, 41,* 576–600.

Nakamura, T. (1987). The communication of dynamics between musicians and listeners through musical performance. *Perception & Psychophysics, 41,* 525–533.

Namba, S., Kuwano, S., Hatoh, T., & Kato, M. (1991). Assessment of musical performance by using the method of continuous judgment by selected description. *Music Perception, 8*(3), 251–275.

Narmour, E. (1990). *The analysis and cognition of basic melodic structures.* Chicago: University of Chicago Press.

Neisser, U. (1976). *Cognition and reality.* San Francisco: Freeman.

Nickerson, J. F. (1948). *A comparison of performances of the same melody played in solo and ensemble with reference to equi-tempered, just, and Pythagorean intonation.* Unpublished doctoral thesis, University of Minnesota, Minneapolis.

Ohgushi, K. (1994). Subjective evaluation of various temperaments. In I. Deliege (Ed.), *Proceedings of the Third International Conference for Music Perception and Cognition* (pp. 289–290). Liege, Belgium: European Society for the Cognitive Sciences of Music.

Oram, N. (1989). *The responsiveness of western adult listeners to pitch distributional information in diatonic and nondiatonic melodic sequences.* Unpublished doctoral dissertation, Department of Psychology, Queen's University, Kingston, Ontario, Canada.

Palmer, C. (1989). Mapping musical thought to musical performance. *Journal of Experimental Psychology: Human Perception and Performance, 15,* 331–346.

Plomp, R., & Levelt, W. (1965). Tonal consonance and critical bandwidth. *Journal of the Acoustical Society of America, 38,* 548–560.

Povel, D.-J. (1981). Interval representation of simple temporal patterns. *Journal of Experimental Psychology: Human Perception & Performance, 7,* 3–18.

Povel, D.-J., & Essens, P. (1985). Perception of temporal patterns. *Music Perception, 2,* 411–440.

Povel, D.-J., & Okkerman, H. (1981). Accents in equitone sequences. *Perception & Psychophysics, 30,* 565–572.

Radocy, R., & Boyle, J. (1979). *Psychological foundations of musical behavior.* Springfield, IL: Thomas.

Repp, B. (1989). Perceptual evaluations of four composers' "pulses." In *Proceedings of the First International Conference on Music Perception and Cognition* (pp. 23–28). Kyoto, Japan: The Japanese Society of Music Perception and Cognition.

Risset, J. C. (1978). Musical acoustics. In E. C. Carterette & M. P. Friedman (Eds.), *Handbook of Perception, Volume IV: Hearing* (pp. 521–564). New York: Academic Press.

Rossing, T. D. (1990). *The science of sound* (2nd ed.). New York: Addison-Wesley.

Rowe, R. (1993). *Interactive music systems: Machine listening and composing.* Cambridge, MA: The MIT Press.

Rowell, L. (1983). *Thinking about music.* Amherst, MA, University of Massachusetts Press.

Saldanha, E. L., & Corso, J. F. (1964). Timbre cues and the identification of musical instruments. *Journal of the Acoustical Society of America, 36*(11), 2021–2026.

Sampat, K. S. (1978). Categorical perception in music and music intervals. *Journal of the Indian Musicological Society, 9,* 32–35.

Sandell, G. J. (1989). Perception of concurrent timbres and implications for orchestration. *Proceedings of the 1989 International Computer Music Conference* (pp. 268–272). San Francisco: Computer Music Association.

Schackford, C. (1961). Some aspects of perception. Part I. *Journal of Music theory, 5,* 162–202.

Schackford, C. (1962). Some aspects of perception. Part II, III. *Journal of Music theory, 6,* 162–202; 295–303.

Schmuckler, M. (1989). Expectation in music: Investigation of melodic and harmonic processes. *Music Perception, 7,* 109–150.

Seashore, C. E. (1938). *The psychology of music.* New York: McGraw-Hill.

Senju, M., & Ohgushi, I. (1987). How are the player's ideas conveyed to the audience? *Music Perception, 4,* 311–323.

Shaffer, L., Clarke, E., & Todd, N. (1986). Meter and rhythm in piano playing. *Cognition, 20,* 61–77.

Shepard, R. N. (1964). Circularity in judgments of relative pitch. *Journal of the Acoustical Society of America, 36,* 2346–2353.

Shepard, R. N. (1982). Structural representations of musical pitch. In D. Deutsch (Ed.). *The psychology of music* (pp. 343–390). New York: Academic Press.

Shower, E. G., & Biddulph, R. (1931). Differential pitch sensitivity of the ear. *Journal of the Acoustical Society of America, 3,* 275–287.

Siegel, J. A., & Siegel, W. (1977a). Absolute identification of notes and intervals by musicians. *Perception & Psychophysics, 21*(2), 143–152.

Siegel, J. A., & Siegel, W. (1977b). Categorical identification of tonal intervals: Musicians can't tell sharp from flat. *Perception & Psychophysics, 21*(5), 399–407.

Sloboda, J. A. (1983). The communication of musical metre in piano performance. *Quarterly Journal of Experimental Psychology, 35A,* 377–396.
Sloboda, J. A. (1985). *The musical mind: The cognitive psychology of music.* Oxford: The Clarendon Press.
Stevens, S. S. (1967). Intensity functions in sensory systems. *International Journal of Neurology, 6,* 202–209.
Strong, W., & Clark, M. (1967). Perturbations of synthetic orchestral wind instrument tones. *Journal of the Acoustical Society of America, 41*(1), 39–52.
Stumpf, C. (1883). *Tonpsychologie I.* Leipzig: Hirzel.
Takeuchi, A., & Hulse, S. (1991). Absolute pitch judgments of black- and white-key pitches. *Music Perception, 9*(1), 27–46.
Takeuchi, A., & Hulse, S. (1993). Absolute pitch. *Psychological Bulletin, 113*(2), 345–361.
Terhardt, E. (1974). Pitch, consonance and harmony. *Journal of the Acoustical Society of America, 55,* 1061–1069.
Terhardt, E., & Seewann, M. (1983). Aural key identification and its relationship to absolute pitch. *Music Perception, 1*(1), 63–83.
Terhardt, E., & Ward, W. D. (1982). Recognition of musical key: Exploratory study. *Journal of the Acoustical Society of America, 72,* 26–33.
Thomson, W. (1991). *Schoenberg's error.* Philadelphia, University of Pennsylvania Press.
Tougas, Y., & Bregman, A. S. (1985). Crossing of auditory streams. *Journal of Experimental Psychology: Human Perception & Performance, 11,* 788–798.
Unyk, A., & Carlsen, J. (1987). The influence of expectancy on melodic perception. *Psychomusicology, 7,* 3–23.
van Noorden, L. P. A. S. (1975). *Temporal coherence in the perception of tone sequences.* Unpublished doctoral dissertation, Eindhoven University of Technology, The Netherlands.
Vos, P. G. (1977). Temporal duration factors in the perception of auditory rhythmic patterns. *Scientific Aesthetics, 1,* 183–199.
Vaughn, K. (1993). The influence of the tambura drone on the perception of proximity among scale types in North Indian classical music. *Contemporary Music Review, 9*(1&2), 21–33.
Wapnick, J., Bourassa, G., & Sampson, J. (1982). The perception of tonal intervals in isolation and in melodic context. *Psychomusicology, 2*(1), 21–36.
Ward, W. D. (1953). Information and absolute pitch. *Journal of the Acoustical Society of America, 25,* 833.
Ward, W. D. (1954). Subjective musical pitch. *Journal of the Acoustical Society of America, 26*(3), 369–380.
Ward, W. D. (1970). Musical perception. In J. Tobias (Ed.), *Foundations of modern auditory theory,* Volume 1 (pp. 407–447). New York: Academic Press.
Warren, R. M. (1993). Perception of acoustic sequences: Global integration versus temporal resolution. In S. McAdams & E. Bigand (Eds.), *Thinking in sound, the cognitive psychology of human audition* (pp. 37–68). Oxford: Clarendon Press.
Warren, R. M., Obusek, C. J., Farmer, R. M., & Warren, R. P. (1969). Auditory sequence: Confusions of patterns other than speech or music. *Science, 164,* 586–587.
Warren, R. M., & Warren, R. P. (1970). Auditory illusions and confusions. *Scientific American, 223,* 30–36.
Weaver, H. A. (1943). A survey of visual processes in reading differently constructed musical selections. *Psychological Monographs, 55*(1), 1–30.
Wedin, L., & Goude, G. (1972). Dimension analysis of the perception of instrumental timbre. *Scandinavian Journal of Psychology, 13*(3), 228–240.
Wessel, D. L. (1973). Psychoacoustics and music: A report from Michigan State University. *PAGE: Bulletin of the Computer Arts Society, 30.*
Wessel, D. L. (1979). Timbre space as a musical control structure. *Computer Music Journal, 3*(2), 45–52.
Yeston, M. (1976). *The stratification of musical rhythm.* New Haven, CT: Yale University Press.

CHAPTER 5

The Perception of Pictures and Pictorial Art

Julian Hochberg

I. INTRODUCTION

In this chapter, I discuss what perceptual theory must learn from, and may contribute to, an understanding of pictorial art.

The importance of perceptual psychology to the applied arts is reasonably clear. The communications industries—film and video, the news media, advertising and packaging, entertainment—must know how to specify and produce given colors, what resolution is needed for presenting pictures and text, what fonts and formats are most legible, and above all how to depict events and scenes (virtual space and movements) that do not in fact exist. A great many similar questions in vision and audition, as well as more cognitive questions about attention and comprehension, need answers each day; these answers come more from intuition, and from trial and error, than from reliable information, although the knowledge base is expanding.

Two areas of communications research are not so straightforward: the interaction of the expressive features of the medium and its substantive content, and the nature of the display or presentation as an intentional interpersonal act. Even the most seemingly transparent and potentially automatic process, like the making of a photograph, entails a selection and a preparation—a directed effort—that implies a presenter with some purpose and that can make each member of the audience a participant in an implicit dyadic communicative act (e.g., Why is he showing that in such unexpected detail?). This is probably an extremely important aspect of

Cognitive Ecology

every artistic presentation, but it is still true that (as I wrote in 1978) we have barely begun to assemble an analytic logic for communicative acts (Grice, 1968; Schmidt, 1975; Searle, 1969; see Clarke, in press).

These points become central in any attempt to distinguish fine from applied art, presumably a distinction based on *aesthetics* (for now, consider aesthetics as disinterested evaluation). Applying a concept of disinterested evaluation to paintings, which can be enormously costly investments, seems somewhat oxymoronic. Where options are available, where a tradition provides the background against which one displays one's own mark and originality, and, above all, where there is a great deal of Veblenesque prestige and gross financial investment at stake in assessing a given artistic presentation as good or bad (and thereby establishing an artist as worthy of investment), simple issues and unequivocal criteria do not exist. In the applied arts, there are usually more or less measurable criteria that can theoretically be drawn on. There, consumer preference measurements (often using techniques borrowed from traditional experimental aesthetics; see Woodworth, 1938) can at least in principle be subject to validation procedures. But with objects or presentations produced "for their own sakes," a great deal of the appreciation of the work depends on the education that enables the viewer to (1) place it in its tradition, or in its line of development; to (2) exercise the expertise that this requires and—perhaps most important in ensuring the stability of the investment—to (3) contemplate the object as an evocative piece of history.

Artistic provenance and tradition are *potentially* susceptible to rigorous pursuit by art historians, and at least some of the many societal functions that art serves are similarly assessible. Various functions (from sensory pleasure through philosophical reference to providing the motor for social change) have been announced, at one time or another, to be *the* basis for evaluating artistic merit. Both philosophers and experts on art disagree within their ranks, and some deny the possibility of an acceptable definition of either art or aesthetics (Kennick, 1958; Weitz, 1960). Some artists maintain that the only opinions worth considering are those of other comparable artists—special pleading that may, nevertheless, be valid where it reflects the artist's actual goal of achieving original and notable solutions to problems posed by the history and demands of the art form.

My own bias is that there is no single realm of art, even within the making of pictures. Many different activities are lumped together that have different purposes and diverse criteria; what is presently called *art* is more a matter of historical and sociological accident (and vested interest) than anything else. This is compounded by the fact that people really cannot, by introspection alone, provide effective analyses of why they seek exposure to what we will call artistic presentations (paintings, dance, motion pictures, music, architecture, etc.) any more than they can coherently and validly explain their choices in more trivial areas, such as fashion design, popular music, automobile styling, and so on, all of which are aesthetically driven applied art.

I will not maintain a distinction between pure and applied art here, except where it is natural to the discussion.[1]

It seems natural to divide the functions and criteria of the perceptual study of artistic presentations in the following three ways.

A. Pictorial Art as Representation and as Communication about the World

Many perceptual treatments of visual art deal solely with its representational function. Such analysis ignores those art forms, such as music, dance, and abstract paintings, that may have no representational or programmatic content, although they may have descended from representational activities. Other art forms, such as architecture, only rarely attempt to represent something. Yet pictorial art, certain acting, models, diagrams, and architectural design *do* aspire to some degree of objective communication. We would surely use pictures for their representational functions regardless of their artistic and other values, as we would continue to use clothes and buildings even if we had no care for their appearances. It is often cause for wonder, among the educated as well as among the unlettered, that some artists could render the likenesses of portrait sitters and their household goods so faithfully.

Making a surrogate that faithfully mimics to the eye the effects of the represented scene or event has lost its aesthetic interest per se (although convincing "special effects" and computer-generated images of dinosaurs and interstellar battles may still invite a little attention and wonder). But ways of making a portrait that is even more like the sitter than is the sitter, remain a matter of interest. Innovation in the manner or style of representation will surely continue, which introduces the two nonrepresentational functions, expression and aesthetic value.

B. Art as Expression: The Communication of the Artist's State, Feelings, or Identity

The ability of art to move the audience emotionally or to express how the artist feels is often taken as the sole touchstone of artistic merit. It is clear, however, that presentations can be expressive, yet not be regarded as good art (even the crudest forms of advertising, propaganda, and entertainment can move the audience), and that much art lacks both representational *and* emotional content aside from what is sometimes termed the aesthetic emotion, which provides the third function.

[1] The teenager immersed in assessing current musical performers is acquiring culture and an appreciation of that art by means that probably draw on the same mechanisms that contribute to a more classical cultural education, although the latter might provide more continuity to society.

C. The Aesthetic Function: Art as Pleasurable, Interesting, or Engaging

Defining the beautiful and pleasurable in terms of physically measurable canons or prescriptions has been attempted since antiquity. The reward value of the pleasurable has long been measured by asking subjects to make preference judgments and, more recently, by recording the subject's tendency to keep looking or listening (the opposite of boredom or habituation), which is more interesting to those theorists concerned with understanding perception as a motivated, constructive process.

These three headings are, of course, not completely separable nor exhaustive. Most writers agree that what makes an artistic presentation more or less successful, more or less sophisticated and deep, is the extent to which these diverse functions (and yet others) can be met in mutually reinforcing ways, using the expressive features of the medium in concert with the content (a desire often expressed in connection with poetry). However, any museum attests, I believe, to the fact that all functions need not be embodied in any one work of art.

We take up a selective survey of perceptual problems and research in each of these three areas.

II. REPRESENTATION AND COMMUNICATION ABOUT THE WORLD

A. Representational Pictures and Perceptual Theories

Perception textbooks that consider art at all treat representational pictures as surrogate objects that act as likenesses because they present the eye with much the same *pattern* of light as would the scene itself. This treatment must engage, unequally, all the perceptual theories.

A glance at the older traditional perceptual theories will distinguish the different kinds of data and analyses that they can bring to art. Associationism—empiricism, the oldest theory, originally assumed that all conscious experience consists of present sensations, of memory images of previous sensations, and the arbitrary linkages between them forged by the individual's mind in its encounters with the structure of the world. *Gestalt* theory rejected these atomistic premises, explaining that what we perceive reflects the characteristics of underlying brain fields. *Ecological realism,* which today remains in this regard much as Gibson developed it, reanalyzes the light reaching the eye to reveal rich information that might (it claims) account completely for veridical perceptions of the world without invoking the mediating processes and memories of associationism, or the organizational processes of Gestalt theory. This approach is uniquely challenged by the fact that things can be recognized from their pictures, as will be shown. Finally, there is the new-old attempt, which is usually referred to Hebb and Piaget, but in fact reaches back to Helmholtz (1909/1924), to view perception as an active process of fitting "mental structures" or "mental representation" (hypothesized objects, scenes, and events) to selected sensory samples, thus building selective attention and schematization, or abstraction, directly into the heart of the perceptual process. The Hebbian approach is clearly dominant today (though sometimes unacknowledged) in the theory and research of cognitive neurophysiology and computer vision.

Within the classical empiricist approach we can distinguish two components—a sensory psychophysical analysis (dealing with local "points" of light), and a cognitive empiricism, as in Helmholtz's doctrine of unconscious inference, which is that *one perceives those objects and events that are most likely, on the basis of past experiences, to fit the present pattern of effective sensory stimulation*. This gives two purely psychophysical approaches to perception (classical and Gibsonian) and two that invoke mental structure (Gestalt theory and those deriving from Helmholtz and Hebb). All approaches have been used to discuss all three functions (representation, expression, and aesthetic value). I consider representation first, and at greatest length.

B. The Psychophysics of Surrogates

To Leonardo da Vinci, the painter must be able to present a likeness of the world, using skills that could be learned in part by tracing the objects to be represented on a pane of glass that is interposed between artist and scene. By studying how the tracings on that picture plane are transformed by different dispositions of the objects in space, the artist learns what we now call the pictorial, or static monocular *depth cues*.

Aided by the geometry of perspective, as introduced or reintroduced by Brunelleschi in 1420, and explicated by Alberti in 1435 (see White, 1967) and by da Vinci's principles for modeling surfaces through shading and shadows (see Braxandall, 1995; Hills, 1987), it was now possible under the proper conditions to present the stationary monocular observer with approximately the same spatial distribution of light as would be given by the represented scene. In addition to trying to imagine how to represent the chosen scene, artists could also consult masters' notebooks and apply the rules (e.g., perspective). Assisted by devices like the camera obscura and the camera lucida (Kemp, 1990; Wheelock, 1979), the achievement of the same input as would perceptual reality (Danto, 1986, p. 12) became a practical goal. With the development of photography, this goal could be achieved "automatically," and in any case offered a great number of substitutes for masters' sketchbooks.

But a picture may be optically correct and yet fail as an effective surrogate. Da Vinci knew the limitations of this method: The viewer must stand in one location and use only one eye, and the picture must be without surface texture of its own, etc., or the flatness of the picture will be betrayed; therefore, *in general, the viewer cannot be fooled* by such simulations. To varying degrees, and for special purposes, these limitations can be overcome. For example, the cues to flatness and texture of the surface may largely be overcome by restricting the position of the viewer and by painting the scene on some surface that is not in the projective picture plane (e.g., using the ceiling of the nave of a church as the surface on which to paint the continuation of the wall produced the convincing illusion of an additional story [Pirenne, 1970], or by drastically restricting the depth being portrayed in a trompe l'oeil painting, etc.).

Once it has been decided *what* is to be portrayed, and where the viewer stands relative to the canvas and to the scene being represented, decisions as to what the *arrangement* of lines and patches of color on the canvas should be becomes merely a task for following the geometry and mixing the pigments—or, with the advent of chemical and electronic photography, using the right lenses and adjustments in the camera and its display. So far in our account, psychology has contributed only the raw materials given by the psychophysics of acuity and color mixing. Geometry contributes the patterns in which the colors are to be distributed, and we have a technology for making representational pictures. After Daguerre replaced the artist's canvas with light-sensitive plates in 1839 (Szarkowski, 1973), the surrogate, *as I have been considering it so far,* does not need the artist. But that is because so far I, like the SMPTE (Society of Motion Picture and Television Engineers) engineer or the vision scientists, have been considering the matter in a very narrow way.

And in any case, this automatic process describes a way to make pictures, not how it is that they are perceived as being like what they represent. As da Vinci said at the outset (White, 1967), the fact that the picture's surface is almost always quite recognizably flat, and usually viewed binocularly from changing positions, makes it desirable to violate projective geometry (see p. 164). And even the matching of the colors in the scene by trying to match wavelengths and their intensities in the light offered by the picture would be quite useless, as pigment limitations, human spectral sensitivity functions, and the phenomena of simultaneous color contrast make virtually certain. Very little is left of the idea of pictures as surrogates, or of *physical* fidelity as a measure of their accuracy of representation.

It is true, of course, that once the pattern on the canvas is established using the geometrical optics of the Renaissance, we can match the appearance of each separate point on the picture's surface to its corresponding point within the real scene, if we can correctly mix the pigments (or set the phosphors). Rules for classifying the possible colors had to be learned by the apprentice; fully developed, they can now be found, with the minimum palettes for additive and subtractive mixtures, in most introductory perception texts. They will not by themselves suffice: the range of pigments' reflectance is much smaller in any scene that includes specular reflections, light sources, and so on; moreover, contrasts effects characteristic of the scene (e.g., the induced hue of shadows in the open air) are not achieved simply by matching *hues* on canvas to those in the scene, point by point. The phenomenon of simultaneous contrast may be used to mitigate this limitation: by juxtaposing shadows, a region's apparent lightness can be enhanced, as notably exploited in the chiaroscuro (patterning of light and shade) of Rembrandt and de La Tour (Hochberg, 1979). A growing attention to induced colors, like those in shadows, that carries forward through Corot to the Impressionists, made the informed use of contrast an important skill.

Given a working knowledge of the laws of color mixing and color contrast, and with relatively few pigments or phosphors, the painter, printer, or display engineer can approximate the apparent colors of a wide variety of objects under a wide range of lighting conditions. In addition, the patterns of color and the afterimages that they produce can generate effects of *vibrancy* and movement, as taken to their extreme in *optical art*.[2] Less obtrusively, these effects, to which all high-contrast contours are prone, were deliberately used by many of the Impressionists, for what I take to be the following purposes.

Pigments can be mixed in various ways (such as physical intermixing, superposition by glazing, etc.) that provide subtractive mixtures by successively interposed filters of pigment; alternatively, colors can be mixed additively by placing small patches next to each other. If these patches fall below the limits of acuity, an additive optical mixture results. This, in effect, is what the pointillist painters came near doing, in their attempts to present the light from the scene scientifically, preserving a range of reflectances and saturations that would otherwise be lost in the process of subtractive mixture. The patches used in most Impressionist paintings, however, are much too large to be below the resolving power of foveal vision at practical distances (e.g., perhaps 70 ft for a Monet, 300 ft for a Van Gogh). How then does color mixture occur? Partly, because assimilation rather than contrast occurs when different patches fall within the larger receptive field associated with the individual retinal cells serving to detect the individual patches (Jameson & Hurvich, 1975). Interacting with this is the factor of the much lower resolving power of parafoveal and peripheral vision as compared to foveal vision. From the right viewing distances, the mixture occurs in some but not all of peripheral vision, producing a vibrancy otherwise offered at contours in daylight illumination (Jameson & Hurvich, 1975), and raising an issue that further complicates any attempt to think of the picture as a surrogate for the scene.

When the viewer's gaze is fixed on some region that is painted in full detail, the lack of detail elsewhere may not be evident. Conversely, the artist may lead the viewer to look predominantly at one place by only giving detail there. Rembrandt and Eakins (among others) have left clear examples of such usage, and when I stand at the correct viewing distance from these paintings and look at the detailed focal region, the pictures look complete, even though they may be blurred blobs elsewhere.

In Impressionist paintings, however, the fovea finds incoherent patches *wherever* it is directed. The viewer quickly discovers that it is precisely that apparently meaningless pattern that when seen by peripheral vision appears as the meaningful, depicted landscape and people. (This contributes, in my opinion, to the felt *rightness* and inalterability of the artist's work, which I take to be a great deal of what is

[2] This was a brief flurry of abstract designs, in the 1950s and 1960s, in which the moiré effects of regular high-contrast patterns, superimposed on their displaced afterimages by the slight tremors and unnoticeable movements of the eye, amplify the tremors and make them grossly visible (Oster, 1977).

FIGURE 1 One of these figures, modeled after a sketch by Seurat, has had a substantial region of the small dots inverted (from Hochberg, 1994).

meant by aesthetic quality in real paintings.) It is like the first unstudied impression of the world itself, only a tiny part of which is seen in detail by the first few glances. The meaningless patches themselves are generally too unrelated to be stored from glance to glance: a set of patches in one of the figures in Figure 1 has been inverted, but even with knowledge it is very hard to tell which. And because a painting's flatness is most evident to foveal vision, and in such paintings no depth cues are given in foveal vision, the immediate conflict between local depth cues and pictorial surface is greatly lessened.

We are now far from a surrogate that can be defined physically, but we can see how, by taking the eye's characteristic response to color and detail into account, the departures from fidelity may serve to surmount the limits of the painted canvas and better approximate the visual impression made by the scene itself. Most of the further massive deviations from projective fidelity that we consider in relation to the representation of space follow similarly from the inherent limits of the flat pictorial surface, making the representation itself an art rather than an exercise in geometry.

C. The Perceptual Issues in Spatial Representation

Let us start again, at what served as the starting point for much of philosophy, art theory, and psychology, and assume that the trompe l'oeil picture and the scene it represents both act on the eye in the same way. An infinite number of arrangements will do the same, by an argument that is surely familiar at least since Bishop Berkeley: one cannot specify three dimensions with only two.[3]

[3] The fact that a flat picture can act as a surrogate for 3-D objects at different distances was early taken to show that we cannot perceive depth directly: Alhazen in 1572, Peckham in 1504, as described by J. White (1967, pp. 126, 129). Developed fully by Berkeley, in 1709, and thence through James Mill, John

Of that infinity of scenes, why does one perceive only *that* scene which the artist intended to represent? Also (and this is why pictures and depth so interest philosophers, psychologists and roboticists, alike), why does one see the world when all that is given is a necessarily ambiguous retinal image, one that can in principle be fooled by an infinity of interchangeable pictures and scenes?

The *classical* answer was that depth is seen in the real world only because humans have learned to associate visual depth cues (da Vinci's, plus others) with the tactual-motor experiences received in the course of dealings with the three-dimensional (3-D) arrangements of the world. Two aspects of this classical answer are important here. The first is that because these cues normally mediate our seeing of the world of space and objects, *it required no new explanation to account for pictures.*

In the second important aspect of this line of thought that evolved from Berkeley to Helmholtz, the observer supplies necessary *mental structure:* This object looks further away *because* the lines in the retinal image seem to converge, and appears larger *because* (with a given visual angle) it appears farther away. *The physical rules of the visual ecology have been incorporated in the perceptual habits of the viewer;* all depth cues are learned, and what makes each effective is its prior association with other depth cues and with the moving, reaching, and touching that the viewer's history has furnished. Although there may be occasional errors in the perceptual structures fitted to such sensory patterns, probablistically they will be right more often than wrong (see Brunswik, 1956).

To Gestalt psychologists, the determinants of perception are the laws of organization, not ecological probabilities (except as the latter have constrained our evolution). According to Gestalt theory, one sees the simplest or most homogeneous image that will fit the pattern of stimulation. In fact, most of the classical pictorial depth cues can as readily be treated as examples of simplicity as of familiarity (Hochberg, 1974b; Hochberg & MacAlister, 1953). Because Gestalt theory sees representation as a result of innate brain-field organization, not of perceptual habit, both the main features of space perception, and of picture perception (following the same laws of organization), should hold for any viewer, regardless of his or her experience with pictures.

In what Gestaltists (and many others since Rubin's observations) take as the basic phenomenon of visual perception, the *figure* is an area that's shape is recognized (marked *f* in Fig. 2B), whereas the ground is shapeless and usually farther back, extending beyond the figure (Koffka, 1935; Woodworth, 1938). What will be perceived as figure (and consequently, what will be perceived in a picture) presumably depends on the so-called laws of organization, and not on familiarity

Stuart Mill, and Hermann von Helmholtz, this argument remains part of the psychologist's somewhat battered infrastructure. Although followers of Gibson sometimes retort that the additional dimension of time restores a third dimension to the retinal image (by introducing movement-based optic flow), one cannot specify four dimensions with three any more than three with two, as moving pictures attest.

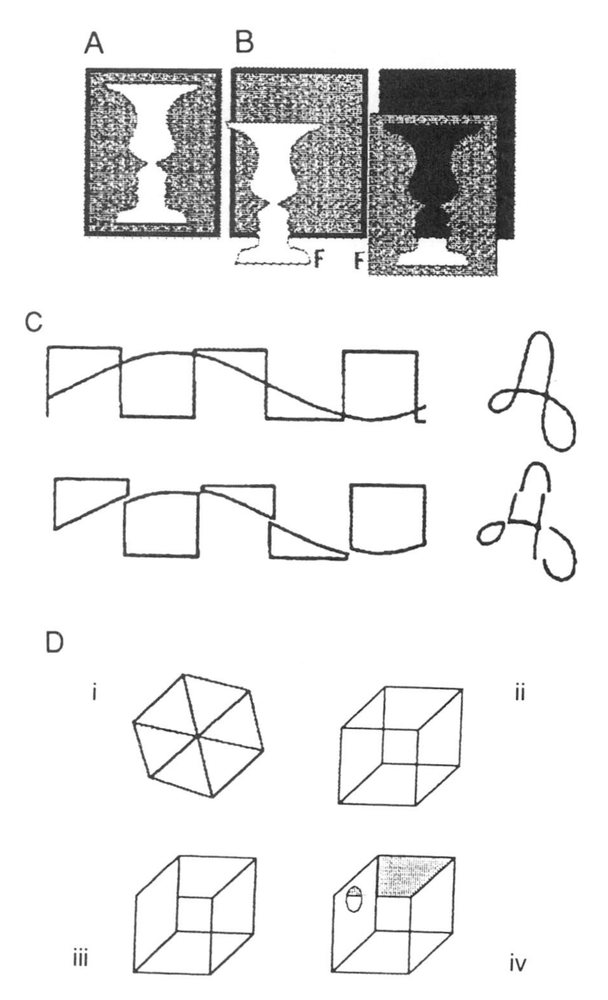

FIGURE 2 (A) At left, a reversible (ambiguous) figure–ground pattern. (B) At right, two alternative shapes that can be seen as figure in A. (C) By the "law" of good continuation, the upper-left pattern should be seen as a sine wave crossing a square wave, not the set of closed shapes that are seen in the lower-left pattern; similarly, no "4" should be visible in the upper-right pattern (as it is at lower right). (D) Figures at i and ii look 2-D and 3-D, respectively, presumably because good continuation (and other Gestalt factors) favor those organizations (Kopfermann, 1930); or perhaps because of a *minimum principle,* that we perceive the simplest overall structure. The latter can be phrased objectively, and also explains de Vinci's depth cues (Hochberg & Brooks, 1964; Hochberg & MacAlister, 1953). But when viewers fixate the ambiguous lower-right intersection in iii, it undergoes spontaneous reversal, in conflict with the upper-left intersection (Hochberg, 1970; Hochberg & Peterson, 1987), or with as many more intersections as we wish (cf. iv), making any object-wide minimum principle, or any other organizational theory which does not define its span of application, currently unviable.

(as Helmholtzians might assume), because meaningless figures may predominate over meaningful ones (e.g., the "4" is concealed in Fig. 2C through the operation of the "law of good continuation"), a point I will shortly question.

Manipulating drawings so as to discover what makes a particular area's shape be seen as a figure, and what causes it to become unrecognizable ground, is therefore a convenient and rapid way to discover what the laws of organization are. It is also a way to help the artist make pictures that will be perceived as desired (e.g., as a visible figure rather than unshaped ground) (Fig. 2A,C), or as 3-D rather than flat (Figs. 2Dii, i, respectively).

It also implies that producing the same pattern of light to the eye as does the scene itself does not ensure that the viewer will see the scene as it really exists. Any snapshot in which a flowerpot seems to be growing out of the subject's head shows how violations of Gestalt laws (again, good continuation, in this case) can make even the most perfect surrogate unveridical or even unintelligible. The Gestalt laws therefore seemed immediately relevant to artists, whose theoretical and graphic concern with the mechanisms of making things be seen one way or another are easy to find (e.g., Matisse's *Dancer,* Tchelitcheff's *Hide and Seek;* Arp's reversible amoebas; and the figure-ground exercises of Escher and Albers). Expression and feeling were held to be just as directly given in the field forces that the configuration engendered in the brain as is the shape itself. Arnheim's important and influential essays and books on the psychology of art, written by a Gestalt psychologist, seemed far closer to what artists were concerned with than any psychophysical treatment had been (see Arnheim, 1954, 1966 for a summing up of that viewpoint).

More objective formulations of the laws of organization were attempted, and usually referred to something like information theory rather than to brain-field organization (cf. Attneave & Frost, 1969; Hochberg & Brooks, 1960; Leeuwenberg, 1971).[4] I will discuss later why both the Gestalt approach and these variants may be useful only as very rough practical approximations, and why it would be a mistake to take them seriously today, either as theoretical explanations of picture perception or as generally viable perceptual approaches (see Fig. 2Diii, iv; cf. Peterson & Hochberg, 1989).

James J. Gibson argued that neither Helmholtzian mental structure, Gestalt organization, nor any other information contributed by the viewer is needed to explain perception in general and depth perception in particular: The information in the proximal stimulation, he argued, is sufficient to account for the direct, correct (veridical) perception of the surfaces and objects of the world. The percep-

[4] Despite the many books for artists that introduce the Gestalt philosophy and demonstrations, a cookbook on their application to pictorial intelligibility remains at once potentially possible, clearly desirable, and essentially unwritten. The possible effects of ground (the space *between* the shapes) as a factor in composition has long been raised in connection with the theory of design (cf. Taylor, 1964) and should theoretically be quantifiable in that regard, but research on this matter has not been done.

tion of such components of the layout as surfaces, edges, corners, and so on, he claimed, is directly specified by higher order variables of stimulation, like texture-density gradients and, above all, by the wealth of optic flow in the light to the eye that results from viewers' movements through a 3-D world. The pictorial depth cues as studied since da Vinci are not the basis of our normal perception of the real world, and the phenomena displayed by artificial preparations such as the scribbles on paper conveying the geometrical illusions and the Gestalt grouping principles, are nondiagnostic for normal perception.

These claims are inconsistent with the nature of pictures, however, because even a high-fidelity picture lacks the motion-produced information that specifies 3-D objects and spaces, whereas it surely provides the motion-produced information that it is a flat pigmented surface, and that the depth cues are thereby specified as being flat markings on that surface. In general, although Gibson has worked at a solution to this problem (Gibson, 1950, 1951, 1954, 1966, 1971, 1979), and others following his approach have made contributions to our understanding of the nature of picture perception (Hagen, 1974, 1976; Kennedy, 1974, 1977, 1993; Sedgwick, 1980, 1983, 1991), I think pictures remain an unresolved and critical problem within his approach. If the cues used in perceiving artwork and illustrations, both ancient and modern, are not also used in perceiving the world, how does it happen that those cues work in pictures?

First, I should note that there is at least some evidence that the static pictorial depth cues do indeed affect perceptions of space and movement even in the case of real, moving objects and viewers (Hochberg, 1987), and then produce illusions of concomitant motion (p. 236n) due to the accompanying parallactic displacement: Gogel & Tietz, 1992; Hochberg & Beer, 1991. Moreover, pictorial depth cues within a single picture may elicit binocular convergence appropriate to the depth they depict (Enright, 1991). Indeed, the classical geometrical illusions, such as the Müller–Lyer, which have traditionally been discussed in terms of line drawings, are also obtained with moving observers and solid objects (see DeLucia & Hochberg, 1991; Hochberg, 1987). These facts contradict the Gibsonian assertions that perception is veridical under normal seeing conditions and that therefore results obtained with drawings are not diagnostic of more general perceptual processes. It also seriously weakens the most common class of explanations of these illusions, which is that the illusions result from misapplied constancy scaling of pictorial depth cues (Gillam, 1978; Gregory, 1970).

Second, and fairly conclusive, there is the central fact that no training at all beyond that given by experience in the world itself is needed to recognize the things that are represented by at least some pictures, including line drawings like those in Figure 3 (Hochberg & Brooks, 1962a), a clear experimental answer for which there is also some anthropological support (Kennedy, 1977). Picture perception is *not* an arbitrary conventional skill, like reading: If its elements are learned at all, they are learned by commerce with the real world. We will return to this point later. Whatever causes viewers to take outlines as equivalent to objects' edges, it is

FIGURE 3 Outline drawings correctly identified by a child who had received no prior pictorial training.

not "symbol learning," and *it cannot be unrelated to how humans see the world itself.* That pictures drawn in outlines (i.e., in ribbons of pigment on paper, Gibson, 1951) are recognized naturally as objects, is a phenomenon that must *reflect an attribute of the viewer, not of the light at the eye.*[5]

Pictorial education does seem to improve the ability to interpret distance and size relations in pictures (Hagen & Jones, 1978; Krampen, 1993; Olson, 1975; Willats, 1977; Yonas & Hagen, 1973), particularly in highly impoverished pictures: As their experience with Western pictures increased, native Africans were better able to perceive spatial arrangements in pictures with sparse and somewhat ambiguous linear perspective (Hudson, 1962, 1967; Kilbride & Robbins, 1968; Mundy-Castle, 1966), although those results have been questioned on various theoretical and empirical grounds (Deregowski, 1968; Hagen, 1974; Hochberg, 1972b, p. 501; Jahoda & McGurk, 1974; Jones & Hagen, 1980; Kennedy, 1977; Omari & Cook, 1972). And an improvement in "reading" outline objects may reflect the older child's greater ability to perceive a line as belonging to more than one shape (Ghent, 1956) and to perceive an object for which only partial outlines are given (Gollin, 1960).

In two other major ways, projective fidelity is insufficient and unnecessary for representing objects and their attributes in pictured space. The first problem concerns viewpoint independence. An optically correct surrogate has, in general, only one viewpoint from which it fits the light from the scene it represents. When pictures are displayed, however, virtually no effort is made to have them viewed from the *one* point that their projective geometry dictates, even though the 3-D layout that can be fit to the two-dimensional (2-D) pattern differs with each viewpoint. Pirenne (1970) argued that the picture remains effective because we compensate for the slant of its surface, of which we are aware because of its frame, its texture, binocular parallax, and so on. That is, we presumably take that slant into

[5] As Hochberg and Brooks noted in 1962, these findings show that *lines must share some stimulus property with edges.* In Hochberg (1962) I thought primarily of luminance differences related to depth and light, but Kennedy (1974, 1993) has carried the problem considerably further.

account when arriving at a perception of the pattern that lies upon it. Something like that compensation does indeed seem to occur, but as will be argued it cannot be the whole story.

The second problem is posed by artists' deliberate violation of projective fidelity: Even while diligently using vanishing-point perspective, artists present certain objects (especially familiar ones) as though their main surfaces were always parallel to the picture plane, regardless of the object's depicted orientation. This distortion, which includes da Vinci's "synthetic perspective" (J. White, 1967, pp. 209–215), is used to overcome a problem that arises in the way we perceive optically correct pictures that are viewed from nearby.

This problem arises because the picture, being flat, is not equidistant from the eye, and the distance differences are significant when the picture is not far off (Fig. 4). Figure 4B is in *incorrect* projection, with each sphere pictured from straight ahead (essentially da Vinci's synthetic perspective), but looks more correct even when viewed from E. There are serious cognitive issues here, which we consider in turn.

Because of the widespread violations of perspective in the viewing and making of pictures, Nelson Goodman (1968) took perspective and pictured depth as arbitrary conventions, learned from pictures: Pictures, in this general view, are a visual language invented by artists (Kepes, 1944), cultural artifacts that in turn determined our nonpictorial vision (Wartofsky, 1979).

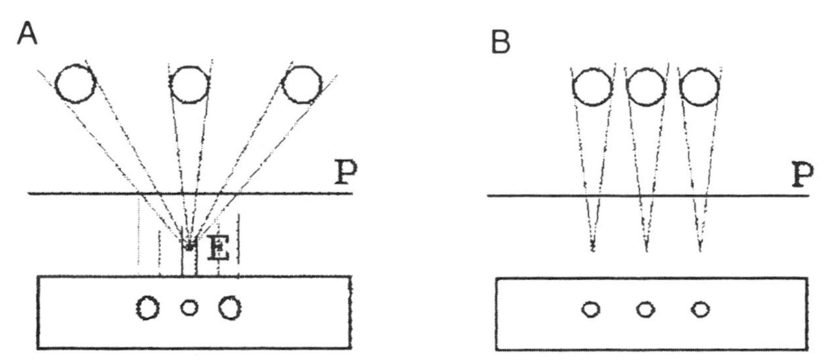

FIGURE 4 (A) Shapes projected on a near picture plane (P). When *correctly* projected as viewed from E, the three spheres at the top are shown as they would be traced on P (all here shown from above); the resulting picture is also to be viewed from E, and is shown in the lower sketch. Note that the pictures of the two outer spheres are larger than that of the center sphere, and somewhat distorted, as they must be if they are to project the same image to the eye. (B) A common distortion in projection that looks more correct than the correct projection at A. In Pirenne's (1970) explanation, viewers compensate for perceived slant to the line of sight in A, thereby correctly seeing that the shapes on the picture's surface are distorted, whereas in B the compensation for the slant yields undistorted pictures of the objects' shapes. (C) A sketch of one of David's large realistic exhibition paintings, made to be viewed from nearby, with all objects painted as in B and with virtually no perspective. (Modified from Hochberg, 1984.) (D) A sketch of a Cézanne painting, showing that at most of the intersections (b–e) good continuation has evidently been used to weaken interposition in foveal viewing.

C

D

FIGURE 4 Continued

This position has many close relatives in modern writings on literature and film. It also has its critics,[6] and it is wrong: Perspective is not arbitrary geometrically (see Gombrich, 1972b; Pirenne, 1970; Sedgwick, 1980, 1983, 1991). As I have just shown, when "realist" artists distort perspective, that distortion itself is usually an attempt to deal with the conflicts between the picture's flatness and the signs of depth it specifies. (Or it may be a specific effort to evoke an eerie, unresolvable, and highly notable conflict, as in de Chirico.)

Moreover, perspective is only one somewhat esoteric pictorial depth cue, among others much older, such as *interposition* and *modeling.* Perspective is global, applying to the entire scene depicted, and it is potentially *metric,* signifying not only that one object is farther than another, but how much farther it is. Other major depth cues are local and nonmetric (in this being like diagonality, a local consequence of perspective: see Gillam, 1978). Interposition, for example, heavily used for centuries in many cultures, is (most simply stated) that the uninterrupted line or contour bounds the nearer surface; there are several related cues as to local spatial structure that are offered by intersecting or interrupted lines or contours (Guzman, 1969; Hochberg, 1994; Kellman & Shipley, 1992; Ratoosh, 1949). It is ordinal rather than metric in the depth it signifies, and is therefore far less vulnerable to changes in viewpoint; and it is local, and therefore does not require the viewer to assess the agreement of distant lines and patterns, which generally seems to require specific effort.[7] Modeling or shading is provided by the light reflected to the eye from diffusing surfaces at varying orientations to the line of sight (for analyses see Horn, 1981; Todd & Mingolla, 1983). Although this cue is good for signifying whether the surface is curved or not curved, the *amount* of curvature perceived is not well represented (Todd, 1989; Todd & Reichel, 1989).

Because of its limitations, many artists have therefore avoided global perspective entirely, as in David's wide paintings (e.g., Fig. 4C). These are highly realistic extended exercises in interposition, in modeling through shading (see footnote 8), and in da Vinci's synthetic perspective (compare Figs. 4B and 4C). Designed to be viewed from nearby and from many different standpoints during commercial exhibition (Brookner, 1980, p. 139), these sidestep the viewpoint problem by using virtually no linear perspective. This was an early and widely used solution (Söström, 1978).

In fact, even interposition has been rendered depthless to foveal vision, as when Cézanne (Fig. 4D, dated 1866 and hanging in the National Gallery, Washington, DC) substitutes a smoothly continuing curve for the abrupt intersection without making the picture's subject unrecognizable (for analysis and comparison to the David painting, see Hochberg, 1984). This method for disarming interposition

[6] For example, Carrol, 1988; Gombrich, 1972; Tormey, 1980; Wollheim, 1987.

[7] The perspective information in a scene may require comparison over large distances, which we will see is not necessarily available to the viewer within a single glance (p. 172f), and which indeed Arnheim (1966) has said is confusing to the uninitiated.

locally is easy to find as well in Matisse, Vuillard, Morisot, and others (Hochberg, 1980, 1984). The use of this method suggests that these artists consider interposition a foveally effective cue that is not essential to recognizing the represented object.

I must take the argument further: Artists in all cultures and at all times have felt free to use *no depth cues at all*. (Except, of course, to the extent to which outlines act as edges, at which one surface occludes the background behind it.) Although the pictorial depth cues are not arbitrary conventions, and can evidently contribute to the perception of both pictured and real depth, there is no reason to assume that the representation of recognizable objects depends in any way on the explicit representation of depth.[8] Remember that in Figure 3, no depth cues were present or needed.

Let us now review these points in the context of perceptual and cognitive theory and research. First, the idea that one compensates for differences in distance or for the slant of the picture's surface, and only then arrives at perceptions of the pictured space from those corrected shapes, implies that viewers use unconscious inference—calculations based on unconscious knowledge about the geometrical couplings of size and distance, slant and shape, in the optics of the physical world (Hochberg, 1974a). The issue therefore has theoretical weight. Is there strong evidence that compensation occurs and is needed? The issue remains open. The geometry of the virtual spaces that would optically fit any given picture certainly changes with viewpoint (see Farber & Rosinski, 1978; Lumsden, 1980; Rosinski, Mulholland, Degelman & Farber, 1980; Sedgwick, 1991). With perfect compensation, the 3-D layout perceived in any picture would not change as the viewpoint changes.

There are laboratory studies that have found that viewpoint has no effect on pictured objects' apparent sizes (Hagen, 1976), on their apparent slant (Rosinski, Mulholland, Degelman, and Farber, 1977), or on their apparent forms (i.e., their rectangularity or nonrectangularity: Perkins, 1973), except at extreme viewing angles. Most of these researchers concluded that some degree of compensation for picture plane must occur. That may be true, but one should note that the thesis that virtual space is mediated by compensation-corrected 2-D pictorial information is

[8] Most analyses by psychologists and computer scientists of how we perceive the world and pictures of it have assumed that objects are seen as surfaces or volumes in 3-D space, defined by their coordinates relative to the viewer (i.e., the distances and orientations at each point in the field of view) (Gibson, 1950; Horn, 1977; Marr, 1982). Admittedly, it is true that viewers can recognize volumetric objects, with no outlines present, defined only by the binocular disparities of the dots in a random-dot stereogram, and by the modeling, shading, and texture in a picture. But although shading effectively distinguishes a flat from a curved surface (Cutting & Millard, 1984), viewer's judgments are unreliable (Stevens & Brooks, 1987; Todd & Akerstrom, 1987), and it seems likely that shading and texture can normally provide only ordinal, nonmetric perceptions of surface orientation (Todd & Reichel, 1989). In such surfaces, viewers use the information given by contours formed at folds or occlusions in the surface (i.e., what amount to outlines) very well. And in any case, however, we know that surface modeling and binocular disparity are not normally essential to the recognition of pictured objects.

not strongly proven. Whatever compensation there may be is certainly not complete over the range of possible viewing conditions: With nearer or farther viewing, as all artists and photographers know, the depth and size perceived within the picture do vary, at least roughly as they should in the virtual space that fits the picture. An anamorphic picture that is recognizable when viewed at the appropriate extreme slant may be unrecognizable when the picture is normally viewed (Clerici, 1954), but this might be true precisely because the "appropriate slant is extreme and compensation fails.[9] Changes toward what one would expect from the geometry, however, are reported anecdotally to occur under much more normal pictorial viewing conditions (e.g., Gombrich, 1972b).

They also have been measured experimentally in laboratory situations: Within a picture's represented layout, the viewer's perceptions of shapes, slants, directions, and distances (Smith & Gruber, 1958) vary at least qualitatively as one would expect from the projective geometry and the viewer's location relative to the picture. Nor is there reason to believe that a given picture is locked to a given virtual space. Thus, when Goldstein (1979, 1987, 1991) had viewers reproduce the layout of poles represented in Figures 5A and B, they showed almost perfect constancy for viewing angle by producing almost identical layouts; but when asked how the poles were aligned relative to the picture plane, viewers' results changed in what Cutting (1988) showed was good agreement with the projective geometry, thus showing little or no compensation. The fact that different questions about the virtual space provide different answers certainly shows that the virtual space that viewers perceive with pictures does not have the invariant properties specified by the optical geometry of the surrogate theory.

That the picture itself (as well as the layout being represented) may appear largely unchanged from different viewpoints does not itself demand Helmholtzian inference (Hochberg, 1971, 1978). The ratios of textural units subtended by the different parts of any shape on a surface that is slanted to the line of sight would remain invariant (Gibson, 1950), and the kinds of distortion that result from viewpoint changes (Farber & Rosinski, 1978) would leave such ratios invariant along any dimension considered separately.

Furthermore, perhaps the distortions are perceived, but go unattended because the objects being depicted remain fully recognizable—identifiable—over a wide range of distortions (Hochberg, 1971, 1978), and thereby define the contents of the picture.

The fact is that any compensation theory in which the perceiver fits a model of physical space that is geometrically dense and consistent has long since been effec-

[9] Anamorphic pictures were traditionally made to be too distorted to recognize when the picture plane is viewed from within a normal range of slants to the line of sight, but to become recognizable when the picture is viewed from the appropriately extreme slant (e.g., the skull in Holbein's *The Ambassadors*). If that slant were then compensated, the anamorph should remain unrecognizable. Familiarity seems to be a factor, and such distortion should provide a psychophysically manipulable variable for studying object perception.

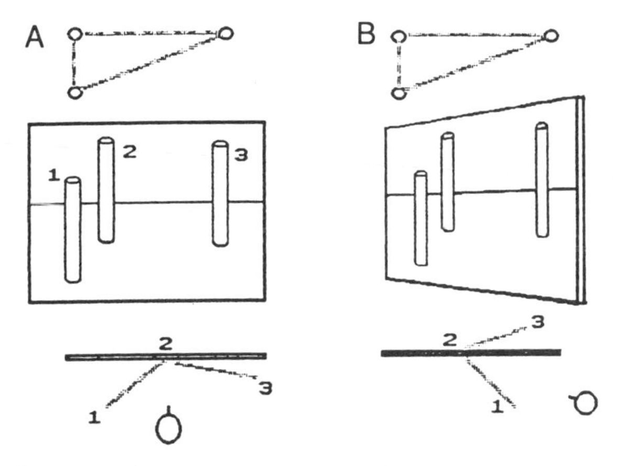

FIGURE 5 (A) a picture of 3 rods in space is looked at from in front, its canonical viewpoint; (B) it is looked at from near its right edge. The triangles over each picture show viewers' judgments of the poles' layouts in depth; the shaded lines connecting 1, 2, and 3 below each picture show viewers' judgments about the rods' alignments with respect to the picture plane (here viewed from above). The two spatial tasks do not yield the same judged 3-D layouts (modified from Goldstein, 1987).

tively ruled out both by artistic practice and laboratory research. Pictorial inconsistencies have been deliberately used, and are tolerated or even unnoticed, since at least the introduction of da Vinci's synthetic perspective. Indeed, it has been reported that viewers judge objects drawn in parallel perspective (which would, of course, be incorrect except when viewed from infinity) to be both more realistic and more accurate than those drawn in the converging perspective that would be correct for their viewing position (see Hagen & Elliott, 1976, a study that has been criticized by Kubovy, 1986, but a conclusion with which I find it hard to disagree).

More direct evidence to the same point is the fact that pictures of inconsistent or impossible objects are not automatically perceived as such. Escher's pictures (and some of Piranesi's), those of Albers, and the demonstrations by Penrose and Penrose (1958) and Hochberg (1968), offered pictured objects that would be physically impossible as 3-D structures but are not clearly seen as much, nor are they seen as complex 2-D drawings, which a Gestalt or informational minimum principle (Fig. 2Di, ii) should argue. Indeed, perfectly possible objects may be perceived as objects that are impossible in this sense (Gillam, 1978; Hochberg, 1970). For example, in Figure 2Diii depth reverses spontaneously when viewers fixate the ambiguous lower right intersection (Hochberg, 1970; Hochberg & Peterson, 1987; Peterson & Hochberg, 1983), so both global minimum principles and likelihood principles are unviable in anything like their present forms (Hochberg, 1982; see also Hochberg, 1987).

Such pictures show that each glance obtains only a limited part of the depth information within the object or picture, and that the overall cognitive structure resulting from those glances does not automatically test for internal consistency. At least two separate levels must therefore be considered to be at work in the perception of objects and scenes: the local features (or local depth cues), such as those offered by the intersection being viewed foveally, and some aspect of the global pattern as currently visible in the low resolution of peripheral vision or as carried in working memory from previous viewing. Without a deliberate effort to *attend* to the mutual spatial relationships specified by the individual glances at difference features, much of the information potentially offered by the layout (spatial and otherwise) is without perceptual consequence (Hochberg, 1968, 1970, 1982). Without good reason to believe otherwise, it seems plausible that these observations are true of pictures quite generally, and apply not only to reversals of near and far but to degree of depth. While attending to a region in which depth differences are small or nonexistent, the differences between picture and the three-dimensional scene it represents may also be small or even rendered nonexistent.

I think that many artists have either by explicit thought or by trial and error taken these considerations into account (see Hochberg, 1980). Avoiding a textured background near the sitter's cheek, using *sfumato* (blurring edge contours) and chiaroscuro to keep attention to such regions as would not offer substantial depth differences in the real object, can provide significant perceptual intervals in which the picture does not force the viewer to note that it lacks the object's solidity. A flat picture may therefore look similar to its three-dimensional subject simply because under many viewing conditions the recognizable features of the latter provide the stationary viewer with little depth information that is important to the object's recognition and appearance (Cutting & Vishton, 1995; Hochberg, 1980). What we perceive is not determined by what distal properties the stimulus information specifies physically. I will go further, in Section IID, and argue that there is simply no support for a surrogate theory of pictures which rests on physical definitions of pictorial fidelity, nor any need for such an account of pictures. There is however need for some account of the process that elects and guides the viewer's glances, and that achieves some integration of their contents.

Let us consider those issues first, and then their relationship to the nonrepresentational functions of visual art.

D. Perception as Purposive Behavior: Schemas, Canonical Forms, and Caricature

When one looks at a picture, one does not (and cannot) direct the eyes everywhere; and even if one did scan a picture in a complete raster (which would take same 200 glances and at least a minute of rapid eye movements to sample an 8 × 10 picture in 4° foveal glimpses), there is no reason to assume that everything would be automatically stored and stitched into a single perceived pattern that contains all the extended information that is physically present in the field of view. The latter, or *optic array,* is therefore not a realistic starting point for discussing the nature of visual processing.

In laboratory studies, subjects look first at regions that are most likely to be informative (by subjects' ratings or by experimenters' definitions [Antes, 1974; Hochberg & Brooks, 1962b; Loftus, 1976; Mackworth & Morandi, 1967; Pollack & Spence, 1968]) and that touch on the main features of the composition as it would be described from a design standpoint (Bouleau, 1963, Buswell, 1935; Molnar, 1968). One can detect large features in peripheral vision while focusing attention on detailed foveal information (Braun & Sagi, 1990), but detailed inquiry must proceed one glance at a time. These acts of looking can therefore be guided by low-resolution peripheral vision, aided by the redundancy of normal scenes and pictures. (A clear demonstration of such redundancy is that a particular object can be located faster when it is in a normal, appropriate scene than when it is in a jumbled or inappropriate one—Biederman, 1972).

One therefore looks at pictures (and at the world) piecemeal. Perceptions must fit that set of limited glances (Gombrich, 1963, 1984; Hochberg, 1968) that are free to continue at length or to terminate after only a very small proportion of the visual field has been brought to detailed foveal vision. *Because glances are purposeful, elective actions, one cannot in advance say what parts of the information in the light offered to the eye by picture or world will in fact be sampled.* Because of the occlusions that are routinely offered in a normally cluttered scene, one must ignore some information that is actually present and add information that is absent in order to achieve a rememberable encoding of what was sampled. Between stimulus description and final perceptual consequence one must therefore posit and investigate mental structures—schematic maps (Hochberg, 1968) or hypothesized objects (Gregory, 1970, 1980)—that serve essential perceptual functions. They motivate successive glances of perceptual inquiry, guide the acts by which the questions are answered, serve as the criteria that terminate the inquiry, and provide for storing the results of the inquiry.[10]

This metatheory is an old one and widely held,[11] but very little is known about the mental structures by which successive glances are integrated, and only a little more is known about what motivates the perceptual inquiry that drives those glances.

The overall view perceived from a series of glances is not constructed by serial integration (i.e., by taking the successive eye movements into account); rather, the

[10] The part that schemas play in remembering verbal narrative, dramatically demonstrated by Bartlett in 1932, is here filled by a structure of contingent visual expectations. An example of such visual expectations might be expressed in words as follows: Is this blurred object a car or a cat? If this is a car, I must look over there to see if it has a headlight. It has a headlight, so it is a car and not a cat, and there is no need to look further: it was a car. In the case of verbal narratives, the listener or reader often consults only local context, rather than the overall story schema (Dosher & Corbett, 1982; Glanzer, Fischer, & Dorfman, 1984; Mckoon & Ratcliff, 1992), even though that structure may be called upon when needed. The visual phenomena discussed in connection with Figure 4 may display the visual counterpart of that looseness.

[11] Really a version of the Helmholtz–Hebb position, and essentially also that of Neisser (1967), it was fairly explicit in J. S. Mill and Helmholtz near the turn of the century.

relative locations of features remain more or less unchanged in the mind's eye despite the shifts in gaze, and probably provide the invariant framework within which each rememberable feature takes its place (Gibson, 1950; Haber, 1985; Irwin, Zacks, & Brown, 1990).

Given that the structures perceived in response even to real or represented objects that permit a perfectly consistent and physically possible construal may nevertheless be quite different, (i.e., physically inconsistent or impossible), it seems safe to assert that the process by which successive glimpses are stored does not follow physical constraints, despite the traditional assumption that perceptual structure reflects physical structure (as in Gregory, 1980; Rock, 1977; Shepard 1984). The perceiver's tolerance of the kind of inconsistencies we have noted, and of many other kinds that can be discovered in any but the most realistic surrogate, argues that the overall structures do not necessarily intrude in the interpretation of each foveal detail. As in verbal story schemes, the overall structure is not automatically consulted at each point (see footnote 10). The same effects occur with real moving objects, showing again that at least some phenomena of picture perception do reveal the nature of more general perceptual process.[12] From this viewpoint, outlines may act like objects' edges because they share the same mechanisms at one or more levels of perceptual processing.

This would surely be adaptive, given the nature of visual inquiry: The eye has the wide field of peripheral vision as a form of on-line storage, a very impoverished reminder of where the fovea had been directed and a preview of what it might find at other loci in the field of view. With their low resolution (no detail, no texture, etc.) peripheral views of the real world and of its pictures must differ little from each other, and the contours at objects' edges, corners, and occlusions must be most important. As noted in Figure 1, the individual glance is limited in the detail it can carry forward; the periphery has an even more limited sensitivity to the detail that has been (or might be) brought to the fovea; and some reduced framework or schema is needed to relate the glances by which we sample both world and picture. Outline drawings seem able to activate such schemas and maps, both in vision and more generally.[13]

[12] For example, a real, 3-D version of Figure 2Diii works in the same way, whether it or the viewer moves (Hochberg & Peterson, 1987), appearing to rotate in illusory motion when it is seen in the wrong spatial arrangement.

[13] As in recognizing an object by touch, which requires haptic exploration and an integrative image (see Klatzky, Loomis, Lederman, Wake, & Fujita, 1993; Lederman & Klatzky, 1987), the foveal exploration of a scene or picture takes time and storage (Hochberg, 1968). But the eye has the wide field of peripheral vision as a form of on-line storage, a very impoverished reminder of where the fovea had been directed and preview of what it might find at other loci in the field of view. With their low resolution (no detail, no texture, etc.) peripheral views of the real world and of its pictures must differ little from each other, and the contours at objects' edges, corners, and occlusions must be most important. As noted in Figure 1, the individual glance is limited in the detail it can carry forward; the periphery has an even more limited sensitivity to the detail that has been (or might be) brought to the fovea; and some reduced framework or schema is needed to relate the glances by which one samples

And perhaps other basic perceptual phenomena may flow from the same conditions: The term *object,* as used here, refers to whatever maintains some relatively fixed structural relationship between its parts, regardless of how its movements (or viewer's movements) place it in the field of view. (Other questions of phenomenal identity arise within short-range apparent motion (see Chapter 6, this volume) but they are sidestepped here.) The viewer needs to know in advance where in peripheral vision there lies an object. Perhaps the figure-ground phenomenon (Fig. 2) reveals where one expects the eyes to find objects' edges, and which side will be the occluding surface: that is, perhaps the Gestalt laws are object cues—cues as to which side of an edge is part of the object, and which parts of the visual field will move together as a unit when moving heads or eyes.[14]

Some object cues are less ambiguous in this way than others (e.g., intersections and corners [Guzman, 1968; Hochberg, 1968; cf. Ratoosh, 1949]), although none absolutely foreclose alternative construal (Chapanis & McCleary, 1953; Dinnerstein & Wertheimer, 1957; see Hochberg, 1994). Most objects display those cues, and the distinctive features that identify those objects are better from one viewpoint than another. Given some arbitrary viewpoint (e.g., a random photograph), it is unlikely that the most informative and characteristic features will be presented as economically and effectively as an artist can choose to combine them, especially *given that the artist is free to change viewpoint within a single object at little cost* (as discussed previously). An object in what may be called its *canonical form* (Which best displays its characteristic features; see Hochberg, 1972a), may offer the viewer a prototype that may help in encoding and storing similar objects in the future (cf. Attneave, 1957a; Gombrich, 1956, 1972b).

If properly constructed, therefore, cartoons and caricatures, despite their drastic

---

both world and picture. Outline drawings seem able to activate such schemas and mental maps, both in vision and more generally: Thus, raised drawings, explored hapticly, work at least to some degree with blind subjects (Kennedy, 1993; Klatzky et al., 1993), who are able to form what amounts to schematic maps of spatial layouts (Arnheim, 1990; Haber, Haber, Levin, Hollyfield, 1993; Landau, 1985).

[14] For example, the law of good continuation is a case of interposition, in the sense that it is extremely unlikely that two different objects, at different distances, will line up within the tolerances of our excellent ability to distinguish misalignments (Hochberg, 1962, 1972b); the law of proximity reflects the fact that things that are close together are more likely to be part of one object (Brunswik & Kamiya, 1953). The first factor should serve both foveally and peripherally, the second would seem more useful in peripheral vision.

Such speculations seem plausible to me, but not enough is now known about the information contributed by peripheral vision to make them testable and applicable. Effective resolution falls off steeply outside of the fovea, as does spatial information (whether through spatial undersampling or perhaps through poor spatial calibration; Hess & Field, 1993). Ginsburg (1980) phrased a speculative description of the Gestalt phenomena in terms of low spatial-frequency filtering, which is one way to achieve a low resolution version of any scene or picture. Still missing is any systematic demonstration that such an account captures the characteristics of what peripheral vision contributes to the succession of combined foveal and peripheral views by which scenes and their pictures are sampled in the course of elective eye movements.

loss of information and fidelity, may better serve to represent the world, clarify visual relationships (cf. Arnheim, 1969), and effect our thoughts (Gombrich, 1956) than pictures of high fidelity. The pioneering work of Ryan and Schwartz (1956) showed that the layouts of at least some objects were better communicated by cartoons that had intentionally been made to help viewers retrieve the information that the task called for than by photographs, shaded drawings, or outline drawings. Some of the generalizations subsequently attributed to that study have been questioned (Biederman & Ju, 1988; Tversky & Baratz, 1985), and experimental research has failed to find caricatures of people superior to their photographs (Tversky & Baratz, 1985), but we do not yet have a principled basis for knowing how to fit the specific caricature to the needs of the specific recognition task.

It would be of great theoretical and applied interest if some finite number of object cues, learned from the real world but applicable to pictures as well, could be found to account for a significant amount of pictorial recognition. One of the problems with the assumption that an object is perceived in terms of the 3-D surface distances specified by its optical projection is that each slight change in viewpoint provides a very different set of such specifications for the same object. The seminal step away from the assumption was taken, I believe, in 1954, when Attneave showed, originally using a guessing-game procedure, that the inflection points in a silhouette or outline carry the (potentially measurable) load of the pictorial information and meaning. It is important to note that the relative placement and topology of these features as they meet the eye in the 2-D optic array (or in a picture) are relatively independent of the 3-D object's size and slant to the line of sight. Note too that in Hebb's (1949) enormously influential synthesis of the classical and Gestalt approaches, embodied today in most connectionist models of computational neurophysiology, the frequently encountered components of objects in the visual environment are taken as the primitives of perception (the *cell-assemblies*). It is by the 2-D arrangement of these components, not by their 3-D structure in depth, that objects and other familiar shapes are recognized.

Proceeding along this line, it would be more effective if viewers (whether humans or devices equipped with computer vision) would analyze outlines into those features that are relatively independent of viewpoint; for example, a rectangle in the field of view is unlikely to be projected there by a nonrectangular trapezoid.[15] Pictorial object recognition based on such features would be relatively unaffected by viewing angle, with no need to assume a compensation process.

For such an approach to have applied and theoretical consequences, we need more specific knowledge about the features by which objects are distinguished and recognized.

Until relatively recently, the framework for most perceptual theories was essen-

[15] Such *nonaccidental* properties are discussed by Biederman, 1985; Binford, 1971; Hoffman and Richards, 1985; Kanade and Kender, 1983; Lowe, 1985; and Richards and Hoffman, 1985. Actual application of these in pictorial perception has not been tested.

tially that of geometrical optics or retinal configurations. If instead (or in addition) object familiarity is taken as an important source of visual primitives, a framework more like category learning (or even the construal processes of discourse analysis) is needed. Although they may contribute somewhat to choosing the best canonical view of an object, the Gestalt laws and the local depth cues do not differentiate between familiar and unfamiliar objects. Biederman (1985) proposed a specific set of primitive components, or *geons,* by which all familiar noun-class objects (telephone, cat, etc.) are recognized very early in the first glance, both in the world and in its pictures. Because such geons are presumably redundant in defining any noun-class object, the occlusions provided by a cluttered environment would often not preclude identifying an object. Although the research that would test the identity and attributes of the proposed geons has not yet been done (or whether a model that relies on a specific set of components is viable),[16] the proposal is admirably researchable, with wide potential extensions to the set of questions on meaning and resonances (like puns and priming in language) that would be important to the concerns of Section III. In any case, it brings meaningful components shared by world and pictures to the fore, and raises the study of objects and their pictures (and perhaps of scenes as well (see Section III.B)) to a new level of relevance and specificity.

Note that in these experiments objects are recognized very rapidly, and that the set of geons that define an object are a product of familiarity. How early familiarity and meaning enter the visual process remains unclear. Even Hebb asserted that figure–ground formation, discussed in connection with Figure 2A, was primary (although he did not spell out what he meant by this, or why). First extract the edges, then identify the shapes they define. That is certainly one logical process for recognizing objects. But I have long questioned whether figure formation is fundamental (e.g., Hochberg, 1972b, 1974b); Peterson (1994) strongly challenged the credo that figure–ground segregation must precede the effects of object familiarity or *denotivity;* and Cavanagh (1987) pointed out that the features needed to identify some shape are often mutually indistinguishable before the object is recognized.

The analysis into geons, even if they work as described, is just a beginning. As noted earlier, where an object has some usual orientation, inverting its picture while leaving its components intact interferes more with identifying that object than it does with an unfamiliar one (Peterson, 1994; Peterson & Gibson, 1991; Peterson et al., 1991). Inversion drastically reduces the recognizability of pictured faces (e.g., Diamond & Carey, 1986; Hochberg & Galper, 1967; Valentine, 1988),

[16] Substantial research by Biederman and his colleagues has now shown that simple outline drawings of noun-class objects are recognized as fast or faster than photographs containing surface information, and that some parts of the outlines are more informative than others. We do not yet know how well the set of geons proposed fits the facts of object recognition (they will not do for the recognition of faces and other individual objects, as we note below), and whether they are independent enough in combination to serve as analytic components.

and of their expressions (Thompson, 1980), which do not yield to an analysis in terms of geons in any case. There are other things to represent than noun-class objects, other components to the recognition process than geons, other attributes or events to recognize and react to than the presence or identity of some object.

Naming responses and explicit recognition are not the only responses viewers can make to objects (see Cooper & Schachter, 1992), although they are by far the easiest to measure in experimental psychology. Objects have what we may call narrative significance, some of which they may share with their linguistic labels, but they have consequence of their own. To the Gestaltists, viewers perceive sadness, glee, menace, and value in objects as directly as they do color or distance (Koffka, 1935; see Sect. IIIa); whether or not such assertions are plausible about objects quite generally, it certainly seems true of those objects that are living creatures. And surely pictured faces are not just pictured 3-D objects: they carry character (Gombrich, 1972a; Secord & Muthard, 1955), and they therefore carry expected actions, as well (Hochberg & Galper, 1974). For that matter, objects themselves are not merely identifiable geometrical structures: their familiar usage, their ritual and symbolic functions, their mutual relationships with other objects in the real or pictured field of view—all support some narrative structures rather than others.

In a simple hard-nosed physicalist view, such *tertiary* properties must emerge late in the order of cognitive processing.[17] How early in the course of perceptual processing such responses are entrained is not known. In any case, one must certainly expect them to be involved in deciding where to glance in the next 250 ms, in how the viewer integrates and stores these glances, and in how the viewer can manipulate the memories of what was seen. For much of human history, most pictures serve as narratives, not merely as surrogates or labels for some layout of surfaces in space. Fine art has, in times past, typically conveyed overt stories,[18] and carried more esoteric references and social commentary as well. Most pictures, whether in galleries or magazines, do the same. Subject matter figures heavily in what people say they want in pictures when they are polled (see footnote 24). I do not see how one could step from perceptual theories dedicated to explaining pictures as specifiers of surface distance and orientation, which was how I opened this section, or explaining them in terms of Gestalt laws of organization, to what people use pictures for, and to what pictures can tell us about the perceptual process

[17] Simple sensory "ideas," such as 2-D locus, brightness, and color, are traditionally termed *primary;* the presumably derived distal properties that we perceive in the world, like depth, reflectance, object size, and so on, were classed as *secondary;* and those properties that were thought to be cognitively derived from the secondary properties (such as the intent expressed by the temporary disposition of some perceived person's face) were therefore taken as tertiary. Most of what is to be communicated by pictures would fall into this category.

[18] See White & White (1965) for a history of how the French National Academy both regulated these and made France preeminent. Historical, religious, and dramatic narrative have in this century essentially lost their importance for fine art; the biography of the artist, and the story and philosophy of what it is to make a picture, have since provided much of the discussable content; cf. Danto, 1986.

(as distinct from explicating the static depth cues). Even if geons should turn out not to work, they are at least a serious step toward an object-oriented theory of picture perception.

Such object-oriented approaches should be able to tackle many questions related to the visual arts far more readily than did the older metatheories, but they are even more closely tied to questions of representation. Representation is what perception as a discipline remains most prepared to discuss. The fact is, however, that most people who are now most interested in art are no longer particularly concerned with the representational function per se (although that was certainly not true when much of the most valued painting was done: see Alpers, 1983).

The two major perceptual yet *non*representational functions that the arts can serve are the main concerns of Section III.

III. NONREPRESENTATIONAL FUNCTIONS OF ARTISTIC PRESENTATIONS: EXPRESSIVE, AESTHETIC, AND ARCHITECTURAL

The expressive and aesthetic functions are far more prominent in writings on art than are questions of the fidelity of representation. The problem in discussing them is that although one can bring agreed-upon and usually measurable criteria to a discussion of representation, that necessary feature of any science vanishes as we proceed further into the nonrepresentational functions of art. To exacerbate the problem, although many different writers offer assertions (and often battle cries) about what is true, no systematic agreement now exists, and most value judgments are made according to principles that are deliberately not made public.

But that does not ensure that there is nothing to pursue. The two major functions here clearly exist, and reflect something about how—and why—people look at (and listen to) works of art, even if one cannot say exactly what that is. *Expressive* refers to feelings, to emotions, to attitudes, and to the self-expression of the artist; *aesthetic* refers (originally) to beauty, to the pleasure provided, and to whatever factors, including those arising from the other two functions (representational and expressive), engage the disinterested[19] evaluative attention of the audience.

A. Expression and Feeling

In one meaning of the term *expression,* artists portray a person's expressions and postures to communicate that person's feelings or character; the issue is then still one of representing that person. But the represented demeanors of the persons portrayed presumably express their feelings, whereas the proper use of the medium

[19] Disinterested in the sense that no extrinsic or *exogenous* motive (Kruglanski, 1975) is evident as the source of the evaluative attention.

expresses the feelings of the artist in a way the spectator can share. When Rembrandt's self portraits reveal the progressive decline of his body over the years (see Zucker, 1963), it is not Rembrandt the subject of the self-portrait but Rembrandt the painter who expresses this unflinching firmness (Sircello, 1965). In another and less representational meaning, a broad, jagged line may be called on to depict a blunt, harsh person, or used in drawing a threatening scene; a thin tremulous line and an unbalanced, tense composition may be used to express anxiety about some event or scene. And the way the artist's medium is used may itself be expressive without representing anything at all, as when one says that music *is* joyous, not that the music is about a joyous event (cf. Beardsley, 1958, 1965; Zink, 1960). In still another and most important sense of the word, expression is what the artist may express himself or herself, trying by choice of subject matter and style to attain a unique identity, one that may carry connotative meaning as well (e.g., being whimsical, excited, or brooding). Most artists use the elements of the medium not only as signatures (i.e., to identify themselves), but also as signals of their characteristic attitudes toward their subject matter as well (and to some degree, as constraints upon their subject matter). There is little research to support such statements and, as I argue later, little reason to undertake such research.

There are innumerable writings to the effect that color and composition in the visual arts (Ball, 1965; Kepes, 1944; Poore, 1903; Taylor, 1964); melodic structure, scale, and rhythm in music (Gutheil, 1948; Meyer, 1956); words and sounds in poetry and prose (Belknap, 1934; Pope, 1949; Wilson, 1931); and movements in dance (Davis, 1972; Kreitler & Kreitler, 1972; Sorell, 1966) are all endowed with expressive meaning. There is also a large but scattered body of experiments to this point, usually designed to show that expressive, or *physiognomic* judgments (Koffka, 1935) can be obtained from subjects who are shown such elements in a research rather than artistic context (Werner, 1948). Reviews of early research of this kind can be found in Hammond (1933) and in Chandler and Barnhart (1938); a large body of later research is referenced in Pickford (1972).

Such analyses are critical in all applied art, but especially so in advertising (and propaganda), where the connotation of words, of visual elements, of layout composition, of mood music, and so on, must relate appropriately and contribute as desired to the audience's image of the product or the person being represented. Relevant research (including what has reputedly been performed in-house and kept as trade secrets) is carefully weighed (albeit with a validity that is essentially untested in any publicly assessible way). If the psychology of expressive art languishes today, it is not for want of belief in its potential economic payoff.

What is missing is a well-developed and testable psychological theory in terms of which such research can be ordered and which it might inform. The Gestaltists, many of whom argued that at least some of these properties are as intrinsic to our experiences of objects as is their color, never offered a principled theory (for most recent discussion, see Epstein & Hatfield, 1994). The classical empiricist approach (e.g., associations shared because of elements in common, the prosodic aspects of the language that are associated with different classes of message, and the abstraction of a form or category to deal with experience) appears in a wide range of proposals like those of Osgood (1976) and Langer (1958). The empathy theory, as in Lipps's attempt to explain both aesthetics and the geometrical illusions (Lipps, 1900) in terms of an emotional or reactive response that is supposedly even made to relatively simple stimuli, is represented more or less directly by Gombrich (1972a) in his hypotheses about portrait perception, by Schillinger (1948) on music and (via identification with what he termed *modal-vectorial bodily functions and rates*) by Gardner (1973). Words, shapes, colors, and feelings may all share a single "isomorphic internal response"—as held by several Gestalt theorists (see Arnheim, 1954; Koffka, 1935), and in a complex way, by Smets (1973). Smets speculated that an aesthetic stimulus elicits those emotional and synesthetic connotations that also evoke the same degree of arousal (where "arousal" would be measured by desynchronization of alpha-wave activity), and reported such equivalences with colors, shapes, and connotative descriptions.[20]

Although writings on art often analyze some work in terms of the effects putatively aroused by its component elements, I know of no theory that provides combining rules. Without such combining rules, discussions about the expressive effects of the components are neither theoretically nor practically useful. We do not know whether or how the effect of a work of art reflects the effects of its parts as measured separately.[21] It seems clear enough that expressiveness can indeed be attributed to larger parts, or even whole works (especially those, like music and dancing, that are the normal means of celebrating such feelings), but very little research has been directed to this point. Subjects will refer to the emotional impact of abstract pictures and mood responses are made with some reliability to musical selections (Berger, 1970). Pickford (1972) and Child (1969) have done general reviews, and Berlyne and Oglivie (1974) and Pickford (1955) have done factor-analytic studies of subjects' responses to works of art. In such research, however, the subject must make some analytic response, as in words or rating scales, and one must question whether these can adequately represent the effect of an artistic presentation (cf. Gardner, 1973), especially with naive subjects.

For this reason, methods in which subjects reliably perform more objective tasks that are designed to reveal their sensitivity to nonrepresentational qualities of the art seem more valid. For example, subjects will reliably match titles (not necessarily those given by the artist) with abstract paintings; will correctly assign tops and bottoms to them (Lindauer, 1970); and are able to match paintings by Klee with

[20] More physiological arousal had previously been found in response to red than to blue (Wilson, 1966), and Smets (1973) reported that subjects matched colors, shapes, and verbal expressive concepts to each other in the same way that their arousal patterns were related (using the duration of desynchronization of alpha waves as the measure of arousal). What alpha-wave desynchronization tells here is of course another question.

[21] That is, of course, the problem which sank Structuralism (meaning the psychological school: Hochberg, 1972a).

the music that presumably inspired them (Minnigerode, Cianco, & Sbarboro, 1976; Peretti, 1972; Wechner, 1966). The method of work-to-work comparison has been used to measure the ability to judge whether works of art are by the same artist, using literary selections (Westland, 1968), musical selections (Gardner, 1972a), and paintings (Smets & Knops, 1976). Such procedures might avoid some of the problems inherent in standard tests of aesthetic judgment (Child, 1969) and provide an effective research tool for the measurement of artists and periods, as well.

Similar procedures might be applied to artistic style, which may be the most important expressive aspect of art.[22] The artist's choice of contents may itself amount only to another aspect of style (Sontag, 1961). Style (what is usually meant by "expressing one's personality," by those who use the phrase), has been an ever-increasing component of the art market and therefore of aesthetic development (cf. Grosser, 1971) since the Renaissance. Without a distinctive and memorable style, no pure artist (and few applied artists, like cartoonists, singers, or dress designers) can have a viable career. If that is true, and I think it is, one cannot really feel that one understands the perception of pictorial art until this aspect of expression is understood. The discussion in Section III.C may be relevant to this point.

Another and much more elusive way to characterize pictures, which is neither a matter of representational adequacy nor really one of expression (as it has been discussed in this section) is traditionally the concern of experimental aesthetics, which is considered next.

B. Art as Pleasurable or Engaging: Experimental Aesthetics and Preference

Experimental aesthetics was founded by Fechner (1876). The central thread is psychophysical: an attempt to predict the aesthetic value of stimuli from their identifiable properties. (Fechner also, of course, founded psychophysics.) Woodworth (1938) presented an admirable discussion of the field to his time; for early bibliographies, see Hammond (1933) and Chandler and Barnhart (1938). Reviews or collections of papers are found in Berlyne (1971, 1972, 1973, 1974), Child (1969), and Pickford (1972). Many of the papers include efforts to relate some aesthetic prescriptions to abstract mathematical formulae, to neurological speculations, or to some variety of stimulus–response motivation theory. Most do not take the aesthetic value of real art (e.g., museum paintings) as their subject; two recent attempts that do (Batovrin, 1993; Stephan, 1990), to which I return briefly in Section III.C, do not attempt to predict measurable hierarchies.

[22] For eloquent defense of this point, see for example, Tolstoy, 1899; Croce, 1915; Collingwood, 1938. There are also some to whom the major function of art education is to teach children to express themselves (Read, 1943; Gardner, 1973). From that viewpoint, the fact that the clearly recognizable individuality of children's drawings declines as they mature is offered as evidence that a decline in artistic ability has occurred, but of course that judgment rests on which definition of artistic function is emphasized.

I raise some of Woodworth's points, add a few, and then discuss why this field might concern perception psychologists. First, I feel that most research in experimental aesthetics really has nothing to do with the perception of beauty or the arousal of an aesthetic experience, and that subjects' preference judgments cannot support any simple interpretation. *For those very reasons,* however, I believe that research in this area is applicable to the appreciation of art (particularly, pure art). Although the criticisms raised in this regard are probably valid, this research area should remain of modest interest both for art and for psychology.

Woodworth pointed out that in experimental aesthetics the object of study is the response to the beautiful, the sublime, the tragic, the comic, or the pathetic. The response should depend on the subject's feeling rather than on intellectual perceptions or judgments. In the laboratory, however, the subject must surely take the questions to mean not How much feeling is aroused in you? but Is this object pleasing or displeasing? In this way the results belong under the heading of judgment or taste rather than feeling. Most research involved having the subject make rankings or choices according to preference, and Woodworth noted that the very fact that nearly everyone was able to select a most pleasing rectangle when Fechner solicited such judgments (in the process of testing claims that had been made about the "golden section," to which we return in a moment) was itself an important psychological result: "A mere rectangle, we might suppose, could have no esthetic effect one way or the other" (Woodworth, 1938, p. 385).

To these comments, I add that psychologists have known for decades that introspection will not serve to reveal directly the inner workings of our minds. The Helmholtzean view, which still was adherents (Gregory, 1993; Rock, 1993), that perception proceeds by *unconsciously* fitting the most probable explanation to the information we receive, should, as in the James-Lange theory of emotions and the attribution theory of social psychology (cf. Bem, 1967; Nisbett & Wilson, 1977; Schachter & Singer, 1962) apply as well to judgments about feelings and attitudes. Even if subjects could consciously observe their preferences, in general they usually seem to do more of what the situation demands than what they are ostensibly asked to do (Orne, 1962).

One might think that the last point can be disregarded because the stimuli used in laboratory studies of experimental aesthetics have often been random polygons or other relatively neutral patterns, with no inherent meanings to contaminate the findings. I do not believe, however, that the aesthetic-preference *task* can be a neutral one: It asks subjects to expose their tastes and their sensibilities, to make themselves vulnerable with regard to a dimension of preference having the strongest of social and intellectual connotations. (In fact, preference tests have also been used as personality tests: cf. Barron & Welsh, 1952.)

Furthermore, nonsense patterns may not be what they seem. First, they are not alone in the subject's field of judgment, because any set of stimuli implies the entire class of stimuli from which they can be inferred to have been generated. This is true of subjects' judgments of the patterns' "goodness" (Garner, 1966; Garner & Clement, 1963), which must surely interact with the subject's familiarity with the canons of culture. Second, the stimuli used are, by and large, random shapes of a sort that no reasonable person would spend a glance on outside of the experiment, and are surely not worth either arousal or preference.[23] What arousal and preference there are must, it seems to me, derive from the challenge of grasping the principles that should guide the choices. I will return to this point after sampling the research area.

Some steps have been taken to move experimental research on art to more meaningful pictures (cf. Lindauer, 1970; Wallach, 1959). As I noted, some factor analyses of similarity judgments or rating scales of various selections of real artworks have been reviewed by Berlyne and Ogilvie (1974) and Pickford (1972). Analytic, evaluative and educational writings about art deal almost exclusively with the artworks in museums and equivalent collections. To the cognoscenti, this provides a large but finite common culture to discuss, to study and to assess. To the perceptual psychologist who seeks generalizable data about visual aesthetics, it should be important to know whether museum and non-museum pictures differ intrinsically in how they affect most average viewers. In 1990 and 1991, Lindauer reported a pair of studies which found virtually no differences in subjects' preferences for (and judgments about) museum art and mass-produced art. Most recently, two Russian artists, Komar and Melamid (1993) commissioned a survey questionnaire on American public attitudes toward various aspects of art, including color, subject matter, style, and so on, which served as a guide for a painting tailored to that taste;[24] this event may arouse some interest among art historians and philosophers, but experimental psychologists cannot take the interviewee's data at face value. Regardless of how reliable or valid data on aesthetic responses (preference, looking time, etc.) may be, they are not informative to perceptual theory unless they allow some kind of stimulus measurement (whether through instruments or through judges) and some degree of generalization.

The bulk of the experimental research remains the work on color preference (summarized in Pickford, 1972), and on how the complexity or typicality of visual and auditory nonsense patterns affects subjects' interest in them and affects the subjects' judgments of preference or pleasingness (a great deal of the work with adults is summarized in Berlyne, 1974, some work with children is summarized in Gardner, 1973, and Pickford, 1972, and work with differential habituation of

[23] To most object-oriented perceptual theories (at least since Hebb), which make some degree of meaning and familiarity part of the earliest visual processing, arbitrary patterns of dots are complex, rather than simple, and findings based on nonsense patterns need to be validated with principled selections of more familiar and meaningful pictures.

[24] This tabulation revealed (among other things) that respondents expressed strong color preferences; that design factors are clearly important in shopping quite generally; that traditional style and landscapes are preferred in pictures; and that more diversity in taste is shown by those claiming greater experience with art. The survey is itself part of an artistic statement, and the questions it raises are more important than the answers it provides, which may nevertheless be useful to the companies that sell the framed pictures one occasionally see for sale in the lobby of the supermarket, and perhaps to advertisers. (In research on advertising, lists of what will attract readers' attention in a picture have long been compiled, based on direct or indirect measures of where they look in a target magazine.)

infants' looking at various kinds of stimuli is summarized in Cohen, 1976, and Olsen, 1976). Given that looking behavior is elective, what keeps viewers looking when no exogenous task requires it should surely be of interest to perceptual psychologists (practical considerations aside, which are not financially trivial). And given that exogenous motives are so numerous, sociologically conditioned, and diverse, progress here would seem to require the identification of some few core endogenous determinants. As Blich (1991) noted, there is a burden of allegedly central concepts in aesthetics too cumbersome to handle, and very few empirical tests. Continuing the theme raised at several points in this chapter (and in its 1978 version), I add that this burden is so cumbersome largely because of different artists' and writers' different agendas, competitive market for selling and investing in valuable pictures, and (absent an authoritative Academy) a lack of agreement on artistic value and how to measure it.

What I have therefore selected from the rich but tangled literature on experimental theoretical aesthetics reflects a path that I feel may be worth pursuing. Except for the Smets (1973) study, the color work does not seem to be of theoretical interest. Works on the effects of complexity and typicality may be of interest, because the theoretical statements hint at ways to incorporate other proposed aesthetic principles and allow connections with each other and with more general theories of visual perception and cognition.

Perhaps the three most famous aesthetic principles for achieving beauty in visual art are these: The *golden section* (that is, supposedly the most pleasing of proportions), which since antiquity was claimed to be the proportion in which the whole is to the larger part as the larger is to the smaller: $1/x = x/1 - x$, or $x = .618$ (in a rectangle, that would require one side to be .618 times the length of the other); Hogarth's *Line of Beauty* (an ogive, or S-curve), used as the main line of myriad works of painting, sculpture, ornamentation, and pottery; and *Polykleitos's Canon,* the Doryphoros, a statue providing a model and set of rules that seem a watershed in Greek statuary. Of these three, the first has been subject to the most research (reviewed by Woodworth, 1938, and Valentine, 1962). Rectangles with that proportion are, by and large, the central tendency of preference judgments. Why?

Witmer (1894) ascribed its preferred status to a pleasing unity of diverse parts; Weber (1931) proposed (in the *Journal of Applied Psychology,* we should note, where a fair amount of such work was published) that any figure sets its viewer the problem of seeing it as a unit and, if it is too easy to do so, interest is quickly lost, whereas too much difficulty spoils the aesthetic effect. These formulations reflect an age-old and developed theme: Beauty or pleasingness is some function of complexity and/or other factors that affect the ease with which the viewer can see it as a unit. The formulation sounds plausible (although we should also note that most recently, Boselie, 1992, found that subjects showed no special preferences for the golden section.) How might we extend it to other measurable stimuli, and what function will predict subjects' preference judgments (i.e., the so-called hedonic tone of the stimuli) from stimulus measures of the objects they are judging?

Birkhoff (1933) proposed that, within any class of objects, the aesthetic value is M

= O/C, where O is some measure of order and C is a measure of complexity. The means obtained from subjects' preferences for polygons he had constructed gave the same order as did his measure of M. There have been several failures to corroborate this model (Davis, 1936; Eysenck, 1968; Eysenck & Castle, 1970). Other quantitative models have been proposed. Adding a "pleasure center" to his nerve-net model for the detection of lines and angles, Rashevsky (1940) provided a good fit to Davis's data. A remarkable effort to provide the mathematical basis for manufacturing music according to his own theoretical formulation was published by Schillinger in 1948—with what effect, I do not know. Information-theory versions of Birkhoff's formula (Moles, 1966) take their somewhat less simplistic accounts of order from *subjective* predictability, or redundance, which should vary with learning and motivation (cf. Moles, 1966; Smets, 1973, measured subjective redundancy in her research by adapting a version of Attneave's guessing technique). Eysenck proposed an inverted-U-shaped function relating preference and complexity (Eysenck, 1968; Eysenck & Castle, 1970). So did Berlyne (1967), on the grounds that arousal (activation of a cortical reward system) increases linearly with complexity, whereas hedonic tone is greatest at an intermediate level of arousal (Hebb, 1955; Lindsley, 1957).

In Berlyne's proposal, the arousal potential of a stimulus pattern depends on several factors, including the pattern's intensity, its association with significant events, and its *collative properties.* These last are formal characteristics such as the pattern's variation along dimensions like familiar-novel, simple-complex, expected-surprising, and so forth. Arousal presumably increases with complexity (among other things), and hedonic tone is greatest at intermediate arousal levels, so hedonic tone should be an inverted-U function of complexity. Judgments of interest versus disinterest, however, and of complexity versus simplicity (and other verbal measures of arousal) should increase with the complexity of the stimulus (often measured in informational terms or uncertainty).

In many cases (Crozier, 1974; Dorfman & McKenna, 1966; Normore, 1974; Vitz, 1966; Walker, 1973; Wolhwill, 1968), the expected relationship between hedonic tone and complexity is found; in others, pleasantness or preference ratings increase monotonically with complexity (Hare, 1974a; Jones, 1964; Reich & Moody, 1970; Vitz, 1964). Reich and Moody (1970) even found pleasantness to decline with complexity, as in Birkhoff's proposal, when they used stimuli to which subjects had been habituated. As Smets (1973) pointed out, however, and demonstrated (using two-element matrix patterns that varied in redundancy as well as in number of elements; cf. also Snodgrass, 1971), given a nonmonotonic function, the part of the curve that one obtains depends on the range tested. The effective order, structure, or redundancy that a subject can discern should depend on the stimulus pattern's familiarity (cf. Goldstein, 1961; Harrison & Zajonc, 1970) and perhaps on the subject's artistic training (Hare, 1974b; Smets, 1973). It seems likely therefore that pleasingness and preference judgments are *not* a monotonic function of complexity in such experiments. And we should remember that where they are, that fact may reflect how viewers handle the preference task when confronting meaningless or nonsense shapes[25] (see pp. 181).

Alternative models can be fitted to these facts. In the McClelland, Atkinson, Clark, and Lowell (1953) "butterfly curve" proposal, a stimulus to which we have become habituated is neither pleasing nor displeasing. As it departs from this adaptation level (Helson, 1964), the stimulus passes through a maximum of pleasingness and finally becomes unpleasant and noxious. An application to stimulus complexity is reasonably straightforward (Terwilliger, 1963) and could serve as a testable amendment to theories that assign beauty or attractiveness to the average or to the prototypical (e.g., in faces; see Langlois & Roggman, 1990; Perett, May, & Yoshikawa, 1994, to support some such formulation; as to the need for amendment, see Alley & Cunningham, 1991; Hochberg, 1978. Recent results reviewed by Blich (1991), however, do not show reliable relationships between judged typicality and judged preference, but that may merely reflect problems with conscious judgments about typicality and perhaps the stimuli used.

In any case, something like the expected curves of preference and distance from norm has been found in some laboratory experiments (Day, 1967; Haber, 1958; Munsinger & Kessen, 1964), and the cycle of unpopularity, popularity, and neutrality through which popular songs and other fashions swing grant it considerable anecdotal plausibility (see also Wohlwill, 1966). We would expect therefore that small departures from some culturally familiar schema help motivate perceptual inquiry. Both the golden section (Fischer, 1969; Lalo, 1908) and Polykleitos's canon (Ruesch, 1977) have been claimed as cultural rather than mathematical norms. In fact, Ruesch argued that Polykleitos's canon embodies the central tendencies of actual early anthropometric measurements from which, once established, subsequent sculptors departed for specific effects. The higher attractiveness of photographically or digitally averaged faces, mentioned above, may reflect the same relationship.

A plausible parallel can thus be drawn between some laboratory findings, and at least some features of the less simplistic world of art. Missing from discussions of both, as I see it, is the question of motivation. In the tasks set by the traditional methods of experimental aesthetics, the subject agrees to judge the relative merits of some members of a stimulus set. That challenge, and not the inherent beauty or interest of the stimuli, must be what maintains the subject's interest. The more complex the stimulus, the more there is for the subject to consider before answering, and the more to sample, integrate, and store. It is not surprising, therefore, that looking and listening time increases with the complexity of the stimulus patterns, as in fact it does (Berlyne, 1974; Crozier, 1974; Faw & Nunnally, 1967; Hochberg & Brooks, 1962b, 1978), and as subjects' ratings of *interestingness* do, as well (reviewed by Berlyne, 1972). The latter ratings, in fact, increase *monotonically* with complex-

[25] As Normore's subjects asked spontaneously in one experiment, "How can a dot be beautiful?" (1974, p. 113)

ity, reflecting, I suggest, the subjects' continued search to find some order or principle in the pattern that will account for the occurrence and placement of most (or at least some) of the elements.

What about the hedonic tone associated with such schema-testing activities? Weber's explanation (p. 183) will do nicely: If a pattern is so simple that it offers no principle to generate and test, or if it is so complex that (given the subject's background and motivation) no discernible schema makes the pattern "right," it is not pleasing, because no perceptual achievement has rewarded the subject's efforts. Note that this makes the grasping and testing of the schema, not the complexity or the arousal per se, the basis of the hedonic tone, and that *a motive to undertake the task is needed: the hedonic tone is not inherent in the stimuli.*

There is much that such a schema-testing account must leave out, especially about aesthetic response to real pictures viewed as art objects.[26] The inexpressible feelings that sometimes arise, for which terms like *beauty, pleasure,* or *preference* seem inappropriate, are not addressed by schemas and their like. Words do not do well either in discussing visual aesthetics, but critics and philosophers of art try to meet the challenge. One strategy is to discuss how the picture affects unconscious, nonverbal, or preconscious levels. Because the affects are not consciously available to the writer or the viewer, some fairly detailed theory is needed about the workings of the unconscious. Variants of psychoanalytic discourse with its repertory of symbolic meanings, related sets of assumptions about the viewer's unconscious childhood memories, and analyses of metaphors (as inherent in the visual means of representation, not as provided by the "text" being represented; cf. Wollheim, 1987, pp. 308–315)—all of these have been used, with good or bad effect, to illuminate or elucidate museum art. Such analyses often start with what the viewer has already perceived explicitly, but point to other relationships entailed in composition, coloring, style, and so on.

In a related vein, Stephan (1990) granted the right hemisphere of the brain a fully developed world of visual associations not available to the linguistic left hemisphere (although recent studies by Biederman & Cooper, 1991, did not find the widely touted hemispheric superiority in recognizing drawn objects on which

[26] For example, Christine and Fred Attneave's verbal response to an earlier (spoken) version of this proposal was: "What about the pleasure of first seeing an intensely blue lake?" I think that such questions can be handled, but this is not the place to try to do so. Another troublesome problem concerns how much a member of the audience expects to fit into a given schema. Some amount of any artwork (particularly one in which the elements are presented over time under the artist's control, like music, dance, literature, or motion pictures) is *texture,* and is needed merely for verisimilitude or for filling. Some of the artwork will have *outcome* and be important to the final structure. In assessing how economical a work of art is, we probably should be attending not to the total complexity, but to that portion that the viewer or listener takes as part of the structure (i.e., the viewer's *subjective outcome structure*). In painting and in drawing, one does not take all of the brush strokes and the cross-hatches as significant elements, yet some of them, in each case, do serve special functions in the artist's design.

this theory rests); perhaps such clouds of partially aroused sensory associations (cf. Titchener's context theory of meaning) could be studied through the priming effects of geons, or other early visual-object components. Similar in some respects is Batovrin's (1993) concern with how art communicates significations which, in Kandinsky's words, are feelings for which we have no words;[27] in Batovrin's system, these arise from the detection of fractal order, not identified consciously, that is normally undetected in the chaos of cognitively unprocessed sensory data.

Although these various and ineffable responses are attributed to the visual stimulus offered by the work of art being discussed, it is not clear how they might be translated into perceptual research. That does not necessarily make them wrong, but it does at present leave them beyond experimental study. In any case, these examples are only a tiny sample of the many different criteria by which philosophers, critics, and artists judge works of art to be of greater or less artistic (or aesthetic) value. That means, I believe, that there are many things we mean by art: in fact, whatever can engage an appreciative and informed following. (It seems significant that the dealer who was "ahead of all others" in recognizing the merits of Cubism when no one else thought it of value, and was instrumental in its immense success in overcoming early opposition, could not tell what was good or bad in other later styles [Assouline, 1990]). And this brings us back to the schema-testing theme. In sophisticated art, an audience needs to be educated in what can be taken as the artist's premises and purposes, and in the tradition against which the artist makes his "statement." Without that education, or without the intention to perceive how the attributes of the picture fit each other and fit the schema provided by the tradition to which the work inescapably refers, there is nothing for the viewer to achieve. If one knows nothing about such art, there is then no way to know what one likes. For that matter, there may then be nothing to get the viewer engaged in the first place.

Although there may be many attributes of pictorial art that a schema-testing approach does not address, therefore, I can reclaim it for the educated aesthetic appreciation of even those works whose individual aesthetic value rests primarily on such evasive attributes. This is so because mobilizing and testing schemas is, in this approach, the process by which visual information is carried past the momentary sample provided by the individual glance. And that process is especially important where (as in much Abstract Expressionist art) there is no economical verbal narrative that can serve.

By this argument, much that is in fact not representational in nature or intent draws on the same cognitive processes by which we apprehend the structure of representational pictures. And that does not stop with pictures, as the last section attempts to show.

[27] Cited in Batovrin, 1993, p. 42.

C. Beyond Pictures: Grasping the Form and Order of Architectural and Environmental Structures (and of Nonspatial Logical Problems)

If pictures work because in some respects they are not that different from the world, then in those same respects the world is not that different from pictures. In Sections I and II, I argued this with respect to pictured and real objects, proposing that they share the same schematic maps. I think the same point can be applied to the setting of those objects, that is, to the structures of those surfaces and edges that lie beyond what one can quickly reach. Objects further than a few meters offer no useful accommodation, convergence, or binocular disparity; nor do moderate depth differences then offer useful parallax with head *rotations* (see Cutting & Vishton, 1995; Hochberg, 1980).

Of course, if the viewer moves laterally while keeping his or her gaze fixed on the edge between a nearer and farther building, or keeps that gaze fixed during a turn of the head so that foveal vision then receives the parallax to which the visual system is highly sensitive, potentially usable depth information then becomes available. As discussed in Section II.C, the invariants in the optical transformation that the viewer receives while moving in space offer information about surfaces and layouts that Gibson had proposed (1966, 1979) would provide for direct and veridical perception of the surfaces, layouts and the movements themselves. Architectural theorists have not missed this point (cf. Benedikt, 1979). Nonetheless, when the viewer's movements are small, the static depth cues may overcome the information theoretically available within the optic flow and once a depth arrangement is misconstrued, instead of being used as depth information the parallax can also be misconstrued as illusory concomitant motion (see p. 236n).

The static monocular (pictorial) cues are therefore mostly what moderately distant architectures and landscapes, in the real world, initially offer the moderately inactive observer. And because the potential field is much wider than in most picture viewing, and extends beyond the limits of peripheral vision, the task of integrating information from successive glances at different parts of the environment must span not only the time from one saccade to the next, but also the much longer times that head and body movements take.

Visual exploration under these wide-field conditions is even more elective than in pictures. When some part of the environment lies beyond peripheral vision, it can offer no invitation to look at it; neither can it offer any expectations (cues) about where desired information can be found, or offer a degraded view of what had been disclosed by previous glances. Structural schemas must be even more important in the perception of architecture (Hochberg, 1983) and of wide-field environments quite generally, and to the perception of their order and their aesthetics.

We use the term *object* for whatever maintains a relatively fixed structural relationship between its parts, despite eye movements or changes in the observer's viewpoint. The environmental or architectural objects considered here are large enough to exceed peripheral vision from a single standpoint: a large building, a complex, a piazza, a scenic landscape. The relevant architectural or environmental object schemas must (at least) act as maps of such structures if they are to guide observer's viewpoint changes, and to store the information then available. Within each glance, we would expect the structural cues (the Gestalt organizational factors and the local depth cues) to apply much as they do in pictures, providing expectations about which side of an edge is the occluding surface and what features are part of the same object. But the scale of the objects relative to the momentary visual field must have consequences.

One consequence is that the large-scale object's structure is more likely to be misconstrued. Given that the objects in question may extend beyond each glance, the "assumption" of good continuation must be particularly important, but is probably subject to error wherever roughly parallel contours of similar contrast (arising from surfaces' edges, occlusions, or corners) fall within successive glances. We have seen that an insensitivity to structural contradictions between local depth cues can be demonstrated even within a single glance (Fig. 2D). Misconstruals of architectural arrangements are probably quite likely, if good continuation is indeed more vulnerable to mistakes and given the greater distances between the local depth cues. (I will not here consider the argument that evolution surely must have assured our ability to perceive environments correctly, which I would be glad to debate in some other setting; cf. footnote 28).

Not all architects wish to avoid illusion (e.g., Eisenman & Gehry, 1991). It seems likely that those who are concerned with having a distinct and identifiable large-scale structure (i.e., one seen as the correct arrangement of surfaces and volumes in 3-D space) must work to avoid such misconstruals: correct construals are not automatic. (Of course, the viewer can usually work out the correct structure with some effort, but that would require first noticing the inconsistency, and then striving to reconstrue the view correctly. Escher's pictures of impossible buildings are suggestions of how much work a much larger structure might require.)

As a second consequence, the viewer may not be able to grasp readily, or to keep in mind, how the different views fit together. To perceive a large construction's overall form (or a region of a landscape or of a city) normally requires at least several glances. Because these successive views may be achieved through quite different behaviors (saccades, head turns, body turns, etc.), it seems unlikely that their information is integrated simply by placing them within a spatial framework or coordinate system. Unless the structure viewed is already a familiar one, which can be identified by some single feature and can summon up an object-centered map, reference points or landmarks are probably used in stitching the glances together. By landmark, I mean some feature or region that is recognizable to peripheral vision in all the glances in which it falls (Hochberg & Gellman, 1977; Lynch, 1960). Other locational indicators may be relative automatic ones (like the slant of sunlight, or the diagonals caused by convergent perspective; but see below), and the viewer can with effort parse the view by careful attention to the extraretinal signals. Although I know of no experimental research here, I can remember situations in which even time and effort were insufficient.[28]

Landmarks and regional recognizability should therefore be important in recognizing the overall form of such large-scale objects. A distinctive tower (Lynch, 1960), or perhaps any large nonrepeating asymmetry (Hochberg & Gellman, 1977), will do as landmarks. Biederman (1987), citing a dissertation by Mezanotte (1981), suggested that *types* of scene (equivalent, I think, to the class-noun objects to which his recognition-by-component proposal is addressed; see p. 175) are recognized very rapidly in terms of clusters of geons (see p. 175), which preserve the aspect ratio and relations of the largest visible geon within each object. If true, this should be important in constructing and evaluating recognizable architectural regions of environments, like neighborhoods.

But location with respect to a landmark or region is not of itself enough to make the form memorable from glance to glance, or from one point to another while walking through the structure or neighborhood. The use of simple Platonic forms (circle, square, etc.) and forms with what we would now term *nonaccidental properties* (Section II) are often stressed as important to architectural aesthetics (Prak, 1968): having well-defined familiar relationships between their parts perhaps informs the viewer where the other parts can be found. Indeed, the visual design of buildings and complexes is usually intended to offer a recognizable form (O'Neill, 1991) that itself serves multiple goals: to communicate the constructions' overall and more local functions, to be individually recognizable, to provide an ambience of time and place, to be interesting, and so on.

With so many goals that the visual form of environmental structures might fulfill, and with many choices probably to be made between them, there is clear occasion for evaluation and aesthetic judgment. Because there now seems to be no agreement about what the aesthetic goals of architecture should be (Shepheard, 1994), the perception psychologist can make a contribution at the center of the enterprise. Some of the less artistic and more popular bases for aesthetic judgments depend only trivially on what perception can offer (e.g., the pomp, the nostalgia, the evocation of earlier or different lives; on the bridge at Concord, or in Rue Danton, it is surely historical imagination and not the architectural form that invites my contemplation; much would be missing from an identical stimulus, an exact replica, in some theme park). Some aesthetic judgments may depend on factors that are clearly perceptual and that can be addressed by perceptual research

[28] Successive glances at a gigantic Jackson Pollock canvas, attempting to decipher a maze, or attempting to detect the inversion in Figure 1 should provide convincing demonstrations, outside the laboratory, that neither extraretinal signals, nor visual invariances under transformation, are enough for easy integration of views without memorable form or landmarks. These are all artificial ("unecological"), but the existence and nature of protective coloration in animals testifies to the general fallibility of the principles on which all visual systems rely (see Hochberg, 1978; Metzger, 1953).

tools (like studies of what makes architectural spaces seem more open or more closed; e.g., Gärling, 1969; Hayward & Franklin, 1974) but that hold no clear importance for current perceptual theory. Some questions of architectural or environmental aesthetics would probably interest both disciplines (e.g., does the smooth linear perspective provided by buildings of equal height, as in many Paris streets, provide vistas of greater perceived depth than the same underlying and invariant but jagged perspective provided by unequal heights, as in most New York streets). Some questions which can probably be raised and answered most naturally in the context of architectural and environmental structure are fundamental to an understanding of visual cognition quite generally.

One such question is that of *affordances* (Gibson, 1979) or *means-end readiness* (Tolman, 1948): the behaviors that some disposition of surfaces will support. Köhler's ape, Sultan, showed that affordances are not automatically invoked by the environment, but it still seems true that something important about the schemas that guide actions must rely on the parameters of environmental structure. Research on how the perception of stairs and doorways relates to their potential ease of use (Warren, 1984; Warren & Wang, 1987) shows that it is possible to give the concepts of affordances in particular and schemas in general more specific content.

A much larger and fundamental question that comes to the fore in the perception of large-scale structures is that of the schematic map, versions of which I held were essential in looking at pictures (Section II) and important in aesthetics (Section III.B). In comprehending architecture and urban environments quite generally, the need for such a concept becomes unavoidable. As long as one talks about stimulus information, which one can do readily with respect to representational pictures and somewhat less readily with nonrepresentational art, the guiding schemas, which are unobservable and difficult to specify and study, can simply be ignored. But those not well schooled in a particular architectural construction or city neighborhood must not only stitch their glances together in order to grasp the form of the structure, but must be able to consult that form, and derive answers from it to guide their behaviors when the overall view is no longer available. Prosthetic devices can help: With map in hand or in mind, the wanderer can traverse an unfamiliar building or a landmarkless neighborhood as though its overall structure were clearly in view and fully grasped.

How well an architectural form is grasped is both practically and aesthetically significant, but it is of cognitive importance as well. Maps and diagrams may guide behaviors in mechanical ways that make it unnecessary to grasp the overall form, but maps and diagrams themselves may be more or less difficult to grasp and to remember (cf. Tufte, 1990). A remembered map substitutes for the schema that the viewer must form to integrate and store successive glances; studying what makes such aids most useful for that purpose should tell much about schemas and their characteristics. For there is no reason to believe that the schematic maps that are used in exploring and regenerating perceived architectural form are utterly different from those used in sampling and storing a picture's representational content;

from those used in discerning the cultural context of that painting (or of any other art object); or those that have proved to be helpful in grasping the meaning of data (Wickens, Merwin, & Lin, 1994) or in reasoning one's way through a difficult problem in logic (Bauer & Johnson-Laird, 1993) and may be used, less formally, in reasoning quite generally. Perhaps the architectural and environmental schema may require more assistance, and be open to such assistance, just because of the scale of time and space over which they must be consulted.

Architecture is obviously not merely a perceptual problem. A building or a piazza or a neighborhood each has its own set of severe economic, mechanical, and historical constraints, and its functions are not normally limited to visually pleasing or impressing the viewer. And the stakes are clearly different, with the architectural structures being more of a salient "statement" by the owners than are paintings. And finally, although there are the odd exceptions, architectural constructions rarely represent the other things and people in our environment, and therefore do not raise the immediate epistemologically flavored questions that pictures do. But on reflection, architecture does raise many other of the problems that picture perception does, and offers a different viewpoint from which to consider the solutions to those problems.

References

Alpers, S. (1983). *Dutch art in the seventeenth century: The art of describing.* Chicago: University of Chicago Press.

Alley, T. R., & Cunningham, M. R. (1991). Averaged faces are attractive, but very attractive faces are not average. *Psychological Science, 2,* 123–125.

Angier, R. P. (1903). The aesthetics of unequal division. *Psychology Review, Monograph Supplement, 4,* 541–561.

Antes, J. R. (1974). The time course of picture viewing. *Journal of Experimental Psychology, 103,* 162–170.

Arnheim, R. (1943). Gestalt and art. *Journal of Aesthetics and Art Criticism, 2,* 71–75.

Arnheim, R. (1954). *Art and visual perception.* Berkeley: University of California Press.

Arnheim, R. (1966). *Toward a psychology of art.* Berkeley: University of California Press.

Arnheim, R. (1969). *Visual thinking.* Berkeley: University of California Press.

Arnheim, R. (1990). Perceptual aspects of art for the blind. *Journal of Aesthetic Education, 24,* 57–65.

Assouline, P. (1990). *An artful life: A biography of D. H. Kahnweiler, 1884–1979.* Translated by C. Ruas. New York: G. Weidenfeld.

Attneave, F. (1957a). Some informational aspects of visual perception. *Psychological Review, 61,* 183–193.

Attneave, F. (1957b). Physical determinants of the judged complexity of shapes. *Journal of Experimental Psychology, 53,* 221–227.

Attneave, F., & Frost, R. (1969). The discrimination of perceived tridimensional orientation by minimum criteria. *Perception & Psychophys Physics, 6,* 391–396.

Ball, U. K. (1965). The aesthetics of color: A review of fifty years of experimentation. *Journal of Aesthetics and Art Criticism, 23,* 441–452.

Barron, F., & Welsh, G. S. (1952). Artistic perception as a factor in personality style: Its measurement by a picture–preference test. *American Journal of Psychology, 33,* 199–203.

Bartlett, F. C. (1932). *Remembering.* Cambridge: Cambridge University.

Batovrin, S. (1993). *The ecology of meaning.* New York: Tsefar.

Bauer, M. I., & Johnson-Laird, P. N. (1993). *Psychological Science, 4,* 372–378.

Baxandall, M. (1995). Shadows and enlightenment. New Haven: Yale University Press.

Beardsley, M. (1958). *Aesthetics: Problems in the philosophy of criticism.* New York: Harcourt Brace.

Beardsley, M. (1965). On the creation of art. *Journal of Aesthetics and Art Criticism, 23,* 291–304.

Belknap, G. N. (1934). *Guide to reading in aesthetics and theory of poetry,* Eugene: University of Oregon Publication, 4, 9.

Bem, D. J. (1967). Self-perception: An alternative interpretation of cognitive dissonance phenomena. *Psychological Review, 74,* 188–200.

Benedikt, M. (1979). To take hold of space: Isovists and isovist fields. *Environment and Planning, B6,* 47–65.

Berger, I. (1970). Affective response to meaningful sound stimuli. *Perceptual and Motor Skills, 30,* 842.

Berlyne, D. (1967). Arousal and reinforcement. In D. Levine (Ed.), *Nebraska Symposium on Motivation* (pp. 1–110). Lincoln, NE: University of Nebraska Press.

Berlyne, D. E. (1971). *Aesthetics and psychobiology.* New York: Appleton.

Berlyne, D. E. (1972). Ends and means of experimental aesthetics. *Canadian Journal of Psychology, 26,* 303–325.

Berlyne, D. E. (1973). The vicissitudes of aplopathematic and thelematoscopic pneumatology (or the hydrography of hedonism). In D. E. Berlyne & K. B. Madsen (Eds.), *Pleasure, reward, preference* (pp. 1–33). New York: Academic Press.

Berlyne, D. E., McDonnell, P., Nicky, R. M., & Parham, L. C. (1967). Effects of auditory pitch and complexity on E.E.G. desynchronization and on verbally expressed judgments. *Canadian Journal of Psychology, 21,* 346–367.

Berlyne, D. E., & Ogilvie, J. C. (1974). Dimensions of perception of paintings. In D. E. Berlyne (Ed.), *Studies in the new experimental aesthetics* (pp. 181–226). Washington, DC: Hemisphere.

Biederman, I. (1972). Perceiving real world scenes. *Science, 177,* 77–80.

Biederman, I. (1985). Human image understanding: Recent research and a theory. *Computer Vision, Graphics, and Image processing, 32,* 29–73.

Biederman, I. (1987). Matching image edges to object memory. *Proceedings of the IEEE First International Conference on Computer Vision* (pp. 384–392). London: Computer Society Press.

Biederman I., & Cooper, e. (1991) Object recognition and laterality: Null effects. *Neuropsychologia, 29,* 685–694.

Biederman, I., & Ju, G. (1988). Surface versus edge-based determinants of visual recognition. *Cognitive Psychology, 20,* 38–64.

Binford, T. (1971). Visual perception by computer. *Proceedings, IIEEE conference on systems science and cybernetics.* Miami, FL:

Birkhoff, G. (1933). *Aesthetic measure.* Cambridge, MA: Harvard University Press.

Blich, B. (1991). Pictorial realism. *Empirical Studies of the Arts, 9,* 175–189.

Boselie, F. (1992). The Golden Section has no special aesthetic attractivity. *Empirical Studies of the Arts, 10,* 1–18.

Bouleau, C. (1963). *The painter's secret geometry: A study of composition in art.* New York: Thames, Hudson & Harcourt.

Braun, J., & Sagi, D. (1990). Vision outside the focus of attention. *Perception & Psychophysics, 48,* 45–58.

Brookner, A. (1980). *Jaques-Louis David.* London: Chatto & Windus.

Brunswik, E. (1956). *Perception and the representative design of psychological experiments* (2nd ed.) Berkeley: University of California Press.

Brunswik, E., & Kamiya, J. (1953).Ecological cue-validity of "proximity" and other Gestalt factors. *American Journal of Psychology, 66,* 20–32.

Buswell, G. T. (1935). *How people look at pictures.* Chicago: University of Chicago Press.

Carroll, N. (1988). *Mystifying movies: Fads and fallacies of contemporary film theory.* New York: Columbia University Press.

Cavanagh, P. (1987). Reconstructing the third dimension: Interaction between color, texture, motion, binocular disparity and shape. *Computer Vision, Graphics and Image Processing, 37,* 171–195.

Chandler, A., & Barnhart, E. (1938). *A bibliography of physiological and experimental esthetics.* Berkeley: University of California Press.

Chapanis, A., & McCleary, R. A. (1953). Interposition as a cue for the perception of relative distance. *Journal of General Psychology, 48,* 113–132.

Child, I. (1969). Esthetics. *Annual Review of Psychology, 23,* 669–694.

Clerici, F. (1954). The grand illusion. *Art News Annual, 23,* 98–180.

Cohen, L. B. (1976). Habituation of infant visual attention. In T. J. Tighe & R. N. Leaton (Eds.), *Habituation* (pp. 207–238). Hillsdale, NJ: Erlbaum.

Collingwood, R. G. (1938). *The principles of art.* Oxford: Clarendon.

Cooper, L. A., & Schachter, D. L. (1992). Dissociations between structural and episodic representation of visual objects. *Current Directions in Psychological Science, 1,* 141–146.

Croce, B. (1915). *Breviary of aesthetic* (Vol. 2). Houston: Rice Institute Pamphlet.

Crozier, J. B. (1974). Verbal and exploratory responses to sound sequences varying in uncertainty level. In D. E. Berlyne (Ed.), *Studies in the new experimental aesthetics.* Washington, DC: Hemisphere.

Cutting. J. E. (1988). Affine distortions of pictorial space: Some predictions for Goldstein (1987) that La Gournier (1859) might have made. *Journal of Experimental Psychology: Human Perception and Performance, 14,* 305–311.

Cutting, J. E., & Millard, R. T. (1984). Three gradients and the perception of flat and curved surfaces. *Journal of Experimental Psychology: General, 113,* 198–216.

Cutting, J. E., & Vishton, P. M. (1995). Perceiving layout and knowing distances: The interaction of relative potency, and contextual use of different information about depth. In W. Epstein & S. J. Rogers (Eds.) *Handbook of Perception and cognition. Vol. 5: Perception of space and motion.* (Chapter 11). San Diego, CA: Academic Press.

Danto, A. (1986). *Philosophical disenfranchisement of art.* New York: Columbia University Press.

Davis, M. (Ed.). (1972). *Research approaches to movement and personality.* New York: Arno.

Davis, R. C. (1936). An evaluation and test of Birkhoff's aesthetic measure formula. *Journal of General Psychology, 15,* 231–240.

Day, H. (1967). Evaluations of subjective complexity, pleasingness and interestingness for a series of random polygons varying in complexity. *Perception & Psychophysics, 2,* 281–286.

DeLucia, P., & Hochberg, J. (1991). Geometrical illusions in solid objects under ordinary viewing conditions. *Perception & Psychophysics, 50,* 547–554.

Deregowski, J. B. (1968). Difficulties in pictorial depth perception in Africa. *British Journal of Psychology, 59,* 195–204.

Diamond, R., & Carey, S. (1986). Why faces are and are not special: An effect of expertise. *Journal of Experimental Psychology: General, 115,* 107–117.

Dinnerstein, D., & Wertheimer, M. (1957). Some determinants of phenomenal overlapping. *American Journal of Psychology, 70,* 21–37.

Dorfman, D., & McKenna, H. (1966). Pattern preference as a function of pattern uncertainty. *Canadian Journal of Psychology, 20,* 143–153.

Dosher, B. A., & Corbett, A. T. (1982). Instrument inferences and verb schemas. *Memory and Cognition, 10,* 531–539.

Eisenman, P., & Gehry, F. (1991). *International Architectural Exhibition, 1991. Venice, Italy.* New York: Rizzoli.

Enright, J. T. (1991). Paradoxical monocular stereopsis and perspective vergence. In S. R. Ellis, M. K. Kaiser, & A. C. Grunwald (Eds.), *Pictorial communication in virtual and real environments* (pp. 567–576). New York: Tayor & Francis.

Epstein, W., & Hatfield, G. (1944). Gestalt psychology and the philosophy of mind. *Philosophical Psychology, 7,* 163–181.

Eysenck, H. J. (1968). An experimental study of aesthetic preference for polygonal figures. *Journal of General Psychology, 79,* 3–17.

Eysenck, H. J., & Castle, M. (1970). Training in art as a factor in the determination of preference judgments for polygons. *British Journal of Psychology, 61,* 65–81.

Farber, J., & Rosinski, R. R. (1978). Geometric transformations of pictured space. *Perception, 7,* 269–282.

Faw, T. T., & Nunnally, J. C. (1967). The effects on eye movements of complexity, novelty and affective tone. *Perception & Psychophysics, 2,* 263–267.

Fechner, G. (1876). *Vorschule der aesthetik* [Elementary aesthetics]. Leipzig: Breitkopf & Hartel.

Fischer, R. (1969). Out on a (phantom) limb. Variations on a theme: Stability of body image and the Golden Section. *Perspectives in Biology & Medicine, 12,* 259–273.

Gardner, H. (1972a). The development of sensitivity to artistic styles. *Journal of Aesthetics and Art Criticism, 29,* 515–527.

Gardner, H. (1972b). Style sensitivity in children. *Human Development, 15,* 325–338.

Gardner, H. (1973). *The arts and human development.* New York: Wiley.

Gärling, T. (1969). Studies in visual perception of architectural spaces and rooms: I. Judgment scales of open and closed spaces; II. Judgments of open and closed space by category rating and magnitude estimation. *Scandinavian Journal of Psychology, 10,* 250–268.

Garner, W. R. (1966). To perceive is to know. *American Psychology, 21,* 11–19.

Garner, W. R., & Clement, D. E. (1963). Goodness of pattern and pattern uncertainty. *Journal of Verbal Learning and Verbal Behavior, 2,* 446–452.

Ghent, L. (1956). Perception of overlapping and imbedded figures by children of different ages. *American Journal of Psychology, 69,* 575–587.

Gibson, E. J. (1969). *Principles of perceptual learning and development.* Englewood Cliffs, NJ: Prentice-Hall.

Gibson, J. J. (1954). A theory of pictorial perception. *Audio-Visual Communications Review, 1,* 3–23.

Gibson, J. J. (1950). *The visual world.* Boston: Houghton Mifflin.

Gibson, J. J. (1951). What is form? *Psychology Review, 58,* 403–412.

Gibson, J. J. (1966). *The senses considered as perceptual systems.* Boston: Houghton Mifflin

Gibson, J. J. (1971). The information available in pictures. *Leonardo, 4,* 27–35.

Gibson, J. (1979). *The ecological approach to visual perception.* Boston: Houghton-Mifflin.

Gillam, B. (1978). A constancy-scaling theory of the Müller-Lyer illusion. In J. P. Sutcliffe (Ed.), *Conceptual analysis and method in psychology: Essays in honor of W. M. O'Neil* (pp. 55–70). Sydney: Sydney University Press.

Gillam, B. (1979). Even a possible figure can look impossible. *Perception, 8,* 229–232.

Ginsburg, A. (1980). Specifying relevant spatial information for image evaluation and display design: An explanation of how we see objects. *Perception & Psychophysics, 21,* 219–228.

Glanzer, M., Fischer, B., & Dorfman, D. (1984). Short-term storage in reading. *Journal of Verbal Learning and Verbal Behavior, 23,* 467–486.

Gogel, W. C., Tietz, J. D. (1992). Absence of computation and reasoning-like processes in the perception of orientation in depth. *Perception & Psychophysics, 51,* 309–318.

Goldstein, A. G. (1961). Familiarity and apparent complexity of random shapes. *Journal of Experimental Psychology, 62,* 594–597.

Goldstein, E. B. (1979). Rotation of objects in pictures viewed at an angle: Evidence for different properties of two types of pictorial space. *Journal of Experimental Psychology: Human Perception and Performance, 5,* 78–87.

Goldstein, E. B. (1987). Spatial layout, orientation relative to the observer, and perceived projection in pictures viewed at an angle. *Journal of Experimental Psychology: Human Perception and Performance, 13,* 256–266.

Goldstein, E. B. (1991). Perceived orientation, spatial layout and the geometry of pictures. In S. R. Ellis, M. K. Kaiser, & A. C. Grunwald (Eds.), *Pictorial communication in virtual and real environments* (pp. 480–485). New York: Taylor & Francis.

Gollin, E. S. (1960). Developmental studies of visual recognition of incomplete objects. *Perceptual and Motor Skills, 11,* 289–298.

Gombrich, E. H. (1956). *Art and illusion.* New York: Pantheon.

Gombrich, E. H. (1963). *Meditations on a hobby-horse.* London: Phaidon.

Gombrich, E. H. (1972a). The mask and the face: The perception of physiognomic likeness in life and art. In E. H. Gombrich, J. Hochberg, & M. Black (Eds.), *Art, perception and reality.* Baltimore: the Johns Hopkins Univ. Press.

Gombrich, E. H. (1972b). The "What" and the "How": Perspective representation and the phenomenal world. In R. Rudner & Israel Sckeffler (Eds.), *Logic and art, essays in honor of Nelson Goodman.* Indianapolis, IN: Bobbs-Merrill.

Gombrich, E. H. (1984). *Sense of order: A study in the psychology of decorative art.* Ithaca, New York: Cornell University Press.

Goodman, N. (1968). *Languages of art: An approach to a theory of symbols.* Indianapolis, IN: Bobbs-Merrill.

Graham, N. (1989). *Visual pattern analyzers.* New York: Oxford University Press.

Gregory, R. L. (1970). *The intelligent eye.* New York: McGraw-Hill.

Gregory, R. L. (1980). Perception as hypotheses. In H. C. Longuet-Higgens & N. S. Sutherland (Eds.), *The psychology of vision* (pp. 137–149). London: The Royal Society.

Gregory, R. L. (1993). Seeing and thinking. *Giornale Italiano di Psicologia, 20,* 749–769.

Grice, H. (1968). Utterer's meaning, sentence-meaning and word-meaning. *Foundations of Language, 4,* 225–242.

Grosser, M. (1971). *Painter's progress.* New York: Potter.

Gutheil, E. (1948). *Music and your emotions.* New York: Liveright.

Guzman, A. (1969). *Computer recognition of three-dimensional objects in a visual scene.* Unpublished Ph.D. dissertation, M.I.T.

Haber, R. N. (1958). Discrepancy from adaptation level as a source of affect. *Journal of Experimental Psychology, 56,* 370–375.

Haber, R. N. (1985). Three frames suffice: Drop the retinotopic frame. *Behavioral and Brain Sciences, 8,* 295–296.

Haber, R. N., Haber, L. R., Levin, C. A., & Hollyfield, R. (1993). Properties of spatial representations: Data from sighted and blind subjects. *Perception & Psychophysics, 54,* 1–13.

Hagen, M. A. (1974). Picture perception: Toward a theoretical model. *Psychological Bulletin, 81,* 471–497.

Hagen, M. A. (1976). Influence of picture surface and station point on the ability to compensate for oblique view in pictorial perception. *Developmental Psychology, 12,* 57–63.

Hagen, M. A., & Jones, R. K. (1978). Differential patterns of preference for modified linear perspective in children and in adults. *Journal of Experimental Child Psychology, 26,* 205–215.

Hagen, M. A., & Elliot, H. B. (1976). An investigation of the relationship between viewing condition and preference for true and modified linear perspective. *Journal of Experimental Psychology: Human Perception and Performance, 2,* 479–490.

Hammond, W. A. (1933). *A bibliography of aesthetics and of the philosophy of the fine arts from 1900 to 1932.* New York: Longmans, Green.

Hare, F. G. (1974a). Verbal responses to visual patterns varying in distributional redundancy and in variety. In D. E. Berlyne (Ed.), *Studies in the new experimental aesthetics* (pp. 159–168). Washington, DC: Hemisphere.

Hare, F. G. (1974b). Artistic training and response to visual auditory patterns varying in uncertainty. In D. E. Berlyne (Ed.), *Studies in the new experimental aesthetics* (pp. 169–173). Washington, DC: Hemisphere.

Harrison, A. A., & Zajonc, R. B. (1970). The effects of frequency and duration of exposure on response competition and affective ratings. *Journal of Psychology, 75,* 163–169.

Hayward, S. C., & Franklin, S. S. (1974). Perceived openness–enclosure of architectural space. *Environment and Behavior, 6,* 37–51.

Hebb, D. (1949). *The organization of behavior.* New York: Wiley.

Hebb, D. (1955). Drives and the C.N.S. *Psychology Review, 62,* 243–254.

Helmholtz, H. L. F. von. (1909/1924). *Treatise on physiological optics.* Vol. III (Trans. from the 3rd German ed., 1909–1911, J. P. C. Southall, Ed. and Trans.) Rochester, NY: Optical Society of America, 1924–1925.

Helson, H. (1964). *Adaptation level theory.* New York: Harper & Row.

Hess, R. F., & Field, D. (1993). Is the increased spatial uncertainty in the normal periphery due to spatial undersampling or uncalibrated disarray? *Vision Research, 33,* 2663–2670.

Hills, P. (1987). *The light of early Italian painting.* New Haven: Yale University Press.

Hochberg, J. (1962). The psychophysics of pictorial perception. *Audio-Visual Communications Review, 10,* 22–54.

Hochberg, J. (1968). In the mind's eye. In R. N. Haber (Ed.), *Contemporary theory and research in visual perception* (pp. 309–331). New York: Holt, Rinehart & Winston.

Hochberg, J. (1970). Attention, organization and consciousness. In D. I. Mostofsky (Ed.), *Attention: Contemporary theory and analysis* (pp. 99–124). New York: Appleton-Century-Crofts.

Hochberg, J. (1971). Pirenne's optics, painting and photography. *Science, 172,* 685–686.

Hochberg, J. (1972a). The representation of things and people. In E. H. Gombrich, J. Hochberg, & M. Black, (Eds.), *Art, perception and reality.* Baltimore: The Johns Hopkins University Press.

Hochberg, J. (1972b). Perception II. Space and movement. In J. W. King & L. A. Riggs (Eds.), *Woodworth & Schlosberg's experimental psychology* (pp. 395–550). New York: Holt, Rinehart & Winston.

Hochberg, J. (1974a). Higher-order stimuli and interresponse coupling in the perception of the visual world. In R. B. Macleod & H. L. Picks (Eds.), *Perception: Essays in honor of James J. Gibson* (pp. 17–39). Ithaca, NY: Cornell Univ. Press.

Hochberg, J. (1974b). Organization and the Gestalt tradition. In E. C. Carterette & M. Friedman (Eds.), *Handbook of perception.* (Vol. I, pp. 179–210). New York: Academic Press.

Hochberg, J. (1978). Art and perception. In E. C. Carterette & M. Friedman (Eds.), *Handbook of perception.* (Vol. I0, 257–304). New York: Academic Press.

Hochberg, J. (1979). Some of the things that paintings are. In C. F. Nodine & D. F. Fisher (Eds.), *Perception and pictorial representation* (pp. 17–41). New York: Praeger.

Hochberg, J. (1980). Pictorial function and perceptual structures. In M. A. Hagen (Ed.), *The perception of pictures.* (Vol. 2, pp. 47–93). New York: Academic Press.

Hochberg, J. (1982). How big is a stimulus? In J. Beck (Ed.), *Organization and representation in perception* (pp. 191–217). Hillsdale, NJ: Erlbaum.

Hochberg, J. (1983). Visual perception in architecture. *Via, 6,* 26–45.

Hochberg, J. (1984). The perception of pictorial representations. *Social Research, 51,* 841–862.

Hochberg, J. (1987). Machines should not see as people do, but must know how people see. *Computer Vision, Graphics, and Image Processing, 37,* 221–237.

Hochberg, J. (1994). Construction of pictorial meaning. In T. A. Sebeok & J. Umiker-Sebeok (Eds.), *Advances in visual semiotics: The semiotic web 1992–93* (pp. 110–162). Berlin: Mouton de Gruyter.

Hochberg, J., & Beer, J. (1991). Illusory rotations from self-produced motions: The Ames Window effect in static objects. *Proceedings of the Eastern Psychological Association, April,* 34 (Abstract).

Hochberg, J., & Brooks, V. (1960). The psychophysics of form: Reversible-perspective drawings of spatial objects. *American Journal of Psychology, 73,* 337–354.

Hochberg, J., & Brooks, V. (1962a). Pictorial recognition as an unlearned ability: A study of one child's performance. *American Journal of Psychology, 75,* 624–628.

Hochberg, J., & Brooks, V. (1962b). The prediction of visual attention to designs and paintings. *American Psychologist, 17,* abstract.

Hochberg, J., & Brooks, V. (1978). Film cutting and visual momentum. In R. Monty & J. Senders (Eds.), *Eye movements and psychological processes, II.* Hillsdale, NJ: Erlbaum.

Hochberg, J., & Galper, R. E. (1967). Recognition of faces: I. An exploratory study. *Psychonomic Science, 9,* 619–620.

Hochberg, J., & Galper, R. E. (1974). Attribution of intention as a function of physiognomy. *Memory & Cognition, 2,* 39–42.

Hochberg, J., & Gellman, L. (1977). The effects of landmark features on mental rotation times. *Memory & Cognition, 5,* 23–26.

Hochberg, J., & MacAlister, E. (1953). A quantitative approach to figural "goodness." *Journal of Experimental Psychology, 46,* 361–364.

Hochberg, J., & Peterson, M. A. (1987). Piecemeal organization and cognitive components in object perception: Perceptually coupled responses to moving objects. *Journal of Experimental Psychology: General, 116,* 370–380.

Hoffman, D. D., & Richards, W. A. (1985). Parts of recognition. *Cognition, 18,* 65–96.

von Hofsten, C., & Lindhagen, K. (1980). Perception of visual occlusion in 4 ½-month-old infants. *Uppsala Psychological Reports, 290.*

Horn, B. K. P. (1977). Understanding image intensities. *Artificial Intelligence, 8,* 201–231.

Horn, B. K. P. (1981). Hill-shading and the reflectance map. *Proceedings of the IEEE, 19,* 14–47.

Hudson, W. (1962). Pictorial depth perception in sub-cultural groups in Africa. *Journal of Social Psychology, 52,* 183–208.

Hudson, W. (1967). The study of the problem of pictorial perception among unculturated groups. *International Journal of Psychology, 2,* 89–107.

Irwin, D. E., Zacks, J. L., & Brown, J. H. (1990). Visual memory and the perception of a stable visual environment. *Perception & Psychophysics, 47,* 35–46.

Jahoda, G., & McGurk, H. (1974). Pictorial depth perception in Scottish and Ghanaian children: A critique of some findings with the Hudson test. *International Journal of Psychology, 9,* 255–267.

Jameson, D., & Hurvich, L. M. (1975). From contrast to assimilation: In art and in the eye. *Leonardo, 8,* 125–131.

Jones, A. (1964). Drive and the incentive variables associated with the statistical properties of sequences of stimuli. *Journal of Experimental Psychology, 67,* 423–431.

Jones, R. K., & Hagen, M. A. (1980). A perspective on cross-cultural picture perception. In M. A. Hagen (Ed.), *The perception of pictures, II* (pp. 193–226). New York: Academic Press.

Kanade, T., & Kender, J. R. (1983). Mapping image properties into shape constraints: Skewed symmetry, affine-transformable patterns, and the shape-from-texture paradigm. In J. Beck, B. Hope, & A. Rosenfeld (Eds.), *Human and machine vision* (pp. 237–257). New York: Academic Press.

Kanizsa, G. (1985). Seeing and thinking. *Acta Psycologica, 59,* 23–33.

Kanizsa, G., & Gerbino, W. (1982). Amodal completion: seeing or thinking? In J. Beck (Ed.), *Organization and representation in perception* (pp. 167–190). Hillsdale, NJ: Erlbaum.

Kellman, P. J., & Shipley, T. F. (1992). Perceiving objects across gaps in space and time. *Current Directions, 1,* 193–199.

Kemp, M. (1990). *Science of art: Optical themes in Western art from Brunellschi to Seurat.* New Haven, CT: Yale University Press.

Kennedy, J. M. (1974). *A psychology of picture perception.* San Francisco: Jossey-Bass.

Kennedy, J. M. (1977). Ancient and modern picture-perception abilities in Africa. *Journal of Aesthetics and Art Criticism, 35,* 293–300.

Kennedy, J. M. (1993). *Drawing and the blind: Pictures to touch.* New Haven, CT: Yale University Press.

Kennick, W. (1958). Does traditional esthetics rest on a mistake? *Mind, 68,* 317–334.

Kepes, G. (1944). *Language of vision.* Chicago: Theobald.

Kilbride, P. L., & Robbins, M. C. (1968). Linear perspective, pictorial depth perception and education among the Baganda. *Perceptual and Motor Skills, 27,* 601–602.

Klatzky, R. L., Loomis, J. M., Lederman, S. J., Wake, H., & Fujita, N. (1993). Haptic identification of objects and their depictions. *Perceptions & Psychophysics, 54,* 170–178.

Koffka, K. (1935). *Principles of Gestalt psychology.* New York: Harcourt Brace.

Komar, V., & Melamid, A. (1993). *American public attitudes towards the visual arts.* Tabular report prepared by Martila & Kiley, Inc. New York: The Nation Institute.

Kopfermann, H. (1930). Psychologische Untersuchungen über die Wirking zweidimensionaler Darstellung körperlicher Gebilde. [Psychological studies on the effectiveness of two dimensional representations of solid structures]. *Psychologische Forschung, 13,* 293–364.

Krampen, M. (1993). *Children's drawings: Iconic coding of the environment.* New York: Plenum.

Krietler, H., & Krietler, S. (1972). *Psychology of the arts.* Durham, NC: Duke University Press.

Kruglanski, A. W. (1975). The endogenous-exogenous partition in attribution theory. *Psychology Review, 82,* 387–406.

Kubovy, M. (1986). *The psychology of linear perspective in Renaissance art.* Cambridge, UK: Cambridge University Press.

Lalo, C. (1908). *L'Esthétique experimentale contemporaine.* [Contemporary experimental aesthetics]. Paris: Alcan.

Landau, B. (1985). *Language and experience: Evidence from the blind child.* Cambridge, MA: Harvard University Press.

Langer, S. K. (1958). *Philosophy in a new key.* New York: Mentor.

Langlois, J. H., & Roggman, L. A. (1990). Attractive faces are only average. *Psychological Science, 1,* 115–121.

Lederman, S. J., & Klatzky, R. L. (1987). Hand movements: A window into haptic object recognition. *Cognitive Psychology, 19,* 342–368.

Leeuwenberg, E. (1971). A perceptual coding language for visual and auditory patterns. *American Journal of Psychology, 84,* 307–349.

Leibowitz, H., & Dichganz, J. (1980). The ambient visual system and spatial organization. In *Proceedings of the AGARD Conference on Spatial Disorientation in Flight.* (AGARD-CP-287), Alexandria, VA: Defense Technical Information Center.

van Lier, R. J., van der Helm, P. A., and Leeuwenberg, E. L. J. (1995). Competing global and local completions in visual occlusion. *Journal of Experimental Psychology: Human Perception and Performance, 21,* 571–583.

Lindauer, M. S. (1970). Psychological aspects of form perception in abstract art. *Scientific Aesthetics, 7,* 19–24.

Lindauer, Martin S. (1990). Reactions to cheap art. *Empirical Studies of the Arts, 8,* 95–110.

Lindauer, M. (1991). Comparisons between museum and mass-produced art. *Empirical Studies of the Arts, 9,* 11–22.

Lindsley, D. (1957). Psychophysiology and motivation. In M. Jones (Ed.), *Nebraska symposium on motivation, 1957* (pp. 36–40). Lincoln: University of Nebraska Press.

Lipps, T. (1900). Aesthetische einfuhlung. *Zeitschrift fur Psychologie und Physiologie der Sinnesorgane, 22,* 415–450.

Loftus, G. R. (1976). A framework for a theory of picture recognition. In R. A. Monty & J. W. Senders (Eds.), *Eye movements and psychological processes.* Hillsdale, NJ: Erlbaum.

Lowe, D. G. (1985). *Perceptual organization and visual recognition.* Boston: Kluwer Academic.

Lumsden, E. A. (1980). Problems of magnification and minification. In Hagen, M. A. (Ed.), *The perception of pictures, 1,* (pp. 91–135). New York: Academic Press.

Lynch, K. (1960). *The image of the city.* Cambridge, MA: MIT Press.

Mackworth, N. H., & Morandi, A. J. (1967). The gaze selects informative details within pictures. *Perception & Psychophysics, 2,* 547–552.

Marr, D. (1982). *Vision.* San Francisco: Freeman.

McClelland, D., Atkinson, J., Clark, R., & Lowell, E. (1953). *The achievement motive.* New York: Appleton-Century-Crofts.

McKoon, G., & Ratcliff, R. (1992). Inference during reading. *Psychological Review, 99,* 440–466.

Metzger, W. (1953). *Gesetze des Sehens* [Laws of vision]. Frankfurt-am-Main: Kramer.

Mezanotte, R. J. (1981). *Accessing visual schemata: Mechanisms invoking world knowledge in the identification of objects in scenes.* Unpublished doctoral dissertation, State University of New York, Buffalo.

Meyer, L. B. (1956). *Emotion and meaning in music.* Chicago: University of Chicago Press.

Minnigerode, F. A., Ciancio, D. W., & Sbaboro, L. A. (1976). Matching music with paintings by Klee. *Perceptual and Motor Skills, 42,* 269–270.

Moles, A. (1966). *Information theory and esthetic perception.* Urbana, IL: University of Illinois Press.

Mundy-Castle, A. C. (1966). Pictorial depth perception in Ghanianaian children. *International Journal of Psychology, 1,* 288–300.

Munsinger, H., & Kessen, W. (1964). Uncertainty, structure and preference. *Psychological monographs, 78,* 586.

Neisser, U. (1967). *Cognitive psychology.* New York: Appleton.

Nisbett, R. E., & Wilson, T. D. (1977). Telling more than we can know: Verbal reports on mental processes. *Psychological Review, 84,* 231–259.

Normore, L. F. (1974). Verbal responses to visual sequences varying in uncertainty level. In D. E. Berlyne (Ed.), *Studies in the new experimental aesthetics.* Washington, DC: Hemisphere.

Olson, G. M. (1976). An information processing analysis of visual memory and habituation in infants. In T. J. Tighe & R. N. Leaton (Eds.), *Habituation* (pp. 207–338). Hillsdale, NJ: Erlbaum.

Olson, R. K. (1975). Children's sensitivity to pictorial depth information. *Perception & Psychophysics, 71,* 59–64.

Omari, I. M., & Cook, H. (1972). Differential cognitive cues in pictorial depth perception. *Journal of Cross-Cultural Psychology, 3,* 321–325.

O'Neill, M. J. (1991). Evaluation of a conceptual model of architectural legibility. *Environment and Behavior, 23,* 259–284.

Orne, M. T. (1962). On the social psychology of the psychological experiment: With particular reference to demand characteristics and their implications. *American Psychologist, 17,* 776–783.

Osgood, C. E. (1976). *Focus on meaning.* The Hague: Mouton.

Oster, G. (1977). Moirée patterns in science and art. *Advances in Biological and Medical Physics, 16,* 333–347.

Penrose, L., & Penrose, P. (1958). Impossible objects: A special type of visual illusion. *British Journal of Psychology, 49,* 31–33.

Perrett, D. I., May, K. A., & Yoshikawa, S. (1994). Facial shape and judgments of female attractiveness. *Nature, 368,* 239–242.

Peretti, P. (1972). A study of student correlations between music and six paintings by Klee. *Journal of Research in Music Education, 20,* 501–504.

Perkins, D. N. (1973). Compensating for distortion in viewing pictures obliquely. *Perception & Psychophysics, 14,* 13–18.

Peterson, M. A. (1994). Object recognition processes can and do operate before figure-ground organization. *Current Directions in Psychological Science, 3,* 105–111.

Peterson, M. A., & Gibson, B. S. (1991). The initial identification of figure–ground relationships: Contributions from shape recognition processes. *Bulletin of the Psychonomic Society, 29,* 199–202.

Peterson, M. A., & Harvey, E. M. H., & Weidenbacher, H. L. (1991). Shape recognition inputs to figure–ground organization: Which route counts? *Journal of Experimental Psychology: Human Perception and Performance, 17,* 1075–1089.

Peterson, M. A., & Hochberg, J. (1983). The opposed-set measurement procedure: The role of local cues and intention in form perception. *Journal of Experimental Psychology: Human Perception and Performance, 9,* 183–193.

Peterson, M. A., & Hochberg, J. (1989). Necessary considerations for a theory of form perception: A theoretical and empirical reply to Boselie and Leeuwenberg. *Perception, 18,* 105–119.

Pickford, R. W. (1955). Factorial studies of aesthetic judgments. In A. A. Roback (Ed.), *Present-day psychology* (pp. 913–929). New York: Philosophical Library.

Pickford, R. W. (1972). *Psychology and visual aesthetics.* London: Hutchinson.

Pillsbury, W., & Schaefer, B. (1937). A note on advancing retreating colors. *American Journal of Psychology, 49,* 126–130.

Pirenne, M. (1970). *Optics, painting and photography.* Cambridge, UK: Cambridge University Press.

Polanyi, M. (1970). Introduction. In. M. Pirenne (Ed.), *Optics, painting and photography* (pp. xv–xxii). Cambridge, UK: Cambridge Univ. Press.

Pollack, I., & Spence, D. (1968). Subjective pictorial information and visual search. *Perception & Psychophysics, 3,* 41–44.

Poore, H. R. (1903). *Pictorial composition and the critical judgment of pictures.* New York: Baker & Taylor.

Pope, A. (1949). *The language of drawing and painting.* Cambridge, MA: Harvard University Press.

Prak, N. L. (1968). *The language of architecture. A contribution to architectural theory.* The Hague: Mouton.

Rashevsky, N. (1940). *Advances and applications of mathematical biology.* Chicago: University of Chicago Press.

Ratoosh, P. (1949). On interposition as a cue for the perception of distance. *Proceedings of the National Academy of Science, 35,* 257–259.

Read, H. (1943). *Education through art.* London: Faber & Faber.

Richards, W. A., & Hoffman, D. D. (1985). Codon constraints on closed 2-D shapes. *Computer Vision, Graphics and Image Processing, 32,* 265–281.

Reich, J., & Moody, C. (1970). Stimulus properties, frequency of exposure, and affective responding. *Perceptual and Motor Skills, 30,* 27–35.

Rock, I. (1977). In defense of unconscious inference. In W. Epstein (Ed.), *Stability and constancy in visual perception* (pp. 321–373). New York: Wiley.

Rock, I. (1993). The logic of 'The logic of perception'. *Italian Journal of Psychology, 20,* 841–867.

Rosinski, R. R., Mulholland, T., Degelman, D., & Farber J. (1980). Pictorial space perception: An analysis of visual compensation. *Perception & Psychophysics, 28,* 521–526.

Ruesch, J. (1977). *Greek statuary of the fifth and fourth centuries B.C.* Unpublished doctoral dissertation. Columbia University, New York.

Ryan, T. A., & Schwartz, C. (1956). Speed of perception as a function of mode of representation. *American Journal of Psychology, 69,* 60–69.

Schacter, S., & Singer, J. E. (1962). Cognitive, social, and physiological determinants of emotional state. *Psychological Review, 69,* 379–399.

Schaie, K. W. (1961). Scaling the association between colors and moodtones. *American Journal of Psychology, 74,* 266–273.

Schillinger, J. (1948). *The mathematical basis of the arts.* New York: Philosophical Library.

Schmidt, C. F. (1976). Understanding human action: Recognizing the plans and motives of other persons. In J. S. Carroll & J. W. Payne (Eds.), *Cognition and social behavior* (pp. 47–67). Hillsdale, NJ: Erlbaum.

Schneider, G. (1969). Two visual systems. *Science, 163,* 895–902.

Scruton, R. (1979). *The aesthetics of architecture.* London: Methuen.

Secord, P., & Muthard, J. (1955). Personality in faces, IV: A descriptive analysis of the perception of womens' faces and the identification of some physiognomic determinants. *Journal of Psychology, 39,* 269–278.

Sedgwick, H. A. (1980). The geometry of spatial layout in pictorial representation. In M. A. Hagen (Ed.), *The perception of pictures. 1,* 33–90. New York: Academic Press.

Sedgwick, H. A. (1983). Environment-centered representation of spatial layout: Available information from texture and perspective. In J. Beck, B. Hope, & A. Rosenfeld (Eds.), *Human and machine vision* (pp. 425–458). New York: Academic Press.

Sedgwick, H. A. (1991). The effects of viewpoint on the virtual space of pictures. In S. R. Ellis, M. K. Kaiser, & A. C. Grunwald (Eds.), *Pictorial communication in virtual and real environments* (pp. 461–479). New York: Tayor & Francis.

Searle, J. R. (1969). *Speech arts: An essay in the philosophy of language.* Cambridge, UK: Cambridge University Press.

Shaw, T. L. (1962). *Hypocrisy about art.* Boston: Stuart Publications.

Shepard, R. N. (1984). Ecological constraints on internal representations: Resonant kinematics of perceiving, imaging, thinking, and dreaming. *Psychological Review, 91,* 417–477.

Shepheard, P. (1994). *What is architecture? An essay on landscapes, buildings, and machines.* Cambridge, MA: MIT Press.

Sircello, G. (1965). Perceptual acts of pictorial art: A defense of expression theory. *Journal of Philosophy, 62,* 669–677.

Smets, G. (1973). *Aesthetic judgment and arousal.* Louvain, Belgium: Leuven University Press.

Smets, G., & Knops, L. (1976). Measuring visual esthetic sensitivity: An alternative procedure. *Perceptual and Motor Skills, 42,* 867–874.

Smith, O. W., & Gruber, H. (1958). Perception of depth in photographs. *Perceptual and Motor Skills, 8,* 307–313.

Smith, P. C., & Smith, O. W. (1961). Ball-throwing responses to photographically portrayed targets. *Journal of Experimental Psychology, 62,* 223–233.

Snodgrass, J. G. (1971). Objective and subjective complexity measures for a new population of patterns. *Perception & Psychophysics, 10,* 217–224.

Sontag, S. (1961). *Against interpretation.* New York: Delta.

Sorrell, W. (Ed.) (1966). *The dance has many faces.* New York: Columbia University Press.

Söström, I. (1978). Quadrataura: Studies in Italian ceiling painting. *Acta Universitatis Stockholmiensis, Stockholm Studies in the History of Art, 30.*

Stephan, M. (1990). *A transformational theory of aesthetics.* London: Routledge.

Stevens, K. A., & Brooks, A. (1987). Probing depth in monocular images. *Biological Cybernetics, 56,* 355–366.

Szarkowski, J. (1973). *From the picture press.* New York: Museum of Modern Art.

Taylor, J. (1964). *Design and expression in the visual arts.* New York: Dover.

Terwilliger, R. F. (1963). Pattern complexity and affective arousal. *Perceptual and Motor Skills, 17,* 387–395.

Thompson, P. (1980). Margaret Thatcher: A new illusion. *Perception, 9,* 483–484.

Todd, J. (1989). Models of static form perception. In J. I. Elkind, S. K. Card, J. Hochberg, & B. M. Huey (Eds.), *Human performance models for computer-aided engineering* (pp. 75–88). New York: Academic Press.

Todd, J. T., & Akerstrom, R. A. (1987). Perception of three-dimensional form patterns of optical texture. *Journal of Experimental Psychology: Human Perception and Performance, 13,* 242–255.

Todd, J., & Mingolla, E. (1983). The perception of surface curvature and direction of illumination from patterns of shading. *Journal of Experimental Psychology: Human Perception and Performance, 9,* 583–595.

Todd, J. T., & Reichel, F. D. (1989). Ordinal structure in the visual perception and cognition of smoothly curved surfaces. *Psychological Review, 96,* 643–657.

Tolman, E. C. (1948). Cognitive maps in rats and men. *Psychological Review, 55,* 189–208.

Tolstoy, L. (1899). *What is art?* (A. Maude, Trans.). London: Oxford University Press.

Tormey, A. (1980). Seeing things: Pictures, paradox and perspective. In J. Fisher (Ed.), *Perceiving artworks* (pp. 59–75). Philadelphia: Temple University Press.

Tufte, E. R. (1990). *Envisioning information.* Chesire, CT: Graphics Press.

Tversky, B., & Baratz, D. (1985). Memory for faces: Are caricatures better than photographs? *Memory & Cognition, 13,* 45–49.

Valentine, C. W. (1962). *The experimental psychology of beauty.* London: Methuen.

Valentine, T. (1988). Upside-down faces: A review of the effect of inversion on face recognition. *British Journal of Psychology,* 7, 471–491.

Vitz, P. (1964). Preferences for rates of information presented by sequence of tones. *Journal of Experimental Psychology, 68,* 176–183.

Vitz, P. (1966). Preferences for different amounts of visual complexity. *Behavioral Science, 11,* 104–114.

Walker, E. L. (1973). Psychological complexity and preference: A hedgehog theory of behavior. In D. E. Berlyne & K. B. Madsen (Eds.), *Pleasure, reward, preference* (pp. 65–97). New York: Academic Press.

Wallach, M. A. (1959). Art, science and representation Toward an experimental psychology of aesthetics. *Journal of Aesthetics and Art Criticism, 18,* 159–173.

Warren, W. H. (1984). Perceiving affordances: Visual guidance of stair climbing. *Journal of Experimental Psychology: Human Perception and Performance, 10,* 683–703.

Warren, W. H., Wang, S. (1987). Visual guidance of walking through apertures: Body-scaled information for affordances. *Journal of Experimental Psychology: Human Perception and Performance, 13,* 371–383.

Wartofsky, M. (1979). Picturing and representing. In C. F. Nodine & D. F. Fisher (Eds.), *Perception and pictorial representation: Making, perceiving and interpreting* (pp. 272–283). New York: Praeger.

Weber, C. O. (1931). Esthetics of rectangles and theories of affect. *Journal of Applied Psychology, 15,* 310–318.

Wechner, W. L. (1966). The relation between six paintings by Klee and selected musical compositions. *Journal of Research in Music Education, 14,* 220–224.

Weitz, M. (1960). The role of theory in esthetics. In M. Rader (Ed.), *A modern book of esthetics.* New York: Holt.

Werner, H. (1948). *Comparative psychology of mental development.* Chicago: Follett.

Westland, G. (1968). The construction of objective tests of a form of aesthetic judgment. *British Journal of Aesthetics, 8,* 387–393.

Wheelock, A. K., Jr. (1979). Perspective and its role in the evolution of Dutch realism. In C. F. Nodine & D. F. Fisher (Eds.), *Perception and pictorial representation: Making, perceiving and interpreting* (pp. 110–133). New York: Praeger.

White, H. C., & White, C. A. (1965). *Canvases and careers: Institutional change in the French painting world.* New York: Wiley.

White, J. (1967). *The birth and rebirth of pictorial space.* Boston: Boston Book and Art Shop.

Wickens, C. D., Merwin, D. H., & Lin, E. L. (1994). Implications of graphics enhancements for the visualization of scientific data: Dimensional integrality, stereopsis, motion and mesh. *Human Factors, 36,* 44–61.

Willats, J. (1977). How children learn to draw realistic pictures. *Quarterly Journal of Experimental Psychology, 29,* 367–382.

Wilson, E. (1931). *Axel's castle. A study in the imaginative literature of 1870–1930.* New York: Scribner's.

Wilson, G. (1966). Arousal properties of red v. green. *Perceptual and Motor Skills, 26,* 947–949.

Witmer, L. (1894). Zur experimentellen aesthetik einfacher raumlicher Formverhaltnisse. In *Philosophische Studien, 1893, 9,* (pp. 96–144, 209–263). Leipzig: Englemann.

Wohlwill, J. F. (1966). The physical environment: A problem for a psychology of stimulation. *Journal of Social Issues, 4,* 29–38.

Wohlwill, J. F. (1968). Amount of stimulus exploration and preference as differential functions of stimulus complexity. *Perception & Psychophysics, 4,* 307–312.

Woodworth, R. S. (1938). *Experimental psychology.* New York: Holt, Rinehart and Winston.

Wollheim, R. (1987). *Painting as an art.* Princeton, NJ: Princeton University Press.

Yonas, A., & Hagen, M. A. (1973). Effects of static and kinetic depth information on the perception of size in children and adults. *Journal of Experimental Child Psychology, 15,* 254–265.

Zink, S. (1960). Is music really sad? *Journal of Aesthetics and Art Criticism, 2,* 197–207.

Zucker, P. (1963). *Styles in painting: A comparative study.* New York: Dover.

CHAPTER 6

The Perception of Motion Pictures

Julian Hochberg
Virginia Brooks

I. INTRODUCTION

Motion pictures, by which we mean television and computer-generated images as well as film, provide an arena in which the major perceptual theories are challenged, and in which most laboratory findings should seek application if only to establish their ecological validity. Still pictures, and Leonardo's analysis of the pictorial depth cues, were the starting point for the major classical problems of perceptual research and theory. Motion pictures promise to be at least as important to theory and application. Not an arbitrary learned code, they draw on the processes that initiate and sustain the successive glimpses by which people sample the world and integrate the information so obtained—processes fundamental to any general theory of perceptual organization and attention. They provide as well an ecologically valid testing ground for theories of cognitive processes and social communication and motivation that might otherwise be untestable.

Although research specifically directed to motion pictures as they are used outside of the laboratory is still sparse, and few attempts have been made to draw lessons from that usage, five major points that strongly suggest the need for significant changes in psychological theory and research will emerge in the course of this chapter. Those points, and the pages in which they are most directly addressed, are as follows:

Cognitive Ecology

1. There are visual sensory mechanisms that respond directly to specific movements within the retinal image or the visual field, but those responses do not account for (i.e., are not necessary or sufficient for) the motions that we perceive (pp. 213–219, 222f, 246f).

2. Cognitive factors also affect the resultant moment-by-moment events that humans perceive, factors that cannot be explained (almost dismissed, as they often are) simply as inferences or hypotheses (pp. 218–233, 241ff).

3. As events on the screen continue to unfold in time, viewers are left only with the cognitive consequences of earlier motions: at some point, psychologists can no longer ignore the contributions of such mental representation (p. 266–270).

4. Mental representations are not simply models of the distal world (inferred or otherwise), but are path-dependent, incomplete, and shifting construals (p. 258).

5. Despite these flirtations with mentalism, it is very important to identify the more sensory (and mandatory) components of motion picture perception (p. 268).

What makes motion pictures, or moving pictures, so important is that they can do at least five things that still pictures cannot and that are of theoretical interest as well as of practical value:

1. They can provide movement-based information about tri-dimensional (3-D) spatial arrangement (motion depth cues) that are unavailable in stills, and that by their absence contradict whatever depth is otherwise portrayed in stills. They also allow us to vary separately components that covary rigidly under normal conditions of seeing, and thereby provide critical tests of theories and models.

2. Scenes or objects that are very much larger than the size of the motion picture screen (or other display device) can be represented through successive views, illustrating dramatically our ability to integrate information over time, and revealing as well how the "what" and the "where" are interdependent in perception whether or not their precursors are separable at some stages of neurophysiological visual processing. Similar in some ways to the normal methods of perceptual inquiry through successive glance, motion pictures permit us to study the processes by which we comprehend, anticipate, and store the visible contents of the world.

3. Motion pictures characteristically permit—in fact, depend on—change per se, making possible levels of visual interest maintenance that cannot be sustained and studied in stills with equivalent subject matter.

4. In motion pictures, scenes and events can be represented piecemeal, juxtaposing parts that were not juxtaposed in fact (in the limit, constructing computer-generated images [CGI] pixel by pixel). Constraints that are essentially impossible to violate in real-world setups are routinely circumvented in motion pictures. At

present levels of special effects, no imagined layout, world, or universe is beyond visual experience. When we add to this the *virtual realities* now being explored in interactive computer displays, viewers can be led to experience almost anything that the entertainment industry, training institution, or research experimenter deems desirable.

5. Redundant sections of actions, periods of time, or extents of space can be elided, and series of events, including human actions, are thereby reduced to their minimal and most effective communicative features. That such omission can go unnoticed, or even improve the viewer's recognition and comprehension, offers a tool with which to investigate the mental representation of such physical and social events.

These options available to motion picture makers are in routine use as part of the medium itself, and in increasing use for the kinds of research they make possible. We first discuss briefly the nature of the medium, and then consider the relevant sensory and perceptual issues (inevitably infringing somewhat on more abstract and nonsensory cognitive processes). There is now a small but growing body of studies on how film construction affects viewers' judgments of actors' characters or mood, and on the effects of film cutting and timing (or their equivalents) on such variables as eye movements and judged suspense, which we discuss in Section III.

On the other hand, only a few experimental studies address directly the points at which major theories of visual sensation and cognition are seriously challenged by long-standing film usage, and even fewer have tried to use perceptual phenomena to make motion pictures (and particularly interactive CGI) more effective. Sections I and II concern these issues.

A. The Medium: What Is a Moving Picture?

1. The Display Considered as a Surrogate: Distance, Size, and Resolution; Routine Violations of Fidelity

First, of course, we are talking about *pictures* that change. Each picture, in *some* sense a *surrogate* for whatever is being portrayed, presents the eye with an optic array in some essential ways the same as that produced by the scene or event itself. That concept invites a reasonably straightforward definition of fidelity, at least in principle (e.g., the display's resolution, contrast and spectral luminosity function, its absence of spatial distortions or anisotropies, etc.). Thus, a motion picture that would be visually indistinguishable from a window open to the represented active scene must meet straightforward considerations of visual acuity, for example, and it should be possible to decide the optimal seating distances, audience volumes (McVey, 1970), and resolution limits as these are imposed by the grain of the scanning raster and pixel size in electronic displays. We should note, however, that resolution and viewing distance are not simple determinants of what scenes and objects the viewer can see (e.g., varying the viewing distance of spatially filtered capital letters, Parish & Sperling, 1991, found that viewing distance had virtually no effect on legibility). Without further research, one cannot estimate the effects of screen size, or simply trade off image resolution for viewing distance, given a constant screen size.

Furthermore, going beyond questions of resolution and visibility, in order to achieve fidelity the correct viewing distance and viewing angle are fixed, for any given viewpoint and focal length of the camera, and for a fixed projector or TV size (we will define this as the *canonical presentation;* Hochberg, 1986). If the viewing distance or angle differs from the canonical presentation, the object being represented must change and distort its shape and its motions through virtual space to fit its moving image upon the screen, and such serious violations of fidelity seem to be made with no compunctions. For example, both film and TV routinely choose to change the effective distance between the scene and the camera (closeup, long shot, etc.), so that fidelity of shape and motion are electively abandoned (see p. 242).

We cannot simply define a motion picture, then, in terms of the fidelity of the motions for which it is physically a surrogate. Let us next address why it appears to move.

2. Why Movies Move: The Persistent "Persistence of Vision" Myth: Stroboscopic Movement and the Related Perceptual Problems

Motion pictures on film consist, as everybody knows, of successive static pictures, usually at a rate of 24 views each second for sound motion pictures; in TV, 30 static pictures each second are traced by a rapidly moving electron beam that lays out alternative lines of the *raster* (the modulated bright lines traced out on the phosphor of the cathode-ray tube). In film, the frequency of view change is on the fringe of the detectable flicker rate, but additional interruptions by a rotary shutter (episcotister) raise the change rate above flicker threshold. In TV, although a new image is only provided at 30 Hz, each overall picture is actually refreshed, by alternate lines, at 60 Hz.[1]

These differences in the underlying display process were considered relevant to the message and its aesthetics (McLuhan, 1964; Zettl, 1973), and indeed the differences in contour stability, resolution, contrast, luminance, moiré patterning, and so on may affect the medium's aesthetic qualities (Hochberg & Brooks, 1973), but both processes are well able to generate apparent motion, and may be irrelevant as a deep aesthetic or communicative issue (Layer, 1974).

What is important to both procedures is that successive small spatial changes in the static patterns of luminance appear as *movements* from one place to another. Correctly termed *stroboscopic motion,* this is the heart of motion pictures, and in

[1] Furthermore, pattern detection probably has a lower contrast threshold as frequency approaches 30 Hz (Keesey, 1972).

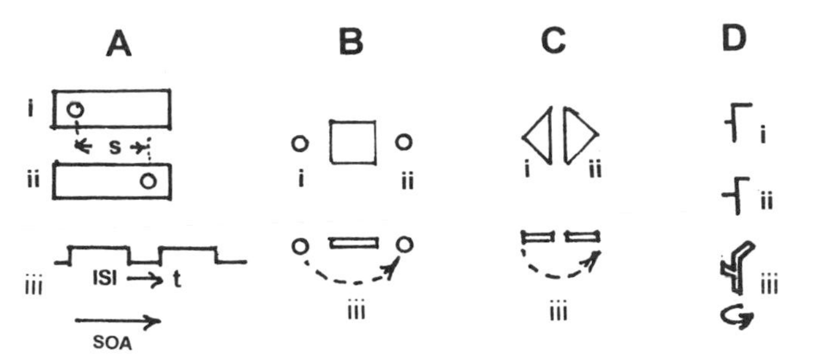

FIGURE 1 Apparent motion. (A) Points of light (i and ii) are successively turned on and off, as graphed at iii. The kind of motion or succession that is perceived depends on the separation in space (s) and time (t); measures of the latter are the interstimulus interval (ISI) and the stimulus-onset asynchrony (SOA), which are not free to vary in motion pictures. (B–D) The perceived motion is not confined to the two-dimensional (2D) plane: At B, with an obstacle between i and ii, and at C in which the two successive shapes are mirror images, the motion appears to curve through the third dimension (as viewed from above at iii). At D, the two successive views i and ii of a meaningless wire shape, each of which looks flat, when viewed in rapid sequence reveal a moving 3-D structure as sketched at iii. This was called the *kinetic depth effect* (KDE) by Wallach and O'Connell, 1953; for a discussion of the limits on this generalization, see Gillam, 1972; Todd, 1982; and Fig. 15 in Hochberg, 1986.

almost all professional film and TV texts it is quite wrongly attributed to "persistence of vision." It is nothing of the sort, because visual persistence alone would merely provide superimposition of the successive frames.[2]

In fact, some kind of successive masking, or an "off" signal, must be invoked to explain why visual persistence does not preserve all of the static intermediate views through which a picture of the moving object passes.[3] But in any case, persistence does not equal movement.

a. The Basic Facts and Problems of Stroboscopic Motion

In what is often taken as the simplest case (Fig. 1A), if two spots of light (i, ii) are separated by a distance s, and by an *interstimulus interval* (ISI) of time (t) during which neither is presented or by an onset-to-onset interval, called *stimulus-onset*

[2] The error started, reasonably enough, in the fact that toys used to combine two rapidly rotating but separate views into one, were also used to provide stroboscopic motion between two locations of some pictured object (Anderson & Anderson, 1993; Cook, 1963). The error was corrected at least as early as Munsterberg (1916/1970), but it apparently cannot be rooted out. Although persistence does *not* explain stroboscopic motion, it does explain why we do not see substantial dark periods in motion picture displays (especially in the electronic displays of TV and CGI, which are always dark save for the one small dot being activated at any moment by the moving electron beam, and the persistent phosphor trail left glowing (at different rates for different displays) on the screen.

[3] For relevant research on masking, see Matin, 1975. Research on off-signals can be found in work in which two successive meaningless, apparently random sets of dots combine to provide a clearly recognizable pattern (such as a word or a complete matrix with one missing dot), which the subject is to identify (Cohene & Bechtoldt, 1975; Eriksen & Collins, 1967; Hogben & Di Lollo, 1972).

asynchrony (SOA), there are some combinations of s and t in which the viewer will report seeing spatial movement. The physical variables seem simple and straightforward. The responses (Wertheimer, 1912) vary from *pure phi* movement (no objects as such are seen, but a strong experience of movement is reported), through *beta movement* (one object appears to move from one location to another), to discrete succession. Under appropriate conditions, beta movement is indistinguishable from real movement (De Silva, 1926; Dimmick & Scahill, 1925; Stratton, 1911; Wertheimer, 1912). The perceived motion is not confined to the two-dimensional (2-D) plane of the display. Like real motion, the apparent motion provides information about depth: With an "obstacle" placed between i and ii, or if the two patterns are mirror images of each other that could only move through s on the retina if they swing through the third dimension, as in Figures 1B and 1C, we see the appropriate movement in depth. (This is not true with the shortest durations and distances; see p. 224.) Indeed, depicted objects that look perfectly flat (2-D) when viewed alone may appear compellingly 3-D when put into apparent motion, a class of phenomena called the *kinetic depth effect* (KDE) (Wallach & O'Connell, 1953; for limitations, see references in the caption). In general, the perceptual system treats a succession of static views, taken with fine enough temporal grain, as being equivalent to the continuous movement from which those views are sampled.

This is, of course, what makes the movement in motion pictures possible: The cameraperson normally needs only to point the camera at the scene, and the apparent motion that results when the film is projected will be the same as what would have been experienced if the camera had somehow produced a continuous (instead of a discontinuous and static) record of the event.

Normal camera recording thus requires little knowledge of the determinants of perceived motion. It is only when normal stroboscopic recording is interrupted, as in cuts or when the camera itself is moving, that knowledge of the determinants of apparent motion becomes important.[4]

b. The Study of Apparent Motion: Its Theoretical Importance in Visual Cognition and Cognitive Neurophysiology and Its Practical Importance in Motion Pictures: The Problem of Identity in Long-Range versus Short-Range Motion

People were paying to view motion pictures by the turn of the century (Pratt, 1973), but the basic phenomenon was not put to major scientific scrutiny until 1912 when Wertheimer named the different perceptual phenomena associated with successive presentations (pure phi, beta movement, etc.; see section I.A.1.a) and used those demonstrations to support the far-reaching conclusions that set Gestalt theories in opposition to the classical associationism. Most important was

[4] Other causes arise occasionally, of course, like the classic interaction in which stroboscopic samples of the real motion produce an apparent backward rotation of a carriage wheel or an airplane propeller, and similar concerns probably apply to other cases like expanded or contracted time.

the argument that pure phi, and the indistinguishability (under the right conditions) of stroboscopic movement from continuous movement (experimentally verified by DeSilva in 1929), showed that the motion experience was a sensation as *direct* as red, green, or any other, with its own underlying physiological basis.

What made this important then, and what keeps apparent motion a lively research topic today, is that motion cannot be defined in terms of a single independent point receptor. Accepting motion as a direct response provides a tool with which to explore the visual nervous system. Let us first very briefly consider the classical system that Wertheimer worked to displace, and its major alternatives, attending selectively to those features that we need for later discussion of motion pictures, and then summarize the few main points about apparent motion that are important to an understanding of motion picture cutting.

At the time motion pictures were introduced (and indeed, through the 1940s), the dominant theory among psychologists and sensory physiologists was that the basic visual system consists of a mosaic of independent point photoreceptors (rods and cones). The elementary photoreceptors within that mosaic were thought to analyze the *proximal* visual stimulus—the pattern of light in the retinal image—into a corresponding pattern of individual independent responses or *sensations*. Such sensations, which (roughly speaking) comprise information about the levels of light at each wavelength at each point in the image (according to the spectral luminosity function of the rod or cone at each respective point in the mosaic) are the starting point of perceptual processing, but are not themselves consciously experienced. Motion, like all other *distal* properties of the world we perceive (i.e., those that are measurable in the physical world but are not present in the point-by-point properties of proximal stimulation: such properties as form, distance, where our gaze is directed or our locomotion is headed, etc.) was thought to be added later in the process. Such perceptions of properties of the world were, presumably, evoked by or associated to the changes in the sensed locations of individual sensations within the mosaic.

That is, the perception of motion (whether stroboscopic or continuous) was thought to follow what we may call Helmholtz's rule of perception in general: that one perceives just that state of affairs that would be most likely to produce the received set of sensations (Helmholtz, 1909/1924; see Hochberg, 1971, 1982). The perception is therefore not a sensory event describable in terms of the retinal image (Fig. 2A) or the optic array, but a *mental representation* (Fig. 2B) of a set of circumstances in the distal world (as sketched in Fig. 2C) that would very probably have provided that particular pattern of sensations. If this view is correct, in order to predict what will be perceived with any given stimulus pattern, the following are needed:

1. A method that generates those 3-D situations that would fit that sensory pattern, physical optics, and some measure of (and assumptions about) retinal sensitivity to the sensory pattern.

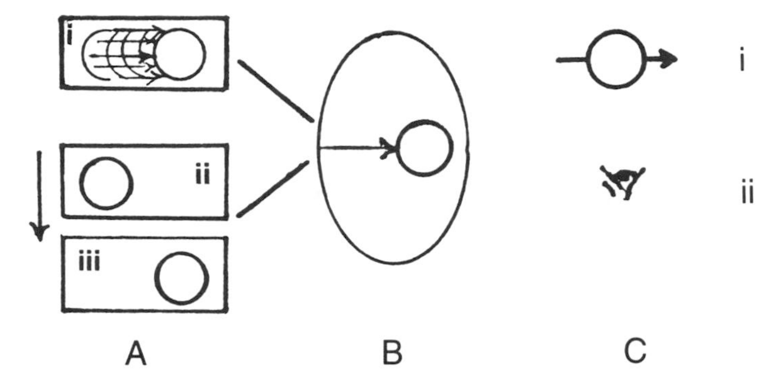

FIGURE 2 Motion as a mental representation. (Ai) A spot moves across the retina. (Aii–iii) Two static spots fall successively on the retina; the downward arrow shows *time* (as in Fig. 1A). Although these displays seem totally different as physical events, they have identical perceptual consequences under the right conditions. In the classical theory, this is so because the most likely event in the world that would produce the same sensory changes in retinal location as either those in i or those in ii–iii would be some object moving in space at some distance from the eye, as sketched at (B); through a cognitive process (often asserted to resemble *inference*) the mind fits the sensory data to the most likely mental representation of object motion, symbolized at (B), which is therefore what is perceived in both cases.

2. Knowledge of how *likely* each of these 3-D situations is, that is, the *ecological probabilities* (see Brunswik, 1956, who thought that ecological surveys could provide such knowledge).

3. One must work out the discriminable proximal stimulus patterns, or *cues,* that each distal situation will characteristically present to the eye (such as occlusion and motion parallax as depth cues in the retinal or pictorial projection of a 3-D scene).

4. A principled way to get from the cues to the percept (i.e., what the characteristics of the mental representation are and the principles by which it is fitted to or determined by those cues).

Helmholtz speculated (1909/1924) that the last is achieved by a process of *unconscious inference,* a process that is essentially the same as that employed in thinking or problem solving, but of which one is not aware. This speculation is still vigorously supported (see Rock, 1977, 1983; also Gregory, 1970, 1980; Shepard, 1981), perhaps because many illusions and other perceptual phenomena can be found that seem amenable to such an explanation, at least qualitatively and retroactively; but one must note in opposition to that easy view that many other perceptual observations do not fit that explanation in any obvious way (Hochberg, 1988, 1994). We can admit that perception involves some process of *construal* that fits but is not determined by stimulus information, without turning to a naive view of how problem solving works (logic does not generally play a large role, as is implied by "inference"), and we can admit the existence of direct sensory response to some aspect of the world, like motion, without ruling out more cognitive paths to the same perceptual consequences (Hochberg, 1974a,b, 1992). Finally, there have been virtually no experimental attempts to compare the notion of inference with alternative ways of getting from receptor response to visual cognition.

One viable alternative approach is Brunswik's proposal (1956) that each cue—each proximal indication as to some distal property—is given weight according to its *ecological validity:* If cue A's presence in the retinal image virtually assures the viewer that the corresponding distal property is present, the validity approaches 1.0, and that cue when present contributes heavily to the resulting perception; if cue B's presence is almost completely uncorrelated with the presence of that same distal property, its validity approaches 0.0, and it contributes little or nothing to the resulting perception.[5] This account needs no assumptions about unconscious inference or reasoning, seems very compatible with recent attempts to model apparent motion endpoints in terms of what amount to multiple cues (cf. Dawson, 1991), and it should be considered whenever notions about inferences and mental representations are entertained in connection with the perception of motion pictures (as in Sections I.A.2.b.3, II.A.1, II.B.2.b, III.C.1.b). Because it makes no provision for direct motion sensations, however, it stands with Helmholtz in the present discussion.

Earlier proposals that motion is directly sensed (e.g., Wohlgemuth, 1911) seemed convincing (Wohlgemuth's view was based largely on the fact that aftereffects of viewing occur with motion as they do with other sensory qualities, an argument still used),[6] but did not start new schools of thought. To Wertheimer, the fundamental theoretical question was whether visual receptors yield only individual punctate sensations of light and color, leaving motion only an inference or an association, or whether the visual system has the kinds of lateral interactions that would be needed to respond directly to motion (as to form). Arguing the latter, he speculated that the neurophysiological mechanism underlying the perceived motion is a cortical "short circuit" between nearby excitations in the visual cortex

[5] Brunswik hoped that ecological surveys would tell us the ecological validities (see Ch. 5, this volume), but that has not been pursued. However, because the system of weights should come to reflect the ecological validities through a form of probability learning (Brunswik, 1956), one could in principle learn about those validities not by studying the contingencies of the visual ecology, but by measuring the different cues' relative weights. Something very like this was part of the motivation for Rosenblatt's pioneering attempt at early parallel distributed-processing (PDP) models, and is discernible in modern attempts to learn what the weights are when a PDP simulation learns some rule in much the same form as humans do.

[6] Prolonged exposure to a pattern moving in one direction (e.g., a waterfall, a rotating spiral) is usually followed by an experience of movement in the opposite direction, one that is not due to eye movements (Sekuler, 1975; Wohlgemuth, 1911) and that has since been taken as diagnostic of differences between different kinds of apparent motion.

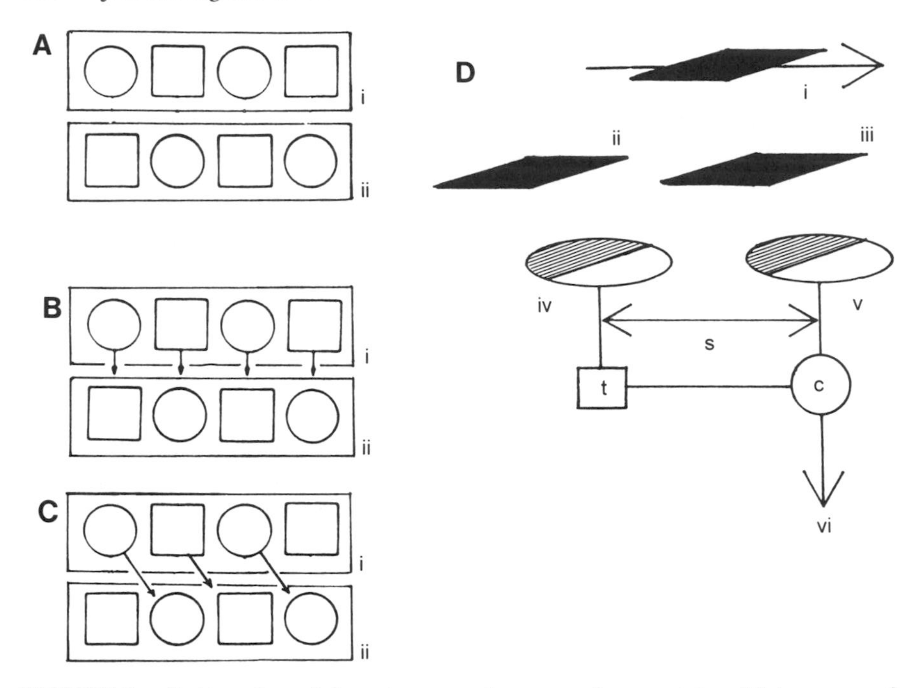

FIGURE 3 Problems for, and alternatives to, motion as mental representation. (A) A sequence of two static views, in which i is replaced by ii. (B) The apparent motion that results, which occurs between the nearest successive shapes rather than between the objects of the same shape, as at (C) (Kolers & Pomerantz, 1971; cf. Orlansky, 1940), even though it is C that one would expect from an inference-based, likelihood-based account (e.g., Fig. 2). Furthermore, such motion (and that in Fig. 1A) provides *aftereffects,* much as direct sensory processes do (e.g., as staring at a red light is followed by a green afterimage), arguing for a direct motion-sensing mechanism. (D) A simplified direct motion-sensing mechanism, with receptive fields, separated by distance (s) and oriented to respond to edges darker on the left. An edge (i) moving rightward with speed (s/t) first stimulates field iv and then field v. After a built-in time delay (t), the input of iv is compared (at c) with that of v; if the delayed and undelayed signals agree, the unit outputs (vi). Two static successive edges (ii, iii) with spatial separation (s) and temporal separation (t) will have the same effect as i. To be useful over the range of motions to which we are sensitive, such mechanisms would need a wide distribution of parameters (edge-orientation, space–time separation, etc.), and recording electrodes have revealed cells that respond appropriately. (For a discussion and advocacy of such motion sensors, see Sperling, 1989.) (E) The direct, sensory explanation in (D), and others like it, is attuned to the ways in which the two displays in Fig. 2A are the same even though they appear totally different; if one's response is indeed based on the ways in which they are the same, no further (mental representation) explanation may be needed. The way in which the two are the same can be built into the stimulus description. To describe a two-dimensional (2-D) stimulus pattern, rather than a one-dimensional point, one specifies its relevant characteristics (e.g., spectral energy at each spatial frequency) in two coordinates, as in the black stripe in X versus Y, which would be best detected by a vertically oriented set of receptive fields. The human nervous system takes relative input times and durations into account, as well as spatial coordinates. That is, the appropriate space–time-sensitive system (perhaps built of units like Fig. 3D) responds to the stimulus distribution over X, Y, and t (Adelson & Bergen, 1985; Watson & Ahumada, 1985); the continuous displacement of the black stripe, and the brief static views shown here, define the same orientation in the XT plane, and should have the sensory consequences.

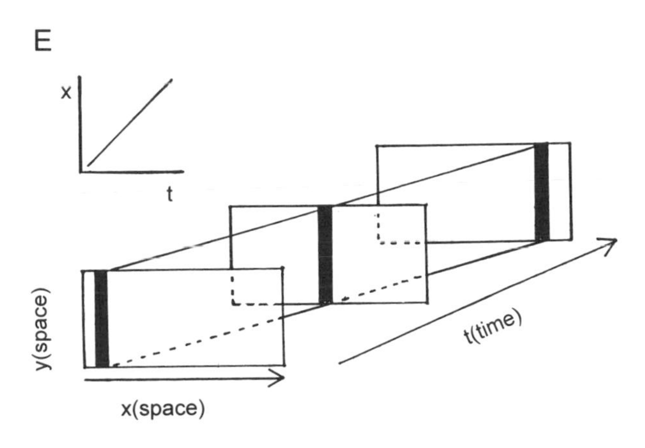

FIGURE 3 Continued

(Wertheimer, 1912), a proposal long superceded by the discovery of cells in the cortex that respond most strongly when specific motions, or their stroboscopic approximations, occur in some region of the retinal image.

i. Receptor mechanisms for retinal displacements. It is now generally believed that there are receptor mechanisms that respond directly to motion, and that almost certainly respond also to the brief intermittent static displacements of stroboscopic motion and motion pictures. The strongest support for this belief comes from the fact that (1) real movement provides an aftereffect (Wohlgemeuth, 1911) that is not attributable to eye movements (Sekuler & Ganz, 1963), as do stroboscopic displacements under the right conditions (see Anstis, 1980); and (2) cells have been identified that are activated by specific velocities of retinal motion within their receptive fields (Hubel & Wiesel, 1968).

The challenge of incorporating the direct-cell evidence of velocity-sensitive response into an understanding of motion perception has driven a great deal of sophisticated analysis and research in the past decade, in which stroboscopic stimuli have played a large part. The first, primitive exploratory psychophysics, testing how luminance, spacing, and timing affect viewers' reports of apparent motion (i.e., whether they report beta movement, pure phi, simultaneity, or discrete succession, etc.), started systematically in 1915 with Korte (1915), Wertheimer's student, using simple stimuli like those in Figure 1A. The attempt to find general relationships between these variables, traditionally known as "Korte's laws"[7] (for reviews, see Boring, 1942; Kaufman, 1974; and Kolers, 1972), have since been replaced by tests of specific hypotheses; the stimuli are now usually much more complex patterns designed according to the hypothesis being tested, and most notably by changing

[7] By (among others) Neuhaus (1930), Sgro (1969), Kolers (1972?) and, as an explicit test of their version of the model in Figure 3D, by Burt and Sperling (1981).

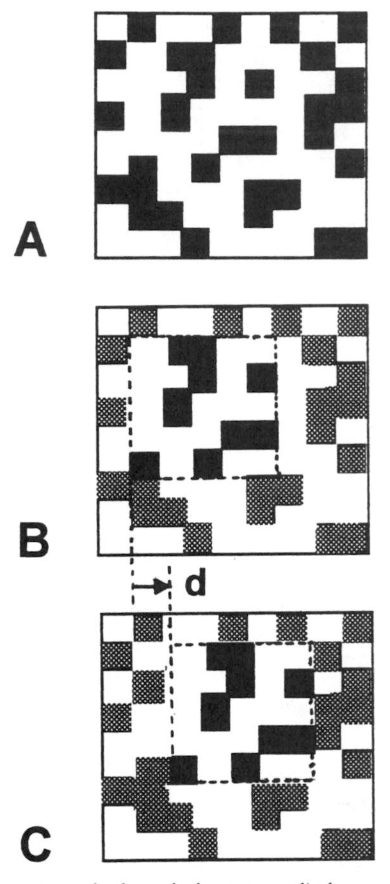

FIGURE 4 A shape that exists only through short-range displacement. (A) A random-dot array. (B, C) A region is displaced coherently (by amount d). In actual use, there are no dotted lines; the dots outside and inside the displaced region are the same shade; and the dots outside the region are changed randomly from frame to frame (and not left the same as is done here for clarity). Although the displaced region is therefore indistinguishable in each static view (as it is at A), it is visible in apparent motion—but only if the displacement (d) is small (Braddick, 1974).

assemblies of "random" dots like those in Figure 4; and the response measures are less often the viewer's subjective judgments about image quality and more often explicit choices, like which of two motions is reported (as in Fig. 3B, C).

By that measure, viewers show no preference for an apparent motion between objects of the same form or color over apparent motion between objects that differ in these attributes as in Figure 3B, C (Kolers, 1972; Kolers & Pomerantz, 1971; Navon, 1976; see also Orlansky, 1940, etc.). The motion occurs between the nearest successive contours, and does not at all preserve phenomenal identity. This is a surprising phenomenon, one that will be important to us in Section III.B.2, and that is more consistent with explanations in which apparent motion perception depends on displacement receptors than with explanations in terms of cognitive inference.[8]

Most current explanations of apparent motion (Adelson & Bergen, 1985; Burt & Sperling, 1981; Watson & Ahumada, 1985) assume the same mechanisms for continuous and stroboscopic motion. Note that the work we survey in the following discussion was not aimed at explaining moving pictures but at using apparent motion to form and test theories about the structure of visual neurophysiology, and ours is consequently a very selective treatment of that work.

The two dominant proposals rest respectively on a sequential change detector like the unit shown in Figure 3D, and on a system of analysis of stimulation to which stroboscopic and continuous motion are (within the appropriate ranges) equivalent. As a simplistic illustration of the former, Figure 3D sketches a mechanism in which the left receptive field would respond to a black edge facing rightward at some moment, and its response is delayed by amount t; if after a delay of the same duration, the right field is exposed to and responds to a black edge facing rightward, and the two responses are multiplied, then the unit will emit a net response. (Note that if only one field is stimulated, or if the interval between the two is not equal to t, the net response will be at its minimum or zero.) The unit should provide equivalent outputs when stimulated by either a black edge moving rightward across the retina with velocity s/t, or by successive stimulation at interval t, within the appropriate ranges; therefore, the real and the apparent motion have the same underlying basis.[9]

The second class of kind of account descends from spatial-frequency channel analyses of 2-D visual patterns into their light-dark ratio, or energy, at each level of the spatial-frequency spectrum (all of these start with Fourier analysis of luminance distributions). When extended to patterns that change over time (i.e., the energy distribution in space–time patterns), the equivalence of stroboscopic and continu-

[8] Virtually any real scene will contain more than one identifiable point in each successive view; how can one then tell in advance which point will move where? The plausible answer is that motion is seen between corresponding objects; that is, each object in one frame moves to the position it occupies in the next (as in Fig. 3A). This would follow from Helmholtz's rule, above, and follow too from those versions of Gestalt theory that hold that one perceives whatever requires the fewest changes (see Section I.A.2.c.). However, starting with Orlansky's finding in 1940 that apparent motion occurs between objects of different shape, with one changing into the other, we now have a great many demonstrations that, at least for short displacements, and over short intervals of time, motion occurs between the nearest luminance-difference contours of the same sign (e.g., Fig. 3B), regardless of color or features of form. Major experiments on the essential irrelevance of form were done by Kolers (1972), Kolers and Pomerantz (1971); on color, by Navon (1976); and essential corroboration of these points by researchers concerned with modeling the early visual response to motion, such as Burt and Sperling (1981), Cavanagh, Arguin, and von Grunau (1989), Dawson (1990), Krumhansl (1984), Ullman (1979), and Victor and Conte (1990).

[9] Such units have various orientations, sizes, and characteristic parameters of (s) and (t).

ous motion becomes evident and calculable (Adelson & Bergen, 1985; Watson & Ahumada, 1985); this is represented in Figure 3E, in which luminance distribution in space (axis S) and time (axis t) are compared for continuous and stroboscopic motion. A network sensitive to the first one should respond also to the second, within the appropriate limits.

Van Santen and Sperling (1985) presented proof in 1985 that the motion–energy explanations are formally equivalent in output to the analysis of the stimulus as performed by a more sophisticated version of the detector in Figure 3D (namely, the revised Reichart detectors proposed by Burt & Sperling, 1981, which pools the responses from two opposed motion directions; see discussions in Chubb & Sperling, 1989, and Sperling, 1989). Both kinds of explanation provide the same responses to continuous movements and stroboscopic displacements, and have led to the successful prediction of apparent motion, otherwise unexpected, in specially devised patterns of repetitive contours, like stripes or plaids (for recent discussions, see Adelson, 1991, and Sperling, 1990). Both will account for the "identity-blind" phenomena like that in Figure 3B. Neither will account for the more intelligent solutions of Figures 1B, 1C, 6, and 7, which are obtained under certain conditions.

In fact, it has been strongly argued (not without opposition) that there are different sources of, and conditions for, apparent motion, and some of the phenomena on which the argument is based are important to our understanding of motion pictures.

ii. On different kinds of apparent motion. It is often proposed that we use both a primitive motion-detection response, and a more detailed form-dependent process. Herewith, some proposals about continuous motion: Exner (1875) proposed that movements are directly sensed at moderate velocities, but at lower velocities they are inferred from position changes. Leibowitz (1955) proposed that only the velocity-sensitive mechanism operates at short exposures (250 ms), so that any more relational phenomena (like the effect of a visible framework, a point important to us in Section I.A.2.c) requires longer views. Gregory (1964) made a similar argument. As to stroboscopic motion, Saucer (1954) argued that it reveals two different neural systems: a primitive response (as undetailed as scotopic vision), signaling only that motion has occurred, and a more detailed response, with the two combined by a superordinate process; and Hochberg and Brooks (1974, 1978a), addressing phenomena that occur in motion picture cutting (see Figs. 7, 17) distinguished at least two processes that could be set to oppose each other, one fast and undetailed, and the other, which allows form to prevail over proximity, which can be found only with longer views.

In the most widely discussed two-system theory, Braddick (1974, 1980) and Anstis (1980) argued that only short-range apparent motions involve the responses of the motion detectors (such as Fig. 3D). Braddick (1974, 1980) found that a region of dots displaced by amount *d* between two successive random-dot patterns (Fig. 4) clearly moves, even though neither view alone bears any discernible

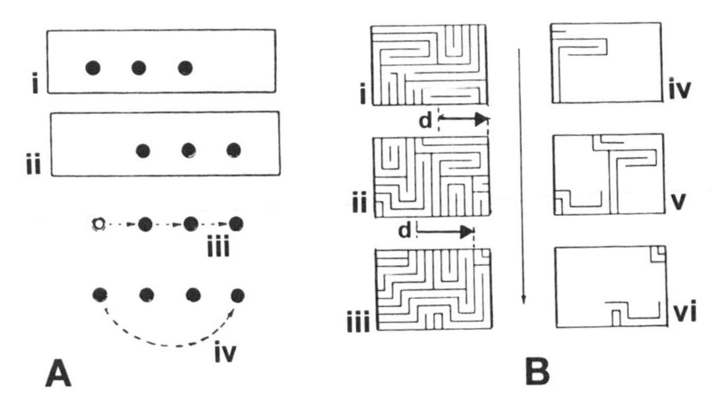

FIGURE 5 Long-range apparent motion. (A) The two successive static views devised by Ternus (1938) often provide the short-range apparent motion shown at iii (following proximity, as in Fig. 3A), but can also be seen as a more long-range movement, iv, especially if the two middle dots are not turned off (i.e., if ISI = 0.0; see Fig. 1A); Petersik (1975), Braddick (1980). (B) The successive static views (i–iii) are correctly perceived to move in the direction of displacement (d) only with small displacements (as in Fig. 4). Removing some of the detail as in iv–vi provides larger units (and lessens the opportunity for motion between noncorresponding elements), allowing correct direction to be perceived with much larger displacements (Hochberg, Brooks, & Roule, 1977; see Fig. 39c in Hochberg, 1986).

shape,[10] but that this only occurred with d ≤ 0.25° and ISIs ≤ 80–100 ms. This value, called D_{max}, is greater with larger moving regions (Baker & Braddick, 1982; Nakayama & Silverman, 1984), probably because it then involves the larger receptive fields, and it increases as well with repetitive displacements separated by brief pauses (Nakayama & Silverman, 1984; Lappin & Fuqua, 1982; Ramachandran & Anstis, 1986; Sperling, 1976). With larger displacements, or when the two successive fields were viewed with separate eyes, the apparent motion does not occur. Only those situations that produce the short-range motion also produced motion aftereffects (Anstis, 1980).[11]

Presumably, short-range motion reflects a point-by-point analysis of the visual field by motion detectors. But long-range apparent motions can also be obtained, which seem to be less identity blind and are subject to what the viewer attends. For example, the long-range motion (Ternus effect) in Figure 5A (Ternus, 1938; Peter-

[10] (This is a *kinematogram;* Julesz, 1971). The dots in the surrounding field were randomly changed from one view to the next.

[11] If aftereffects occur only with receptor-based motion, Anstis's finding argues that short-range motions are receptor based. For intriguing examples of phenomena that the distinction may explain, see Anstis (1980) and Ramachandran and Anstis (1986); for a comparative review of research on such differences, see Petersik, 1989.

sik, 1975);[12] to account for such motion, Braddick (1980) and Anstis (1980) invoke a second motion system.

The appeal to separate neural mechanisms has been strongly questioned. Cavanagh and Mather (1989) argued that the behavioral differences attributed to differences in detection systems resulted from the specific stimuli chosen. For example, the very existence of a maximum displacement results from false targets that activate detectors with the largest receptive fields. Note that removing clutter results in apparent motion at very large displacements (see Fig. 5B). The basic motion detectors of Figure 3D, they proposed, are supplemented only by an inference system that is more cognitive in nature.[13]

Regardless of how this resolves, however, we must note that these issues follow a somewhat different line in motion pictures. An uninterrupted continuous sequence of images is what we will call a *shot*—roughly, what we would record while the camera is kept running. A discontinuity between such sequences is a *cut*. We will discuss these from various viewpoints. For now, however, we want to point out that apparent motion is normally short-range *within* a shot, and therefore presumably reflects the simple action of the motion-detector system. The recording of most motions within a shot is therefore essentially automatic and need not concern the filmmaker;[14] but across a cut, apparent motion is likely to be mostly long-range in

[12] In a more recent long-range example, when motion can be seen in either of two directions viewers tend to see the pictured persons or objects move in the direction they face (McBeath et al., 1992), and where one and the same object can be perceived to face in either of two directions (as in Figure 6), the direction of apparent motion is biased in accordance with that intentional construal (McBeath et al., 1992; Hochberg & Peterson, 1993). Because there is already evidence that the viewer's set can affect the outcome of long-range apparent motion experiments (Ramachandran & Anstis, 1986; Sigman & Rock, 1974; Ternus, 1938), one might speculate that such effects are mediated by how the viewer chooses to attend to the field of potential correspondences.

[13] There are other bases for proposing separable systems of motion perception. Sperling and his associates propose two processing systems (discussed at length in Sperling, 1989), one that responds to luminance distributions within small local areas, or modules, and therefore responds differently to edges of opposite contrast (i.e., which differ in phase) and a "non-Fourier" system that discards such phase information (i.e., responding in the same way to both of the above edges) and coordinates computations made over different modules. Luminance-based responses of the first-order (linear, Fourier) system, responses of the motion detectors, provide most of what is available for any later processing. The second-order non-Fourier system (which can detect motion in patterns so balanced that there is no motion–energy difference from one frame to the next), is effective only at large spatial scale and in central vision (Chubb & Sperling, 1989). Going beyond apparent motion to what we call the motion's *perceptual consequences* (pp. 258), Dosher, Landy, and Sperling (1989) showed that the non-Fourier system was ineffectual in perceptual tasks needing motion-direction information in more than one local region at a time, although it served to discriminate the motion direction of a small moving patch of random dots.

[14] Within a shot, normal motions fall below D_{max}; rapid events (a batter's swing), or rapid camera moves (like a *swish pan*) might exceed that, but such rapid motions tend to be avoided because the viewer cannot grasp what is being presented at such speed. As to the first point, data obtained by Nakayama and Silverman (1984) suggested that with continuing displacements, and pauses of about 33

extent, is usually unwanted and unforseen by the filmmaker, and is very different—and far from simple—in its characteristics.

Between shots, the displays will have changed to contain different objects, or to contain the same objects in different places, unless the views were so matched that the film remains essentially continuous; some few contours may fall within D_{max} of contours in the preceding view, but most will not, and even those that do may not belong to the same object. Cuts (and "slide shows") are central to the use and understanding of motion pictures, so we must be concerned with long-range motion, regardless of whether or not it implicates a second system.

Merely appealing to cognitive inference or perceptual logic will not do. Central to any idea of inference-based motion lies the concept of phenomenal identity or correspondence (see Attneave, 1974; Dawson, 1991; Ramachandran & Anstis, 1986; Ullman, 1979) and of position change or displacement. The apparent motions sketched in Figures 3B, C show that we may be surprised by what *identity* means (see footnote 8), so that the notion of inference cannot simply be applied without an explicit listing of the premises. As in the impossible pictures designed by Penrose and Penrose (1958), and the fact that we misperceive perfectly possible objects as being structured in ways that are impossible (Gillam, 1972, 1979; Hochberg, 1968, 1988), the too-ready explanation of perceptual intelligence fails because we do not know, in advance of the data, what counts as intelligence (see Section II.A.1).[15] This point will recur as we proceed.

The identity problem (or correspondence problem) arises when there is more than one object in the second view that might be seen as the end of the movement of some object in the first view. It is often held that we perceive just the motion that involves the least relative displacement or lowest relative velocities (Dawson, 1987, 1991; Johansson, 1950; Metzger, 1953; Restle, 1979; Ullman, 1979). Ullman proposed that the resulting motion pattern is one that minimizes both the motion lengths ("nearest neighbor" principle) and the splitting or combining of elements from one view to the next. Dawson's model (1991) assigns weights to these factors, and to a third factor that minimizes the relative velocities between neighboring elements of the display (Dawson, 1987, 1991), accounting thereby for most of the traditional body of apparent motion displays. (These same factors seem

ms (as in TV), D_{max} would be about 1° per frame, which would mean that an object could move at ca. 30°/s and remain within short-range motion. Translating from screen dimensions to real motions being recorded, that would be roughly 4, 6, and 32 ft/s in close-up, medium shot, and long shot, respectively. As to the second point, Smith and Gulick (1956, 1962) found that the contour of a small moving square could not be seen clearly at velocities greater than ca. 13°/s, but the upper limit to discern the contours clearly went to about 40°/s if the stationary stimulus was presented prior to the movement (see also Kaufman et al., 1971).

[15] There is a second and more metatheoretical aspect to the concept of inference: it views one percept as the causal consequence of another, a variety of mentalism that used to be anathema, and may still be found to provide too glib an explanation.

important to the partitioning of motions, discussed in Section I.A.2.c.) Although such explicit and quantitative treatment is clearly preferable to explanations that invoke unspecified mental representations and inferences, evidence of the latter would be important if it requires a completely different way of thinking about the processes that underlie apparent motion and moving pictures. And, as we will see, it looks as though it will.

Mental representation, and possibly inference, seems most required for apparent motion when trying to account for the effects of form, depth, and viewers' attention, intention and expectations or knowledge. And, as is seen next, these factors have measurable effects on apparent motion in the laboratory and seem to have very important effects in the viewing of motion pictures.

iii. Beyond the motion-detection systems: Forms, depths, displacement, intention, expectation and attention. Although form, attention, and expectation are much more difficult to specify than variables of nearness and velocity, they are important to any understanding of motion picture cuts (and perhaps to apparent motion quite generally, even though they are manifest only in long-range motion; see Sect. I.A.2.c).

Most recently, McBeath, Morikawa, and Kaiser (1992) reported that where motion can be seen in either of two directions we tend to see the depicted persons or objects move in the directions they face. And where the same object can be perceived as facing in either of two directions (as in Jastrow's, 1899, duck–rabbit, Fig. 6) the direction of apparent motion is biased according to the viewer's construal (McBeath et al., 1992). Any discussion of such apparent motion must therefore include the object's shape, the viewer's attention to that shape, and some process that chooses between alternatives not simultaneously present in the stimulation (therefore, between "mental representations").

Where such meaningful, long-range factors are in conflict with short-range determinants, the latter seem in general to prevail. Within no more than 80–100 ms after an abrupt change of view,[16] apparent motions occur between nearby successive contours of the same sign (or contrast direction). Detailed shapes themselves take longer to be recognized: the very earliest response seems to be to "blobs," that is, regions defined only by their *low spatial frequency* components and lacking in detail (see Breitmeyer & Ganz, 1976; Breitmeyer & Julesz, 1975; Breitmeyer, Love, & Wepman, 1974; Williamson, Kaufman, & Brenner, 1977). As noted here in a different context, by using large displacements, nearness between objects in successive views can be put in conflict with actual identity, as in Figures 7 and 8, producing apparent motions that can be opposed in direction to the true displacement (S). (We believe, with Vorkapich, 1972, that many problems in film

[16] From Braddick's original data (1974, 1980); those of the Nakayama and Silverman experiment (see above) suggest much shorter times.

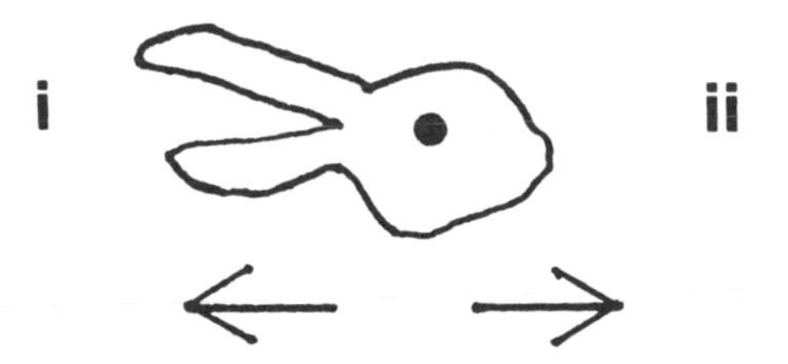

FIGURE 6 Construal and apparent motion. A version of Jastrow's duck–rabbit. The same pattern can be seen as a duck facing left (i) or a rabbit facing right (ii); which animal is perceived is affected by which animal the viewer is set to perceive, and affects the predominant direction of apparent motion (McBeath et al., 1992).

cutting are built around undesired apparent motion between noncorresponding objects; see Section III.B.2.a.)

At about four views per sec, therefore, the set of geometrical shapes in Figure 7 appears to scroll leftward (the F arrow) while actually being displaced rightwards; their shapes are confused even though they are distinct in longer views. Similarly,

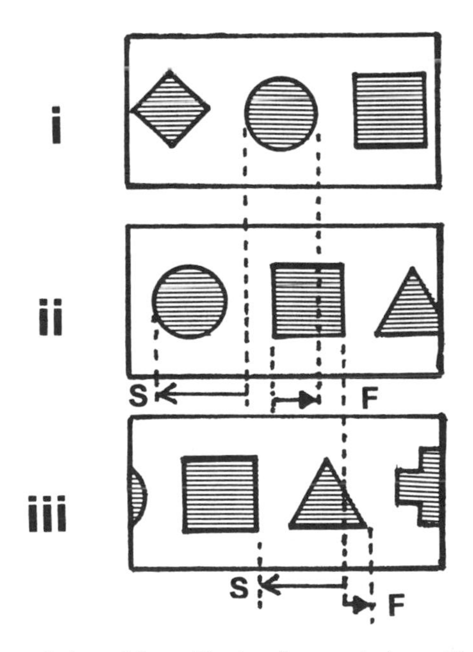

FIGURE 7 Fast proximity and slower identity of geometric shapes. Three successive static views that appear to move rightward (F) in accord with proximity at fast rates (i.e., at short SOAs); they appear to move leftward (S), in accord with identity of shape, at much slower rates (i.e., at longer SOAs) (Hochberg & Brooks, 1974; see Hochberg, 1986).

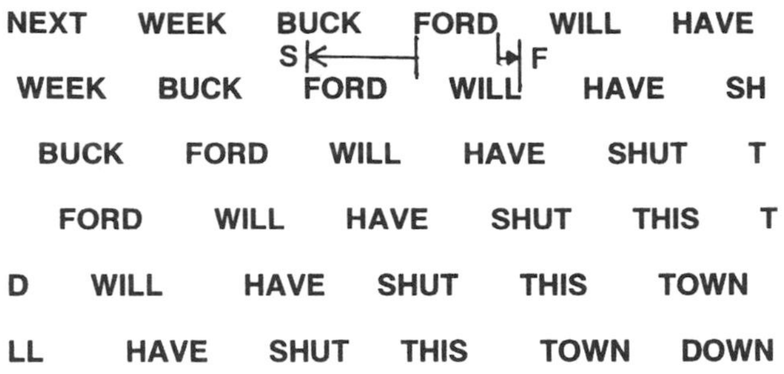

NEXT WEEK BUCK FORD WILL HAVE
S F
WEEK BUCK FORD WILL HAVE SH
BUCK FORD WILL HAVE SHUT T
FORD WILL HAVE SHUT THIS T
D WILL HAVE SHUT THIS TOWN
LL HAVE SHUT THIS TOWN DOWN

FIGURE 8 Fast proximity, slower identity of text. As with Figure 7, this shows static successive views from a continuously scrolling sentence: lines are seen one at a time, replacing each other at from 250 to 3000 ms. At fast rates, text strongly appears to move rightward (F), and is illegible; at much slower rates, the leftward (S) displacements are discerned and the sentences can be read (Hochberg, 1984; see footnote 18).

even though words are normally recognized very rapidly, the sentence in Figure 8 appears to scroll rightward and is illegible. At short exposures, the nearest neighbor wins out over identifiable shape. But with longer views (500–3000 ms: Hochberg, 1986; Hochberg & Brooks, 1974),[17] the sequences that gave leftward motion in Figures 7 and 8 look successively displaced to the right (the *S* arrows), following a very brief and disorienting leftward "jump," and the sentences in Figure 8 are now legible.

The slow-acting, long-range, correct motions (those indicated as S) are, plausibly, due not to a second motion-detection system, but to a knowledge-based recognition of objects (e.g., the words in Fig. 8) and their displacements. Palmer (1986) showed that viewers' discriminations that a displacement has occurred may reflect two different mechanisms, one for delays of 200 ms or less, and one for delays of 500 ms or more. Kinchla had earlier argued that such discrimination draws on a comparison with a decaying memory of the previous display (Kinchla, 1971; see also p. 211). The slowness with which correct direction is obtained in Figure 8 when word recognition itself can be extremely rapid suggests that some additional recovery or reconstrual time is needed. Perhaps the slow response in Figures 7 and 8, and in the recoveries from the motion picture cuts of Section III.B.2.a, rest on the viewer's recognition at some point in the sequence that one or more remembered objects have been displaced leftward: the rest of the sequence may then involve testing that expectation by appropriately shifting attention with each cut.

The phenomena in Figures 6–8 call for some degree of mental representation

[17] As MacKay (1980) noted, one must admit to multiple systems when considering that pursuit movements made when the eye follows a moving object may keep its image reasonably stationary on the retina (Dodge, 1907), but the object nevertheless appears to move.

(at least for memories across views), but have no need for inference-like processes. Spatial relations based on depth cues offer one of the major arenas in which inference has been invoked, historically, and space from depth cues intersects our present concerns by way of several studies showing that it is the locations in represented 3-D depth, rather than nearness in the 2-D display, that appear to determine apparent motion. Depending on the conditions, such findings may imply an inference-like process (e.g., taking depth cues into account to arrive at distance from view, taking distance from viewer into account to arrive at separation between objects, and basing perceived motion on that inferred separation). They may imply only that the information about the third dimension must be extracted before the processes that provide apparent motion, with no inference needed. Or they may be independent consequences of the same stimulus patterns that provide the depth information.

Most recently, Green and Odom (1986) showed that apparent motion occurs between objects that appear to lie in the same depth planes in binocular space rather than between objects at different depths but equidistant in 2-D layout. Although this implies that binocular segregation can occur before apparent motion, it does not necessarily imply that apparent motion is determined with a mental representation of the 3-D space, because the effect might rest on disparity-specific connections. Earlier examples to the same point had not used binocularity to achieve depth (Attneave & Block, 1973; Corbin, 1929, Rock & Ebenholtz, 1962; Shepard & Judd, 1976), but they face contrary data: Ullman (1978) found that when presented within a perspective drawing that represented different 3-D distances, successively presented lines appeared in the apparent motion appropriate to their 2-D rather than their 3-D separations, and Mutch, Smith, and Yonas (1983) reached similar conclusions. Given the theoretical importance of the issue, work replicating both sets of findings, if possible, resolving their differences, and studying the parameters of the depth effect, if one is vindicated, is urgently needed.

In summary, one is left with a plausible argument: that short-range stroboscopic motion reflects the activities of the same motion-detection mechanisms that detect real or continuous motion, and that the movement in motion pictures within uninterrupted sequences (within shots) consists predominantly of such short-range apparent motion, with the direction of motion determined primarily by the nearness of same-sign contours in successive views. Stroboscopic motions also occur over longer distances, however, and in particular across cuts in motion pictures, and the processes then involved, which are more shape-dependent and attention-dependent, are less understood but of greater concern.

Because short-range apparent motion prevails over long-range and meaning-determined factors of apparent motion when they are placed in conflict, and because the short-range factors are easier to model, one can afford to set the latter aside, and thereby keep one's table relatively clean. That is true, however, only if no claims are being made to *explain* motion perception. That we can perceive motion without the involvement of the short-range system argues for another source of apparent motion, and for a superordinate system that accepts input from either.[18] This in turn means that the existence of motion detectors that respond directly to velocities in the retinal image is an interesting finding for neurophysiological concern, and must be considered as a possible perceptual component but does not "explain motion perception" and amends but does not demolish the classical account in Figure 2. This point becomes even stronger when we consider the perceptual consequence of camera movements on the perception of motion pictures.

c. The Partitioning of Relative Movements: Induced Motion, "Separation of Systems," and the Parsing of Complex Movements

In this section, we describe how one set of movements on the retina, or on the screen, can be made to look like a set of totally different movements, and how this fact is central to an understanding of motion picture perception and practice.

As we noted in Section I.A.2.b.ii, the threshold for relative movement is lower than that for absolute movement (Aubert, 1886; Brown & Conklin, 1954). This suggests (but does not prove) that there is a range within which viewers will be certain that movement has occurred, but will not be certain which objects have moved and, consequently, what the direction of movement is.

With a stationary dot (sd) surrounded by a moving frame (mf) in an otherwise dark or featureless space, the frame appears stationary *despite its stimulation of the relevant motion detectors,* and the dot appears to move (Duncker, 1929)[19] (see Fig. 9A). With two dots (Fig. 9B), the smaller object usually appears to move. What is perceived is not transparent to the action of motion detectors as they have been discussed; one might without too much difficulty imagine receptive fields to explain this phenomenon, but not the following: Objects that are characteristically mobile (cars, airplanes, people) are more prone to be perceived as moving (Brosgole & Whalen, 1967; Comally, Werner, & Wapner, 1957; Duncker, 1929;

[18] Starting with the geometric shapes in Figure 7, we have been varying this demonstration since 1973, with the general parameters given; this last procedure, using sentences, was shown during an American Psychological Association address (Hochberg, 1984) and to many classes since. In the latter, by a show of hands the fast effect was always unanimous, with a few stragglers failing to record the last sentences completely at sequences of 3 s per view.

[19] The only visual test to distinguish these is to change the gaze in some known direction (e.g., to find out whether one must move his or her gaze to keep one or the other foveally fixated). In fact, knowledge of the extent and even the direction of the pursuit movements the eyes make is quite poor (Festinger & Easton, 1974; Mack & Herman, 1973; Stoper, 1973), and, of course, the precision with which we can identify the visual effects of our bodily changes, or even of the vehicle in which we sit (witness the moving train illusion, etc.) are much worse. We need other reasons, therefore, for attributing the relative motion to one or to both objects in Figure 9A. These uncertainties probably help make possible the most characteristic feature of cinema (that is, the expansion of space beyond the screen; cf. Sect. II.B), but may also create incomprehensibilities that the filmmaker did not intend (cf. Vorkapich, 1972).

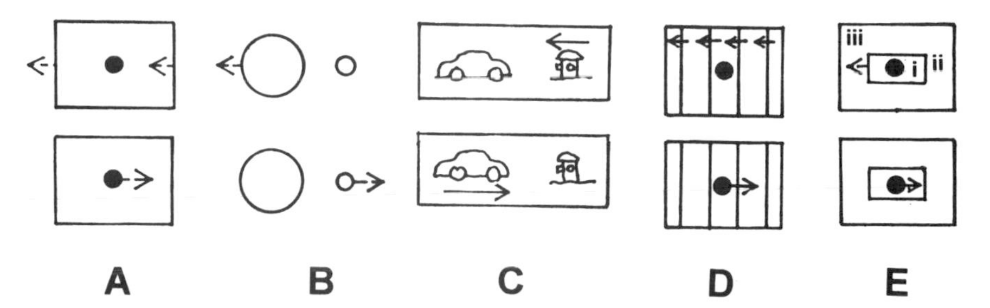

FIGURE 9 Induced motions: the movements on the screen versus those we see. In A–E, the real movements are shown by the arrows in the upper row (e.g., at A, the frame moves leftward and the dot is stationary). The motions that are actually perceived are shown by the arrows in the lower row. At D, the real motion of the background in the upper row, and the perceived motion of the dot (which in reality is stationary on the screen), as it is shown in the lower row, can go on indefinitely, a phenomenon that must underlie all but the shortest actions in motion pictures. This is important to most camera work, particularly to that in Figure 13C2, D.

Jensen, 1960), although the difference can be eradicated if the stationary member of the pair is enclosed in a moving frame (Brosgole & Whalen, 1967).

Given a stationary object within a stationary frame, and a textured field moving continuously behind the object, the object appears to move in the direction opposite to that of the background. Because the object is motionless with respect to the outer frame, but appears to be moving, the logical (and, to some very slight extent, the perceptual) implication is that both the stationary outer frame and the room that contains the frame and the viewer are keeping pace with the moving object. This apparently paradoxical phenomenon is *absolutely essential to the cinematic representation of movements that continue for any appreciable duration* (because an object that actually moved relative to the screen would soon run off it); see the sketch in Figure 9C. How does this phenomenon fit the rule of induced movement?

At least five possible explanations for this important cinematic effect are apparent. First, the fact that the object is immediately adjacent to its moving background may provide a purely local relative motion, and may override the effects of the frame, which lies at a greater distance from the object (cf. Gogel, 1977).[20] Second, the background's movement covers much of the surrounding peripheral retinal image, and may therefore[21] misinform the viewer that his or her gaze direction is changing. Third, those who take perception as inference might point out that the

[20] As with depth in the impossible figures, according to one theory (Hochberg, 1968, 1986), the partition of motion may be primarily a local matter, and the effects of the ground on the object may not interact with the more distant frame.

[21] There is some question about whether or not peripheral vision has a special status in determining self-movement.

viewer accepts the moving background within the frame as being the implied surrounding (as the scene viewed through a moving train's window is seen as part of the general, stationary environment within which train and viewer are moving; cf. Koffka, 1935). Fourth, in an account that we favor, the viewer's primary construal can be of the object as moving in relation to its background, with a secondary or weaker construal that the background is being viewed through a window that is itself moving relative to the background but is stationary with respect to the object (and the viewer). The secondary construal is available with effort but otherwise does not have cognitive consequences in comprehending or remembering the motion picture. Finally, and not in conflict with the previous explanation, Wallach (1959) proposed a general noninferential rule, a "separation of systems," in which apparent movement is always determined by the immediately enclosing frame: Thus, in Figure 9E (a version of a demonstration described by Wallach) the objective movement of the inner rectangle (ii) induces point (i) to appear to move in the opposite direction (iii), even though (i) is stationary with respect to (iii).[22]

Any or all of these may explain the motion picture situation sketched in Figure 9D. Research is sparse but not difficult in principle, and the various explanations have different cinematic implications. For example, according to the first explanation, the effect would decrease as the visual angle subtended by the screen decreased, whereas the last two explanations do not lead to this prediction. But research in this area, as with the apparent motion of Figures 3–8, deals in ambiguous displays, offering alternative motions. As is common with ambiguous stimuli, viewers' may also report perceiving different alternatives with slight changes in conditions, instructions or intentions, or even report spontaneous alternation within a single encounter. Methods exist to measure how strongly the alternatives are perceived and not merely reported (Hochberg & Peterson, 1987; Peterson & Hochberg, 1983), but these have not been used with the phenomena now under discussion, and without such measures one cannot know whether these phenomena are relevant to the very robust-appearing motion picture effects they appear to mimic.

The previous examples are usually discussed as *induced motion;* for a recent review, see Mack (1986). In discovering and exploring a closely related body of phenomena, Johansson (1950, 1974) proposed an account that is simple yet appears to be potentially powerful, stimulus-based, and objectively specifiable: (1) the visual system extracts a motion that is common to all of the moving parts of the field of view; (2) this motion itself becomes the framework against which the residual movement is seen, performing the vector analysis shown in the example of Figure 10. In effect, the diagonal motion of the object (d) in Figure 10A, which is surely

[22] Research by Farber and McConkie (1977) suggested that the effect illustrated in Figure 9E may not be reliable enough to be the basis of the robust, fundamental cinematic phenomenon sketched in Figure 9D, but we think the point sufficiently important to deserve more work.

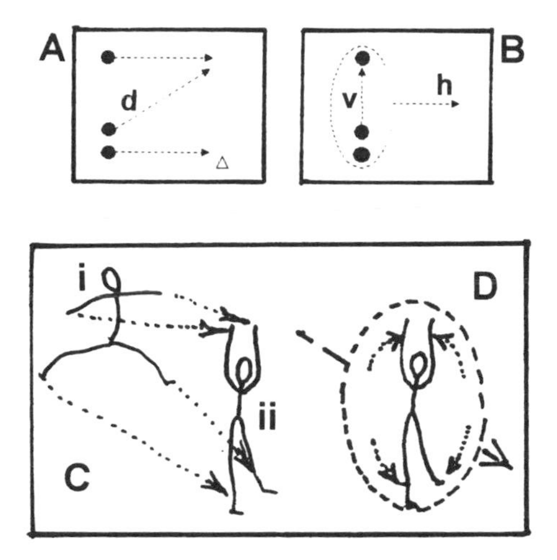

FIGURE 10 Parsing events into common and relative motions. (A, B) An example of the Johansson (1950) phenomenon: At A, Three lights moving, in an otherwise dark field, in horizontal and diagonal (d) paths. These are not what is perceived. At B is sketched what is instead perceived: an in-line motion between the upper and lower dots (i.e., the vertical motion [v] embedded in a horizontal drift [h]. (C, D) What seems very likely to be the same phenomenon pervades motion pictures (and perhaps the perception of complex motions quite generally): At C, a dancer is held aloft at i (although the partner is not depicted here), and the movement paths she takes on the screen as she is set down at ii are shown by dotted arrows. But the sketch at D is closer to what is actually seen: arms brought up and legs brought down, both within a diagonal drift.

picked up by whatever the relevant motion detectors may be (Sect. A.2.b.i), is simply not perceived; instead, it is clearly seen as the vertical motion indicated in Figure 10B, while the whole configuration enclosed by the dotted oval may or may not be discerned as moving horizontally.

This is a very robust phenomenon, and it seems as though it should be important to the understanding of motion perception in general and motion pictures in particular. It is not an arcane or abstract issue. In Figure 10C the motions perceived in life or on the screen are those at 10D, not those actually made in our visual fields. And whatever process explains Figure 10B should, it would seem, thereby explain the important phenomenon of Figure 9D, as well.

One extremely simple account would be that the viewer's gaze follows the main common motion (h in Fig. 10B), leaving only the vertical (v) component on the retina, and that the pursuit movement itself is essentially ignored (Stoper, 1973), and similarly in the kinds of displays with moving cameras on which motion pictures rely so heavily (Section II.A.2). But the effect is obtained almost equally well if a fixation point (the open triangle in 10A) keeps the viewer's gaze stationary (Hochberg & Fallon, 1976), and what seems to be object-relative motion can be obtained in several directions at once. Would two dancers (like the one in Fig.

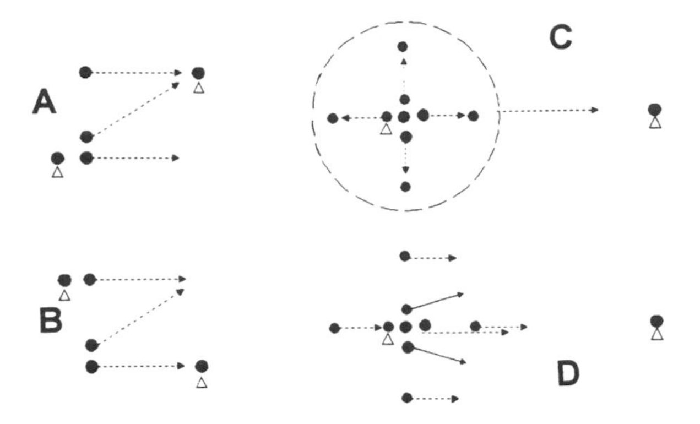

FIGURE 11 Fixation, minimum motion, and viewer's intention in event perception. Points marked by open triangles are stationary, and are potential fixation points. At A, with either fixation point attended, perception of the diagonal motion (d, in Fig. 10A) predominates; at B, in-line motion (v, in Fig. 10B) predominates; in both cases, what viewers report they see is influenced as well (but not determined) by which motion they *try* to see. Events C and D are both perceptions of the movements shown at D. When the rightmost stationary (triangle-marked) point is fixated, the in-line radial motions of 11C strongly predominate; with fixation at the leftmost stationary point, the diagonals of 11D are more often reported (Hochberg & Beer, 1990). Because the paths perceived and those on the screen (and retina) are so different, the filmmaker must know when such divergence may arise (Brooks, 1989), and how to recruit attention to support the desired perception (see Fig. 20). Theories in visual science must make explicit provisions for the effects of fixation and attention if they are intended to explain the perception of motion.

10C,D)—say, one rising and one falling, be differently perceived? We have no research to this point, but in Figure 11D, the Johansson pattern of Figure 10A has essentially been orthogonally duplicated, with two stationary points (marked here as open triangles), and a viewer who fixates the rightmost stationary point perceives not the diagonal motions of 11D but the verticals and horizontals at 11C. Cancellation by pursuit eye movements is therefore not the explanation of the Johansson effect, or of Duncker's induced-movement phenomena.

But the vector-analysis rubric has its own problems. First, Johansson's phenomena are much more robust than his explanations suggest, and are perceived even when surrounded by visible stationary frameworks (the projection or video screen, the screening room, etc.), which are evidently not counted in the vector analysis (Hochberg & Beer, 1990; Hochberg, Fallon, & Brooks, 1977). *How then can we tell in advance what elements in the stimulus display should be included in the analysis?*[23]

[23] The stationary points in Figure 11A & B should appear to move leftward if the common "framework" motion (shown as *h* in Figure 10B) is subtracted according to the vector-analysis explanation, and the Johansson effect should not appear at all if the stationary points look stationary; the situation is like that of the motion picture screen sketched in Figure 9C. The perceptual system must first decide what is being parsed before applying any system of vector analysis or simplicity assessment, which is also evident from the research by Proffitt, Cutting, and Stier (1979) and by Proffitt and Cutting (1980) in their closely related studies of the apparent paths followed by lights, mounted on rolling wheels, when only the lights can be seen.

Second, that principle often fails to tell us which the common motion will be, especially where multiple levels allow repeated extraction (cf. Fig. 11C).[24] Third, changes in configuration that provide ready reference points for simple motions within the whole pattern (Fig. 11A vs. 11B) increase the strengths of those motions, and their effectiveness is greater if the viewer attends to those points (Hochberg & Beer, 1990). Although viewers see only the in-line motions (i) in Figure 11C when attending the leftmost and rightmost point, they see the diagonal motions much of the time when asked to fixate the leftmost stationary point. In Figure 11B, viewers see only the in-line motion (that is, as in Fig. 10B) when asked to fixate either stationary point, but they report seeing more of the diagonal in Figure 11A when fixating either stationary point. (Subjects generally report seeing the diagonals in such patterns more when asked to try to see them, which may reflect shifts in what they take as reference points.)

From most of the examples in this section (Figs. 9–11), it seems that any account of the motions that we actually see must include factors of proximity, of relative velocities, and of what the viewer elects to choose as the origin(s) relative to which the motion is taken. (Dawson's, 1991, model as to which of several alternative apparent motions will be seen, and Palmer's, 1986, formal account of displacement discrimination, mentioned in Section A.2.b.ii and A.2.b.iii, would seem good starting points for such an account.) The effects of familiarity on induced motion, and the effect of set on the Johansson phenomenon, might both reduce to the last point, that is, to what the viewer chooses as the central motion and its origin. And that opens a very different set of questions and implies a very different level of causation.

An account like Johansson's or like Dawson's implies the use of some physically specifiable function, even though it might in practice prove fairly complicated to formulate its measure. Nothing about a viewer's expectations, intentions or knowledge need be invoked, *nor is there a specified role for such discussion.* According to a totally different approach, viewers might start with a repertory of simple, familiar, or expected events, against which they test the low-level input for an acceptable projective fit. Such mental representations seem completely gratuitous in the present (and indeed the previous) examples. But now consider human actors on a stage, seen only by means of small lights at their major joints, in otherwise dark surroundings. When static, only a flat and meaningless array of dots is seen, yet when the humans move, their actions (Johansson, 1973), genders (Cutting, Profitt, & Kozlowski, 1978), and even their identities (Cutting & Kozlowski, 1977) are, at least under some circumstances, discernible. We do not know how much recognizable human motions contribute to the ways in which such moving-dot patterns are parsed (e.g., are they parsed in the same ways with the images inverted?). Note that vector analysis or anything like it could *at most* replace the moving pattern of

[24] Several writers (see Cutting & Proffitt, 1982; Hochberg, 1964; Restle, 1979) suggested these motions arise according to a minimum principle: one perceives the motion pattern that best reduces the number of separate motions; such formulations face the other problems described previously.

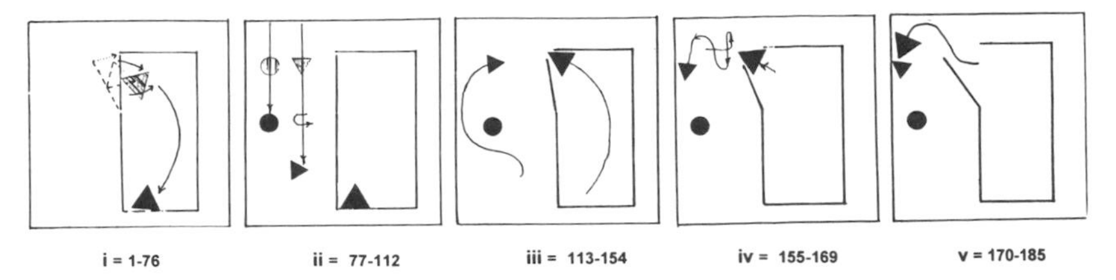

FIGURE 12 A story acted by moving dots. The opening events (ca. 7.5 s) of a narrative animated film (ca. 1 min, 14 s) by Heider and Simmel (1944). Intervals of time covered by each panel are given in frames of film; each frame comprises 1/24 s. The movements that the dots make are clearly and unambiguously perceived, but *as geometrical paths* they are incoherent and unmemorable. As elements of a story, the actions that the dots perform are obvious, coherent, and rememberable (this segment of it is described in Fig. 21). It is actions and events that motion pictures represent, and that govern their use and function; we consider these research areas in Section III.C.1.b.

individual dots by the virtual stick figures' limbs that join those dots, *saying absolutely nothing about action, gender, or identity.*

The fact is that purposive motions can be portrayed even more minimally: In the Heider and Simmel (1944) animation, part of which is sketched in Figure 12, the moving dots reprove, defy, and flee each other in a small story that audiences virtually all recognize and recall (Greenberg & Strickland, 1973; Massad *et al.*, 1979; Shor, 1957). It may be argued that it is only in memory, or at other nonperceptual levels, that the motions are organized as purposive actions, but counterarguments have been offered (Massad *et al.*, 1978), and in any case it is through memory, or other after-the-fact processes, that all our responses to moving stimuli must eventually exercise their cognitive and behavioral consequences. Geometrical motions themselves, unless they form some simple and predictable pattern, are quickly lost, and no explanation of how we perceive motion pictures (or the world more generally) can claim success if it stops with the detectors, or the vector analyses discussed so far.

What one perceives in the case of the moving dots of the Heider–Simmel film is, in our opinion, much more important to understanding the perception of motion pictures (and perhaps perception in general) than any of the apparent motion phenomena described up to this point. We will return to this example and its challenge in Section III.C.1.b.i.

In summary, no current account of early visual response to movement on the retina or on the motion picture screen will, by itself, tell us what motions will be perceived on the screen (or in the world that motion pictures represent). Nevertheless, the partitioning of motions on the screen appears to be lawful, is largely (but not crucially) dependent on measurable features of the stimulus patterning over space and time, and seems close to a general solution. The fact that *nonstimulus* factors, namely familiarity, attention, intention, and set, partially determine whether and how a retinal displacement will be perceived as a motion, has been largely ignored beyond being demonstrated; it is both harder to ignore and easier to accommodate in the context of motion picture use.

This and the previous section have addressed attempts to explain perceived motions on the picture screen. We have not considered how we perceive the screen's pictorial content (i.e., the objects and events in time and space that are conveyed by the motion picture). Many psychologists hold that such perceptions as 3-D depth and layout, and the viewer's motion within these (heading, time-to-touchdown, etc.), are at least as direct a response to the changing pattern of visual information that is received over time by a moving observer, as are the 2-D motions we have discussed. Given that the motion picture itself is a 2-D array changing in time, we refer to the perception of depth, of solidity, and of space and time beyond the screen, as problems of representation, which we consider in the next section.

3. The Motion Picture as It Represents and Reconstructs Events in Space and Time

Having discussed the perceptual processes by which motion pictures move, and by which a moving picture can provide a visual substitute for looking at the event itself, we consider next the great power that motion pictures have to improve on direct recording, and to provide surrogates for events that never occurred—that is, for virtual events.

a. Virtual Events in Our Visual Ecology

The attempt to make pictures that move is ancient (cf. Cook, 1963; Pratt, 1973). Relatively recent devices abound: the zootrope, the praxinoscope, and Muybridge's line of cameras, each of which took a picture as the moving subject tripped a wire. These devices have been replaced, of course, by the linear sequence of photographs on successive frames of film (and by the magnetic or optical storage of an electrical signal, analog or digital, that modulates the intensity of the flying video spot or a scanned liquid crystal display [LCD]), but the goal long preceded the present means of achieving it. The motion picture is thus not a modern device that has imposed an arbitrary and artificial visual language on humanity. Rather it is an evolving discovery, starting with still pictures (see Chapter 5, this volume), that taps into the processes, inherited and acquired, by which we perceive and think about the world itself.

It is clear what added steps are needed to make motion pictures more an interaction with the world itself, and they are technical rather than syntactic or conventional. Motion pictures that change in response to viewers' action as their real-world counterparts would do, now called *virtual realities,* are still largely training devices, games, and laboratory tools (see Section II.A.2.a.iii. for brief discussion). They could in principle be indistinguishable from real action in a real world (a form of mechanical idealism), and although they depend on artificial devices, they emphasize how far from being arbitrary are the motion pictures on which they are based. They also emphasize that motion pictures are not just recordings, a point that the now ubiquitous use of completely convincing special effects should already have made clear.

The fact that we can substitute very different displays and still perceive the same world is an unmatched opportunity for research into our own nature. It is also a powerful tool for studying events in the world with more care and over greater distances than would otherwise be possible: Given the enormous effect that the economics of entertainment and communication have had on motion picture practice, it is worth remembering that motion picture technology was also developed for scientific, technical, and educational reasons.[25] One particularly important reason is the way that motion pictures remove viewer's observational constraints in space and time.

b. The Expansion and Contraction of Time

Muybridge (1882) first set up a string of cameras, recording successive images of a galloping horse to determine the sequence of leg movements—a question that unaided vision cannot resolve. Conversely, by taking pictures at slow rates, events too slow to perceive (such as plants growing) are made perceptible. Such changes in time scale have many functions, but another method for changing time has wider use: The filmmaker may simply record an event or part of an event (say, the planting of a seed), stop the camera, and resume filming only later (say, after the plant has sprouted), with no attempt at achieving smooth apparent motion across the discontinuity or *cut*. Given a moment to recover from the abrupt change in view, the viewer readily accepts the elision of the (redundant) intervening time. With this ability (and with relatively easy combining of CGI with images from optical devices), filmmakers now have almost complete control of space and time.[26]

c. The Construction of Space and Time: Parallel, Repeated, and Multiple Images

The ability to elide stretches of time at will, and therefore stretches of motion through space, allows the filmmaker an economy that is the essence of an art form,

[25] Most writing on film considers it as an artistic development, using theories and methods of discourse drawn from the humanities and, at most, from semiotics, psychoanalysis, neo-Marxism, and feminism (Bordwell, 1992, p. x); Gomery (1992) showed how much that is the subject of such humanistic discourse derives from the economics of the motion picture business. But if, as seems clear, film is not an arbitrary visual language, constraints and directions must derive as well from the nature of the human perceptual and cognitive system.

[26] With the development of trompe l'oeil pictures in the Renaissance, painters could provide the eye of a static viewer with an image very faithful to that which would be provided by some object that never existed save in the painter's imagination; pictorial fidelity therefore could only be assessed in comparison with some virtual reality. Because viewers are rarely perfectly stationary, fidelity was usually compromised to allow for multiple viewpoints, and viewer motion and interaction with the world still distinguished pictures from reality. Viewer motion and interaction were therefore given critical importance in the definitions of natural (as opposed to artificial) perception stressed by psychologists attracted to doctrines of ecological realism, but the developing field of interactive CGI known as virtual reality now promises to void that distinction.

and that imposes its own requirements in a medium that the viewer knows to be a record of past events.[27]

In assembling a representation of a real or virtual event, the filmmaker can use brief portions of scenes, fragments that no one before Porter's epoch-making construction of *The Life of an American Fireman* in 1903 (Pratt, 1973) would have thought to be comprehensible. Many film devices need no instructions to the viewer, some need a little. For example, to understand that two successively presented events are occurring simultaneously in the story, a clue to that effect is needed at least the first time the device is used in the film, because otherwise viewers assume that time moves inexorably forward (Arnheim, 1960, p. 21). But from that point such *parallel action,* like the alternating subchapters of Dickens or Edgar Rice Burroughs, is made clear by prior context. The increasing use of multiple-image presentations, especially for educational purposes (Allen & Cooney, 1964; Goldstein, 1975; Perrin, 1969) offers the viewer an opportunity for more dense, simultaneous comparison. But questions of parallel action and flashback are matters of narrative, and do not tell us anything about visual perception or about moving pictures, as such.

Motion pictures have two major perceptually unique features: first, that the camera moves relative to the scene, whereas the viewer is stationary before the display; second, that massive use is made of discontinuous cuts between places that are separated in space, time, or both. We discuss these in turn.

II. THE MOVING CAMERA AND THE REPRESENTATION OF SPACE

The camera's freedom to move and change its viewpoint has four main consequences, which we consider in turn.

A. The Varieties of Continuous Changes of View and Their Differences in Spatial Informativeness: Movement Perspective, Dynamic Occlusion, and Transformational Invariants

As Leonardo da Vinci noted in his original analysis of pictorial depth cues, in real life we normally use depth cues that are missing in or contradicted by still pictures. Those motion-dependent depth cues have received a great deal of attention in recent years, for several very different and almost unrelated reasons: (1) they are needed for motion pictures and interactive devices used in training and in entertainment; (2, 3) they seem to offer a path to the perception of 3-D objects and layouts that are otherwise very difficult for computer vision—and for neuro-

[27] The live TV presentation of an event may be viewed with a tolerance that a prepared work would not receive. The "rhythmic requirements of the re-created event are quite different from those of the live event. The condensing of the event to its 'highlights' is often a necessary reenergizing procedure when the event is replayed at a later date" (Zettl, 1973, p. 265). There may be reason to distinguish, with Bordwell (1985) and de Wied (1991), between elision and compression, depending on whether omissions occur between or within breakpoints.

physiological models of the nervous system—to achieve through other routes (e.g., through the traditional depth cues, that first require the achievement of meaningful pattern vision); and (4) because it was argued by Gibson and by other *ecological realists* that under natural (not manmade) conditions such motion-produced visual information is the inherently direct and error-free source of our perceptions of our environments and their content. What gave this view added interest was that only Gibson's formulation (Gibson, 1947, 1950) addressed the issues that motion pictures raised.

We consider first how motion-based information brings its own contributions to the pictures that also include the pictorial (or static) depth cues.

1. Motion Information Versus Pictorial Depth Cues

Much of the research on motion-based information about depth uses *only* motion-based information, with displays that look flat when static but are perceived as 3-D volumes or surfaces when moving, in what was called the *kinetic depth effect* (see Fig. 1D; Wallach & O'Connell, 1953) and the parallax-related changes, or *dynamic occlusion,* that is often provided at objects' edges. Most such studies (Braunstein, 1962, 1968; Green, 1961; Johansson & Jansson, 1968; Kaplan, 1969; Metzger, 1953, pp. 328–337; Thompson, Mutch & Berzins, 1985; Todd, 1985; Todd & Norman, 1991; Wallach & O'Connell, 1953; Yonas, Sen & Knill, 1993) aim at identifying what in the changing stimulation leads viewers to perceive the moving 3-D object from the succession of views, each flat when seen alone. Machine-vision scientists have proposed analytical procedures by which computers decide what 3-D structure in the world is generating a particular pattern of moving dots (i.e., they extract *structure from motion*) (Ullman, 1979). But human perceptual systems do not work that way. For example, by these proofs three distinct views are needed to recover 3-D structure, yet humans perceive that structure with sequences of only two views (Braunstein, Hoffman, & Pollick, 1990; Norman & Todd, 1993; Todd & Norman, 1991). And all these proofs, as well as some psychological theories (Gibson, 1979; Johansson, 1982) explicitly or implicitly assume a *rigidity principle:* that we normally perceive that rigid 3-D layout or object (if one exists) that will fit the moving 2-D pattern (we perceive the invariant under transformation). The fact is that viewers fail to extract structure-from-motion in accordance with the rigidity principle (Norman & Todd, 1993) when there are no pictorial depth cues, and when there are pictorial depth cues in conflict with motion-produced information one even sees perfectly rigid moving objects as different, deforming, nonrigid objects (for reviews, see Hochberg, 1987, 1988).[28]

[28] As in the spectacular Ames windows, which have been known for some 43 years to be seen incorrectly when they move, despite the rigidity and simplicity of the trapezoidal object itself; for misperception of a rotating Ames window, see Ames, 1951, Ittelson, 1952; for a stationary object that is incorrectly seen by a moving observer to be in illusory concomitant motion consistent with the misleading static depth cue, see Gogel & Tietz, 1992; Hochberg & Beer, 1990). For reviews of failures of stimulus information specifications to account for what people perceive, see Hochberg, 1988, and Todd, 1990.

One cannot therefore simply regard the pictorial depth cues as being irrelevant in motion pictures, as Gibson and Johansson both argued, on the grounds that motion information is present. *In fact, motion picture makers have for at least 80 years chosen to use procedures that ignore or contradict the motion-produced information* (as seen in Figs. 13B,D), a practice that should be of much more concern to perceptual and cognitive theory than it has been.[29]

Motion does not, therefore, totally supercede the ways in which still pictures work. But it does introduce totally new information about the relative motion of viewer (or camera), and about the layout of space being represented. We consider these next, first in terms of motion picture practice and then in terms of the theoretical and research issues.

2. The Moving Eye: The Information about Viewer (or Camera) Motion in Motion Pictures

By 1914, camera movement (along with extreme long shot, close-ups, parallel action, and cross-cutting) were accepted motion picture usage, and zoom lenses moved from Air Force use into camera filming in the 1950s. Although Helmholtz had written briefly about motion parallax, it was Gibson in 1950 who looked seriously at the visual information received by the moving viewer, or recorded by the moving camera. Because Gibson thought that such information is directly picked up by the visual system, he was concerned to show how that information could directly specify the distal, 3-D world, and did so by considering the effects of three of the four basic camera movements as these correspond naturally to the view received by a moving observer.

a. Dolly, Track, Pan, and Zoom: Potential Movement Information about Three-Dimensional Space

i. Camera change with and without differential optic flow. Camera movements are of two classes: those that change things' sizes on the screen (i.e., the dolly and zoom shots of Figs. 13A and 13B, respectively) and those that move the scene laterally on the screen—*tracking* (or *trucking*) and *pan* shots (as shown in Figs. 13C1, 13C2, and 13D, respectively). (We use these terms in slightly specialized ways; for one thing, all horizontal camera displacements are normally called dollies and all vertical displacements are *crane* shots, but we here reserve dolly for approach or retreat along the sight line.) The arrows placed below the figures show camera or viewer movement (zoom only changes the lens's focal length), and those within the displays indicate changes in view from one moment to the next.

Changes that are correlated with both the motion in depth and the spatial layout within the scene, and that therefore offer potential information about both space and velocity, normally abound in any version of the dolly shot. That information is offered as well in the transformations of the optic array received by an advancing

[29] Being in constant evidence in motion pictures, this point should have been redundant when we raised it in 1978, but only Gibson (1979), among the many ecological realists, has noted it as a concern.

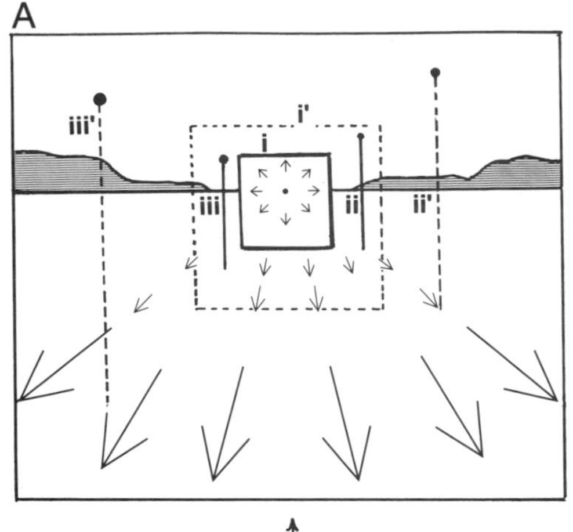
A
i'
iii'
i
iii
ii
ii'

LS
DM

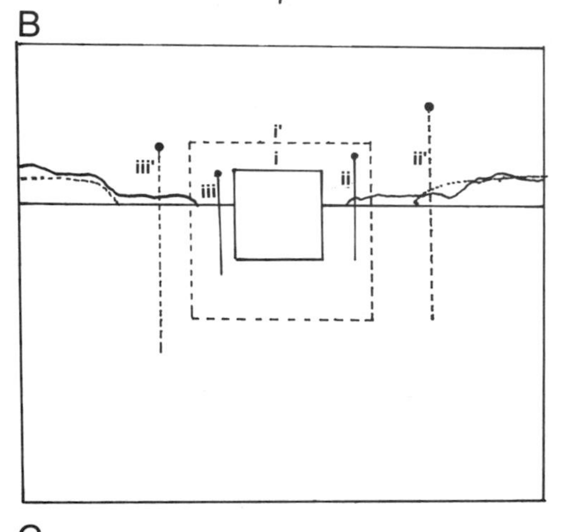
B
i'
i
iii'
iii
ii
ii'

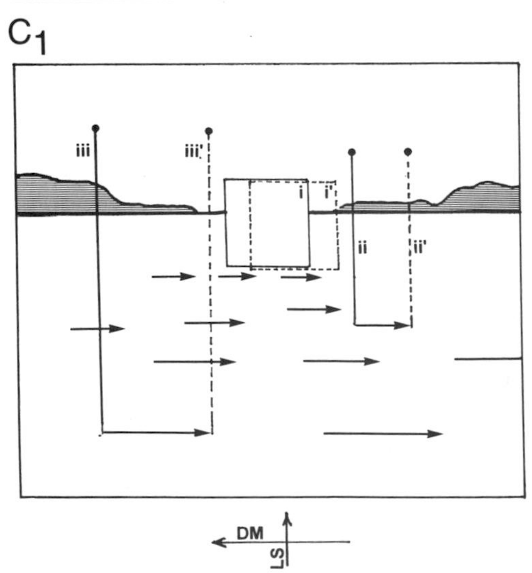
C_1
iii
iii'
i
i'
ii
ii'
DM
LS

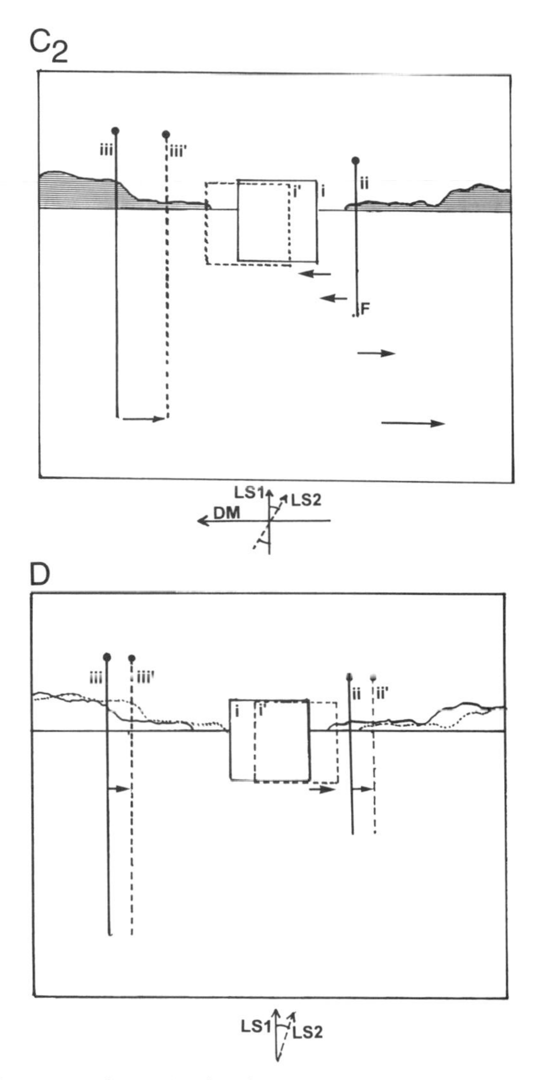

FIGURE 13 Component changes in what the camera records. (A) Dolly. Direction of movement (DM) and line of sight (LS) of the camera (or eye) are the same; here, both are directed into the layout, heading for the *point of contact,* or point of convergence of the optical flow pattern (arrows). The movement DM has continued here until object i doubled in size (i′); note different effects on poles ii and iii. (B) Zoom. A stationary camera and an increased focal length merely magnifies the image (enough to double the size of i). Unlike the dolly, objects i′, ii′, and iii′ retain the proportions of i, ii, and iii. (C_1) Trucking shot. LS is at some fixed angle (here, 90°) to DM. Note the motion perspective in the flow pattern (arrows) and the different displacements of i, ii, and iii. (C_2) Tracking shot. LS changes to keep some point stationary. Here, the fixation (F) keeps pole ii fixed on the screen. Note reversal of motion gradient beyond that point. (D) Pan. LS changes, camera swivels around one point in space (in principle, around the nodal point of the lens). Note that proportions of i′–iii′ are the same as those of i–iii.

(or retreating) viewer, on foot, in a plane, or in a car. As anyone who has seen a long dolly shot on a reasonably large screen knows, convincing feelings of self-movement are likely to accompany it.[30]

There are many different indications of motion and layout in any such motion picture or optic array (Cutting, 1986; Hochberg & Smith, 1955). To know which of those changes is actually used, and how, they must first be specified and isolated. Gibson (1950) introduced the study of the optical expansion pattern (Calvert, 1950, made a similar proposal), and the study of optic flow more generally, and he called attention (Gibson, 1954, 1957) to the several features that have since received more detailed and quantitative theoretical and empirical attention.

There has since been substantial mathematical analysis of the heading, distance, and surface-orientation information offered in the flow pattern (Koenderink, 1986; Koenderink & van Doorn, 1976, 1981; Longuet-Higgens & Prasdny, 1980; Prasdny, 1980). Recent research (Warren & Hannon, 1988; Warren, Mestre, Blackwell, & Morris, 1991) showed that viewers of such displays can indeed judge the *heading* (where their movement, or the camera's, is directed) within the 1 to 2° of allowable error that Cutting (1986, p. 276) estimates is needed for what he calls safe *wayfinding* when running, driving, skiing, and so on.[31] The same pattern provides information, potentially usable to viewers, about how long it will take to collide with the object toward which the movement is headed. That information depends on the rate of expansion of that object within the optic flow pattern, and might be used by the viewer to help direct steering and braking behavior (Lee, 1976, 1980).

The optical flow analyses are interesting, but they are not the only ones possible. As to measures derived from object-expansion rate, we have no actual studies on

[30] There has long been evidence that illusions of self-motion or *vection* can be induced in the stationary viewer by the optic flow patterns in a motion picture (downhill auto chases and views from the roller coaster have counted on such viewer involvement for many decades).

Earlier thought to be determined mostly by peripheral vision (Brandt, Dichgans, & Koening, 1973; Held, Dichgans, & Bauer, 1975; Johansson, 1977, 1982), the effects now seem more dependent on total area, the motion information about depth, and viewer posture (for subsets of these, see Andersen & Braunstein, 1985; Delorme & Martin, 1986; Howard & Heckman, 1989; Post, 1988; Telford & Frost, 1993). The point is that optic flow patterns themselves generally differ in center versus periphery. Thus, the Warren and Kurtz (1992) proposal that central vision and peripheral vision are more sensitive to radial flow and lamellar flow respectively, may reflect the informational properties of the stimulus patterns more than any retinal specialization (Crowell & Banks, 1993).

In any case, one should expect that, other things equal, smaller screens offer less self-motion than larger ones, and the loss in effectiveness and comprehensibility of sequences that depend on such self-motion when these are transferred from film in a theater to small TV in a home remain substantial. Presumably those who make film for TV have learned this point, but there have been no studies on the matter of which we know. The issue is not only of aesthetic importance, given that it adversely effects coordination and performance (Previc, Kenyon, Boer, & Johnson, 1993). In general, the actual study of how well optic flow information is realized in spatial behaviors has only started recently (see Warren, 1990).

[31] But there have also been studies with heading errors that far exceed this limit: Johnston (1972); Regan and Beverly (1982); Warren (1976).

whether these variables determine, say, how motorists brake. (For what we do not know and cannot assume, here, see Kaiser & Phatak, 1993). More important for perception of motion pictures, what research we have shows again that camera motion has not made the pictorial depth cues irrelevant or ineffective: in a brief dolly shot with two approaching objects, if one is large enough that it projects a larger 2-D image even though it is the more distant, it will be judged to have a shorter time-to-collision, in accordance with the static pictorial depth cue of relative size (DeLucia, 1991), and despite the fact that the rates of optical expansion of the two targets should reveal their true distal sizes and times to collision.

Nor is the pattern of the optic array necessarily the only (or even the main) interface between the visual system and the 3-D world in which it is moving. The optic array (essentially, the sheaf of rays with which the world confronts the viewer at a particular station point) does not itself change as the viewer's gaze changes its direction while the head is stationary, and is therefore far easier to analyze than the retinal image. The optic array and the retinal image can be quite different if the eye pursues some point that is not in the direction of motion (see Cutting, 1986; Hochberg, 1986, for discussions). On the other hand, the retinal consequences of eye movements provide a less passive route to obtaining the information needed for wayfinding, as Cutting (1986) pointed out: With a fixation to the right of the heading direction, objects shear rightward through the line of sight faster than those moving left, and vice versa, so that recursive fixations will converge on the direction of movement (as in Fig. 13Ai–iii). Simulating such fixations in computer-generated motion pictures, Cutting, Springer, Braren, and Johnson (1992) showed that subjects do such wayfinding with the required accuracy, although we do not now know how much these very different sources of movement-based visual information, or others not yet considered, are used in guiding real behavior.

The tracking and trucking shots (Figs. $13C_2,C_1$) also result from camera movement. In the former, the moving camera is kept directed at some point in the field of view as the camera itself is moved past ($13C_2$); in the latter, the camera's angle to its direction of motion does not change ($13C_1$). Both views change in ways correlated with both movement and layout, so that they provide potentially usable information about depth and motion. There has been far less analysis and experimental research here than there has been with the dolly shot, but it is clear that motion parallax—with gradients of displacements that Gibson dubbed motion perspective—normally is present and is usable.

The pan shot (Fig. 13D), in its pure case, contains no motion parallax or other motion-dependent information: the successive frames show an accretion at one edge and a deletion at the other, but they are otherwise unchanged by the camera movement. And in the zoom (Fig. 13B), only the degree of magnification changes.

Track and dolly shots provide motion-based depth information about the layout of the environment, whereas pan and zoom do not. But it is sometimes impossible, and almost always more expensive, to obtain track and dolly shots. Accordingly, to provide the same overall view of the layout, and to provide visual movement on the

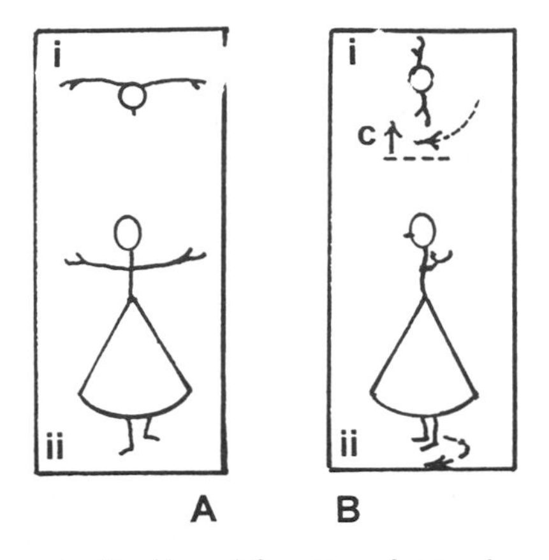

FIGURE 14 The (unnoticed) rubbery deformation of pictured people and things. (A) A 3-D figure (dancer) viewed in telephoto (i.e., from nearer than the camera's position), sketched from above (i) and from in front (ii). (B) The dancer rotates into profile at ii. With the viewer nearer or further than the canonical position (i.e., that which preserves perspective ratios), the depth in the object that would fit the image on the screen is decreased or increased, respectively. Thus, as sketched at Bi, the arms that would fit the telephoto image at *ii* are compressed relative to A*i* by amount c, which nobody seems to notice and which may in fact be unnoticeable. In a wide-angle shot (long shot), the arms would instead be stretched.

screen when the actors and the objects are themselves relatively static, filmmakers routinely substitute pan and zoom for track and dolly.

For those who claim that viewers respond directly to the spatial information in a motion picture, making the depth cues irrelevant, the pan and zoom procedures should be worse, if anything, than providing a static picture (see footnote 29); this is clearest in the zoom. A changing focal length, by its nature, means that if there are rigid objects and surfaces in the environment being represented, the information they project must be that of nonrigidly deforming structure and noninvariant events. The image on the screen provided by a dancer with outstretched arms in Figure 14A can only be fit by a prodigy whose arms can be lengthened and shortened while pirouetting, as at Figure 14B (Hochberg & Brooks, 1989). In fact, the ubiquitous use of different focal lengths (the long shots, medium shots, and close-ups discussed in Section III.A.1.c), and the wide range of seating that prevails, means that the viewer is very seldom at the *canonical* viewpoint that alone would provide optic arrays that could be fitted by rigid objects (Cutting, 1987; Hochberg, 1986), because only at that point (by definition) does the motion picture preserve the same relative visual angles as the camera itself receives.

ii. What movie use and viewing arrangements reveal about the fundamental nature of object and event perception. The nonrigidity sketched in Figure 14B, which applies to every animate and inanimate object in a moving picture sequence that is viewed noncanonically, is the counterpart of the related fact (see Section II.A.1, above) that perfectly rigid moving objects may, due to misleading pictorial depth cues, appear as deforming nonrigid objects. The point made here is that sequences are routinely made and viewed in such ways that their optical information specifies not the rigid objects and events that are in fact being recorded, but grossly nonrigid deformations of structures and motions.

Movie usage thus tells us clearly that objects and events are not perceived only (or even preferentially) by responding to the invariant 3-D structure undergoing motion-produced optical transformations. Indeed, there is good reason to believe that only a very limited depth, little more than ordinal in nature, is involved in much of one's normal perception of the world (see Hochberg, 1980, 1994; Chapter 5, this volume).

Although therefore not fundamental to perceiving and recognizing objects and events, the kinds of information sketched in Figures 13A, C1, C2 almost certainly do have important consequences and applications in the growing use of moving pictures to guide spatial behaviors, as noted next.

iii. Virtual realities, vection and the sense of presence. Moving pictures are now used not only for the entertainment and education of passive spectators, but in *virtual realities,* which are interactive sensory environments in which the viewpoint changes with the head and body motions made by the viewer, along with the auditory and haptic inputs appropriate to those actions. These permit remote or augmented task performance (e.g., telerobotics, such as a unit which provides scaled up vision and appropriately scaled tool movements and haptic feedback in order to perform remote eye surgery; Hunter, Tilemachos, Lafontaine, Charette, Jones, Sager, Mallinson, & Hunter, 1993); have been debated for use in psychotherapy (Begelman, 1991; Tart, 1991); and are in increasing use as training devices and in games. (For an overview of this actively developing field, see Durlach & Mavor, 1994; for a careful analysis of terminology, see Schloerb, 1995.)

In fact, some movies (especially those made to be viewed on a very wide screen, with binocular stereopsis, and with the audience seated only near the canonical viewpoint) might be considered virtual realities for constrained viewers. Many shots in ordinary movies evidently intend the methods of Figures 13A and 13C to elicit feelings of self-motion, or vection (see footnote 30) in members of the audience.

Measuring the viewer's sense of *presence* within the virtual environments is not simple (see Held & Durlach, 1991) but plausible procedures have been offered that may make this a useful construct with which to predict performance in such environments (Barfield & Weghorst, 1993). Regardless of the usefulness of measures of presence in telerobotics and simulators, they may provide an important new tool with which to study moving pictures made mainly to engage the viewer. In such movies, setting the level of the sense of presence may be an important but previously unmeasurable factor in the choice of shot style. More than mere economy may be at stake when selecting a pan rather than track, a zoom rather than a dolly. The filmmaker may wish to achieve a passive, alienated viewing with one method and a strong, involved sense of presence with another. Presence measures may reveal major differences between the different camera movements, focal lengths, and viewer seating arrangements.

Theoretically motivated research on how these shots actually differ in effect would seem to remain important to the study of visual cognition.[32] In any case, there are other functions served by these camera procedures compared to which the loss of compelling depth information may be trivial, once we set aside the assumptions of ecological realism. Two such purposes concern us next. First, the construction of space over successive presentations, and second, the motivation of visual attention.

B. In the Mind's Eye: Space beyond the Screen

What should most concern the psychologist is the motion picture's ability, most salient in pan and tracking shots, to provide the visual information for a scene or layout that is many times larger than the screen (cf. Burch, 1973; Hochberg, 1968). By successive piecemeal presentations of limited portions of a scene or object, visual information about a relatively unlimited region is offered to the viewer over a period of time.

In some ways, this mimics a task that is performed many times in each minute of each day (Hochberg, 1968), and it now seems more likely than ever that an understanding of the nature and limits of such *space beyond the screen* will help explain how humans perform that common, essential, and mysterious task. The fovea is only a very small portion of the field of view, and using the fovea (with its surround of undetailed peripheral vision) to sample the visual world, as the eyes move in the head and the head moves in space, requires one to combine the visual input from successive partial glimpses received.

What principles govern how one combines successive glances?

1. Neither Visual nor Nonvisual Compensation Is Necessary or Sufficient: What Motion Pictures Tell Us about the Classic Problem of Combining Successive Glances

One class of explanation, debated since Helmholtz and James, is that either the efferent signals (von Holst, 1954), which direct the muscles that control the eyes, or the afferent signals from the muscles themselves, are taken into account in interpreting the changing visual information. Relatively specific forms of these theories of "compensation for visual direction" have been proposed (see Matin, 1986, and

[32] In Ford's *Red River*, a 360° pan is accepted by uninitiated psychology students as fully 3-D, as is the lengthy zoom that imitates a crane shot in the Library of Congress in *All the President's Men* (after the first few feet, only the focal length changes). One must repeatedly point out in detail the absence of occlusion to convince the students that the shots differ from the trucking and crane shots they so closely resemble.

Bridgeman, van der Heijden, & Velichkovsky, 1994), but motion picture usage shows clearly that these nonvisual mechanisms are *neither necessary nor sufficient to account for the integration of successive glances.*

That is, the fact that one can comprehend the visual information as recorded by the moving camera (e.g., Figs. 13C,D), although the images obtained from the changing viewpoint are themselves viewed on a stationary screen, shows that neither efferent nor afferent information about the eye direction is needed to solve the perceptual problem. Although the camera has virtually unlimited freedom, the motion picture itself is normally restricted to a small region of space: the projection screen or the TV tube. The camera may point in different directions at different times, but the scenes are still all viewed only from some one place (such as the specific theater seat, living room couch, or helmet-mounted display).

And at most, therefore, nonvisual information about gaze direction could suffice to tell the viewer only that the successive views in a motion picture are all shown at the same place in space. Viewers need that in order be able to keep looking at the screen, of course, but they need something more than that in order to explain the integration of successive visual information in motion pictures. It seems plausible that when researchers know what that is, it will help the understanding of how one integrates the successive glimpses of the world obtained in normal circumstances as well.

There are two accounts of how this ability might work. One can be formulated fairly precisely, whereas the other is still in a less coherent and more formative stage. We consider them in that order.

2. Visual (Optical) Kinesthesis as an Alternative to Nonvisual Proprioception

Gibson (1954, 1966, 1979) proposed that one responds to views of the world, which undergo continuous transformations as one moves and as viewpoints change over time, by extracting the structure within that changing array of light that remains invariant under those transformations, and which therefore specifies the structure of the invariant surfaces of the world. We noted this proposal in connection with the perception of movement (p. 237); we now consider whether it provides a viable alternative to nonvisual information about direction of view when (as in motion pictures) nonvisual information cannot supply the answer in any case.

There are really three questions:

1. Could it account for how one constructs the space beyond the screen—the space in the mind's eye (Hochberg, 1968)—which is the primary domain in which motion pictures (and perhaps at least some thoughts) run their course?

2. Does any direct or indirect evidence support this proposition?

3. To the extent that it does account for the space beyond the screen, does the process remain one of direct perception, as proposed? That is, is it an autonomous discrimination of the informative higher order variables of stimulation

that provide potentially direct information about the movement and about the underlying structure?

We consider the first question here, and return to the other two in Section III.B.2.b, after surveying experimental findings and relevant observations about motion picture usage.

As to whether this account could explain what looks like the construction of the space beyond the screen, note that the optic array is invariant under the changing glance when the eye moves, which is why those who follow this approach can ignore all questions of gaze direction (see pp. 252f).[33] Because the transformation provided by an eye movement or by a pan (Fig. 13D) is uniform, it specifies that it is not the world that has moved but the observer's eye (or the camera lens).

Gibson's proposal is simple to state: Successive views *specify* the space beyond the screen. Because the viewer perceives directly what the stimulus information specifies, no inference, compensation, or other mental events are needed.[34] This gives the space beyond the screen and its contents the same properties as those of the physical world, which is a point we consider later. But even if this proposal may work and must be considered, when stimulus information is present, it is hard to see how it can explain the extraction of information from a pictured space that is no longer present in the optic array,[35] about which we can answer space-dependent questions and through which we can even take shortcuts.

In a recent study of offscreen space (Beer, 1993 in press), viewers saw a long shot of a particular far-off layout of poles (interpole distance S varied over trials) as in Figure 15A and were then shown a trucking shot, in close up (second panel), with the same set of poles being brought successively to the screen by the ground plane moving at a velocity (V) (the V was not the same from trial to trial). By varying the relevant parameters and measuring when pole 3, still offscreen, seemed just about to appear (e.g., T′), the nature of that offscreen space could be explored in detail

[33] This whole proposal ignores the fact that sensitivity drops enormously outside of the center of vision, and that only a small sample of the detail within the array can actually be obtained within a few glances.

[34] This is essentially a behaviorist (or realist) attempt to make it unnecessary to appeal to mental maps, to remembered or imagined layouts, or to any other mental structures in explaining the facts of perception-guided performance. And just as the latent learning and place learning experiments pressed the overt animal behaviorism of the 1940s past its limits, making it necessary to invoke what can, at least as a verbal shortcut, be called mental representations of the animals' environments, the same strategies now force the same changes in how one accounts for the human perception of motion pictures.

[35] Although not directly addressed to the issue of the space beyond the screen, one should remember that mental rotation (the term is explained in Sect. III.B.2; see Shepard and Cooper, 1982, and particularly Cooper, 1976) shows that subjects can behave as though remembered objects were rotated at a specific rate, and in a specified direction, while out of view. Such research, which has recently come close to measuring how events proceed during cutaways (e.g., Cooper & Munger, 1993), (discussed in Section III.B.2), seems to be a potentially important paradigm for the study of the offscreen events.

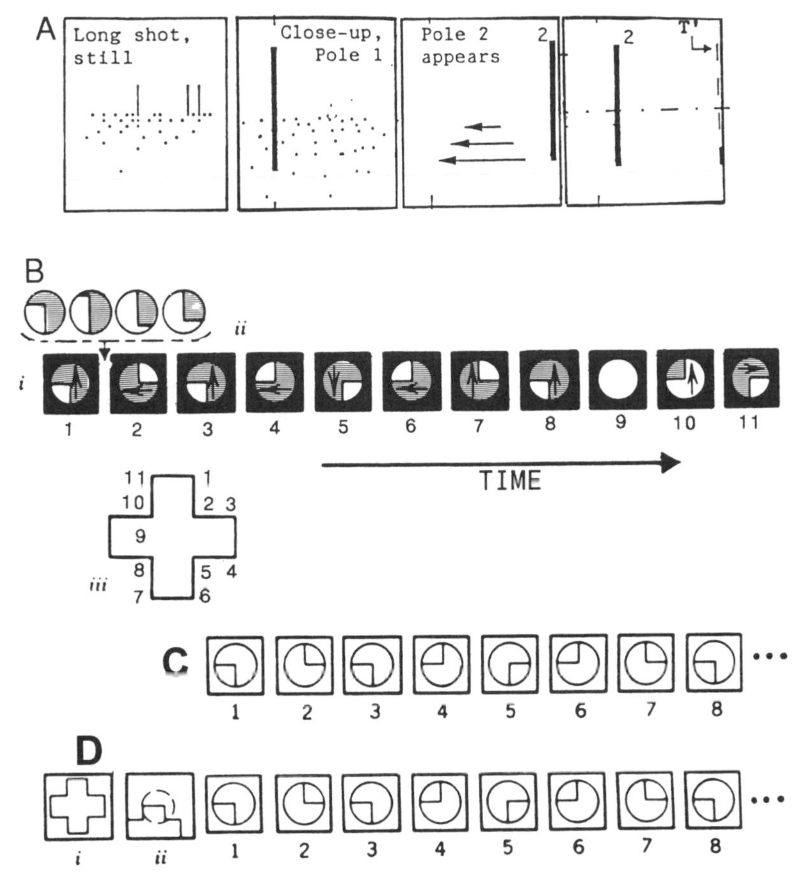

FIGURE 15 Achieving "what" from "where": Space beyond the screen and layouts in the mind's eye; perceptual consequences of perceived motion. Not all motion experiences have robust and evident spatial consequence (e.g., the aftereffects mentioned in Sect. I.A.2.b, Fig. 1A, and the object that remains in the same place in Fig. 9D while it clearly appears to move), but in general motion pictures use represented motion to communicate perceived spatial location, to depict the layout and form of things and surfaces that are not presented all at one time. This, despite the evidence of neurophysiological separation of systems. (A) *Passing information from statically depicted to transformationally specified space across a change of scale.* In sequences used by Beer (1993), which consisted of a static long shot of three poles followed by a trucking shot in closeup, viewers correctly used the information about velocity magnitude (symbolized by the arrows) and duration to judge offscreen separation of the poles, and to press a key just as pole 3 is about to enter (T′) from offscreen at right. (B) *Constructing a moving shape from partial views through a stationary window.* (i) A cross (or other such figure) presented within an aperture one corner at a time; apparent movement, as indicated here by the arrows, is provided by intervening frames like those at ii. Given at least 250 ms/corner, viewers see a cross (iii) moving around behind the window, and see frames 8–10 as skipping the left arm of the cross; the cross being shortcutted in this way of course exists only in the mind's eye (Hochberg, 1968). (C, D) *Constructing a shape from nonoverlapping, static partial views through a stationary window.* Removing the animation (e.g., deleting the frames at ii in A, above) leaves the sequence at C incomprehensible. Adding a long shot and an intermediate frame as at i and ii in D restores comprehensibility, as indicated by correctly locating shortcuts and by other measures (Hochberg, 1984; described in Hochberg, 1988).

(note that because V was varied, this viewer could not be responding merely on the basis of time, but on something like V × T).[36] What is most important here is that the spatial information given in the long shot survived a change of scale when being used to anticipate the offscreen motion. We need more work to know how its characteristics vary with filmic circumstances, but it is certainly not reducible to information in the stimulus sequence.

Similarly, in the case of shortcuts, if a pictured cross is presented piecemeal, one angle at a time (Fig. 15D), and the camera shortcuts one of the arms in doing so, the audience identifies the shortcut immediately for what it is (especially if the sequence has been preceded by a long shot (Hochberg, 1968, 1986). It is easy to demonstrate this kind of phenomenon, and hard to talk about it without including something like a cognitive map in the discussion. That map may be envisioned as something spatial and formlike, with properties isomorphic at some level with those of physical space, or as a specified feature list (Pylyshyn, 1973) and attentional tags (Pylyshyn, 1989), or as a set of contingent expectations or "sensorimotor plans"—what the viewer would see if he looked at this or that point on the object or in the scene (Hochberg, 1968, 1970).

Given the rotations, shortcuts, and changes of scale that can be demonstrated in connection with the space beyond the screen, it seems clear that optical kinesthesis cannot account directly, with no appeal to mental representation, for the construction of that space. We will return to this point, the nature of mental content (and how it is achieved), after we consider the motion picture technique that best raises the issue, that is, the motion picture cut. That issue also allows us to address the third question, whether the process is to be viewed as "automatic pickup" of information or whether the viewer's intentions and mental structure must enter into the explanation (see Section II.A.2.a.ii). There is at present little basis for choice as long as we consider only motion picture sequences in which all changes are continuous.

[36] The mentally represented distance S′ from pole 2 to pole 3, beyond the screen, could then be measured as S′ = V × T, where T was the time between the appearance of pole 2 and the subjects' signal. Note that the close-up layout had never appeared in the optic array. Nevertheless, S′ varied linearly with S as long as the space beyond the screen did not exceed the width of the screen itself (after that, space was somewhat compressed).

In both this research, and the demonstration in Figure 15D, "where" and "what" are closely linked. Although our language about objects and their locations strongly suggests an underlying disparity in how these are mentally represented (Landau & Jackendoff, 1993), and although it has long been held that these are separate cognitive processes, with their own channels (Held, 1968; Ingle, Schneider, Trevarthen, & Held, 1967; Mishkin et al., 1983; Livingstone & Hubel, 1988; Trevarthen, 1968), and perhaps those pathways do not converge even as late as the frontal lobes (Maunsell & Gibson, 1992), one must nonetheless steadfastly remember, with Attneave (1974), what motion pictures assure about the what–where connection: "Where" clearly can have perceptual consequences about object identity or scene structure, and must evidently must mean different things to different neurons.

III. DISCONTINUOUS CUTS AND THEIR CONTRIBUTION TO MENTAL STRUCTURE AND VISUAL MOMENTUM

A. The Varieties of Discontinuous Transitions (Cuts) and Their Cinematic Uses

A cut is the transition between the end of one shot and the beginning of the next. Before discussing the perception of cuts, we classify the shots between which cuts are made, describing something of their usage and setting the terminology.

1. Classification and Terminology

A *shot* is a single run of the camera, and the segment of motion picture resulting from it. As units of motion picture construction, in this sense of contributing different views of the same spatial layout, shots differ in two major ways.

a. The Proportion of the Frame Area That Is Occupied by the Subject

This is a central variable; the three usual terms are *long shot, medium shot,* and *closeup* (see Fig. 16). This is usually determined by the focal length of the lens; the various "distortions of perspective," well known in still pictures, appear here and are accompanied by corresponding "distortions of velocity,"[37] and violations of rigidity that necessarily accompany any departure from the one viewpoint that provides the eye with the same optic array as the scene provided to the camera (Fig. 14B; see Sect. II.A.2.a).

Despite the numerous challenges that these techniques offer to the more direct perceptual theories, and the opportunities for inquiry in visual cognition they suggest, they have received virtually no experimental research.[38]

Differences in narrative or affective consequence are often described in discussions of film, but are not considered in this chapter.

b. Angle of Shot

When the camera cuts between two different directions of viewing some scene or object, it provides an *angle shot.* A camera angle change of almost 180° is a *reverse*

[37] A movement in depth, or laterally on the screen, makes less of a size change and covers less of the screen in long focal length (long shot, telephoto) and in medium shot than in closeup (wide angle). That is, speed on screen is slowed in long shot and speeded in closeup. These are not merely theoretical possibilities, but affect how viewers judge a performer's speed of movement (Brooks, 1977). If unintended, and the filmmaker does not take them into account, such effects can be bewildering or ludicrous; where changing velocities are the very subject of the motion picture, as in filming dance, careless changes in focal length can invalidate the reproduction.

[38] Although we know of no research directly to this point, Kraft and Green (1989), using stills, did show that viewer's judgments of distance throughout the pictorial depth plane changed with focal length.

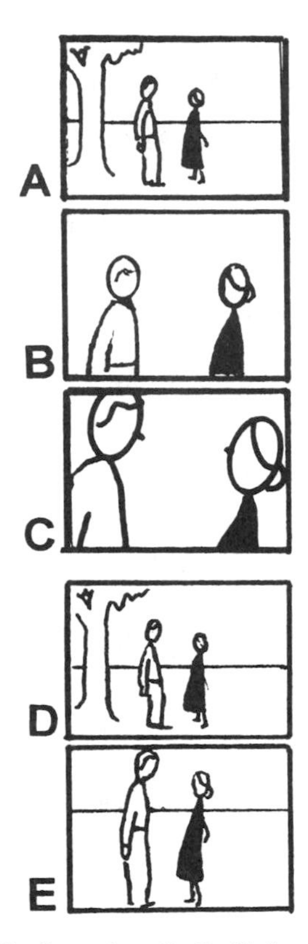

FIGURE 16 Setting the field of attention. (A, B, C) Long shot, medium shot, and close-up, respectively. (D, E) Writers on film warn against a change in screen size that is not large enough (e.g., Reizs & Millar, 1968); we believe that small changes are undesirable because contours are then close enough to provide undesired apparent motion (Hochberg & Brooks, 1978a), what d'Ydewalle & Vanderbeeken (1990) called a *distance jump.*

angle shot (Fig. 17A). Changes greater than 180° are known as "crossing the camera axis"; if unmitigated, they tend to produce confusion for reasons given on p. 224, but they can be variously counteracted, as by the embedding motion of Figure 17C (sketched from *Boomtown:* see Hochberg, 1986).

Vertical angle (*camera level*) has its main effects on field of view, and on judgments of the characters or actions represented. The high angle with its wide field of view is critical in providing literal information about the layout or choreography of a scene (a football game, a ballet corps in ensemble pattern), the lower angles display the individual actions that may be obscured by the higher angles (Brooks, 1984).

And of course, angle determines the all-important composition that satisfies and holds the gaze. It also comments on the location of the viewer (think of the

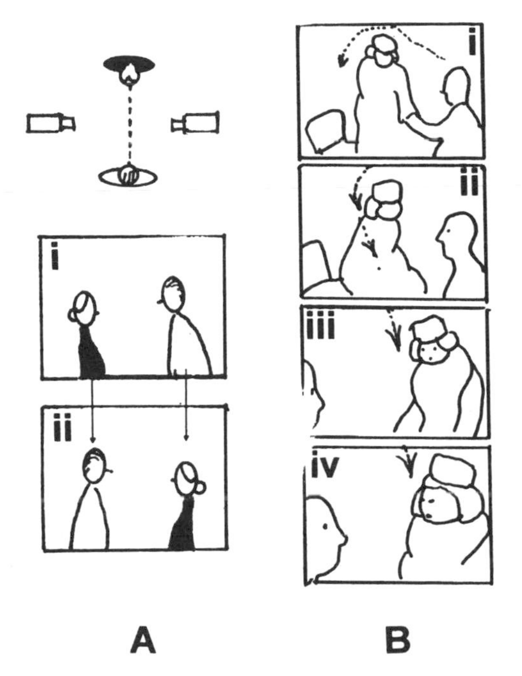

FIGURE 17 A major sin. Crossing the axis of the field of action and the line between salient people or things. A complete change in camera direction would be one of 180°; one of almost that much is a *reverse-angle shot;* a direction change greater than 180° *crosses the axis,* and filmmakers are warned against it. Characteristically, it reverses locations on the screen. More specifically, in a cut between two shots made on different sides of the line that joins two actors, as shown in (A), a cut from the left camera's view, at i, to the right camera's view, at ii, has crossed the axis and confused the viewer. To some degree, the cut is confusing because the viewer has changed orientation beyond what could normally be done by changing gaze direction (Vorkapich, 1972). We believe that a factor that is faster and more disrupting is the undesired apparent motion that occurs between the woman in i and the man in ii, and vice versa (indicated by the arrows). (B) Aware of the danger, the filmmaker can take steps to make even these cuts seamless. In this sequence sketched from Boomtown, views i and ii are parts of one shot with a moving camera (arrows indicate the path of motion on the screen); views iii and iv are the next shot, with the motion continuing as shown. The cut between ii and iii does indeed cross the axis, but the actress's motion (both on the screen and in depicted space) preserves her identity and makes the cut completely comprehensible (and virtually unnoticeable as such).

helicopter shot tracking the courageous isolated group crossing the desert in *Hair*), and on the relative stature of audience and performer (think of the low-angle shots that allow the fiends to tower over us in every horror show).

2. Uses of Cutting

Cuts between shots have many purposes: to present a longer scene or event than can be shown at one time; to represent by use of selected fragments a scene or event that really does not exist in one place or time; to convey some sequential message that would be lost in continuous presentation; to obtain a rhythm of change; and to maintain visual interest. Cuts are also made for technical and adventitious reasons that have nothing to do with our present concerns.

Because the view sequence was not provided by the viewer's own inquiring gaze, an *establishing shot* is often used to orient the viewer within the overall layout, place, time, and so on. A long shot establishes the relationship between objects later shown in more detail (Fig. 16A–C). A cut to a medium shot and then to a closeup channels attention and permits detail that could not be discerned in a long shot (especially in TV, with its relatively small screen and lower resolution). Reestablishing shots (again, usually long shots) may be needed after the sequence has gone on for a while, and especially after a *cut-away* shot has disclosed other events going on elsewhere (usually in *parallel action*) or at other times. A sequence of short shots, often having a definite rhythm and accelerating or decelerating pace, is called a *montage.*

B. The Integration of Discontinuous Successive Views

1. The Relationship to Saccadic Viewing

Change in camera viewpoint is often small, with substantial overlap between successive views (Fig. 18A). Overlap is also substantial for most small saccades (i.e., saccadic, or jumping, eye movements); we might therefore apply those theories of saccadic integration that rely only on visual information (Sect. II.B.2). If the rigid translation (symbolized by the arrows in Fig. 18A) in such successive overlapping views is applied to the screen and space coordinates of the *non*overlapping portions, those portions will automatically be specified at the appropriate locations in the overall layout. In Johanssons's formulation, the perceptual system extracts the shared vector (which becomes the framework, i.e., the translational movement of the gaze). We might now argue that if there are no residual vectors, the scene appears stationary despite its shift in the retinal image; this is a simple application of the argument in Figure 10A,B (Johansson, 1950, 1974).[39] It seems reasonable, therefore, to consider the suggestion that we combine both overlapping saccadic views and overlapping successive shots by recognizing the invariant layout undergoing the translation, but we see next that at best the formulation is incomplete in major ways.

As for saccadic view changes, it is difficult in real eye movements to disentangle the contribution that may come from extraretinal (nonvisual) information about gaze direction, especially with larger saccades; simulated saccades are often used,

[39] The Johansson phenomenon obtains not only with continuous motion but also with discrete static views (Hochberg, Fallon, & Brooks, 1977), which is what both saccadic glances and discontinuous cuts amount to. (But this was not tested more slowly than six views per second, whereas normal saccadic rates are somewhat slower—about four per second—and even the shortest expository shots would be slower still, say 0.5 s each.)

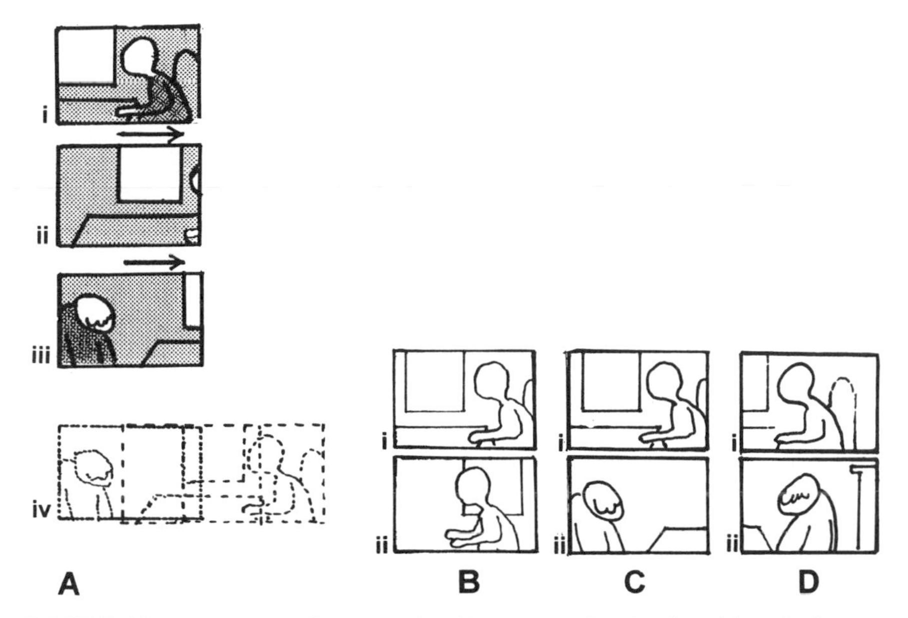

FIGURE 18 Discontinuous but comprehensible cuts. (A) There is substantial overlap between views i and ii, ii and iii, and the rigid translation that relates the overlapping regions could in theory serve to specify the relative loci of the nonoverlapping regions (iv). (B) A cut from one view of one actor (or object), as at i, to another at a small screen distance away (as at ii) invites apparent motion between the two images. An increase in screen separation is one remedy. (C) To avoid the confusions in Figure 17A and 18B, main objects should be widely separated in successive views (i, ii). Cues as to the relationship between views are then needed, and actors' gaze direction (as here and Fig. 19) is a strong cue. (D) If different main objects in successive views are made to coincide successively on the screen (i, ii), the cut may be seamless because notable apparent motion is avoided, but comprehension may be impeded (see text) and "cutting tone" reduced (see Sect. III.C.2).

therefore, comparing their effects to true saccades. In a review comparing localization by the moving and stationary eye, Sperling (1990) concluded that subjects have no special memory for images that are related by simple translation; so do Irwin (1991) and Pollatsek, Rayner, and Henderson (1990). It is even held that the information integrated across saccades is in postcategorical, semantic form (O'Regan & Levy-Schoen, 1983), an argument with which we are sympathetic but that cannot be applied until the principles of a semantics of form and layout are achieved. We know that at least some simple shapes and scenes can be built out of successive small views, none of which contains the overall form (Fig. 15D; Hochberg, 1968), and simple shape judgments can be based on parts viewed before and after an interposed saccade (Hayhoe, Lachter, and Feldman, 1991).

We have no reason, therefore, to assume that saccadic glances are integrated by extending the picture presented to the viewer (that is what taking the optic array rather than the retinal image amounts to, unless we assume, quite invalidly, that all glances gain equivalent information; see footnote 33).

The notion of a direct response to the structure shared by overlapping shots in motion pictures seems even less plausible once we acknowledge that overlap as such does not tell us what viewers perceive. The main point is that we seem not to treat all information equally: certain aspects act faster (or more strongly) than others;[40] and viewers' tasks and expectations determine how space beyond the screen, or overall spatial layout information, is obtained from successive overlapping views (Hochberg, 1992; Hochberg & Peterson, 1993; see p. 257).

These points will be easier to appreciate when we consider why some cuts between overlapping shots seem to work far better than others.

2. What Are Bad Cuts? The Limits of a Transformational Analysis

There is a considerable lore, of course, on what makes cuts comprehensible. Let us first note some of the major injunctions (cf. Reisz & Millar, 1968; Richards, 1992; Vorkapich, 1972), then discuss available research, and finally consider what explanations fit these observations.

Prescriptions fall into four categories:

1. The viewer should expect to be shown what the new shot offers (Reisz & Millar, 1968).
2. The cut should not come in the midst of a movement, but should be made at its inception or terminus (Reisz & Millar, 1968).
3. A shift from one focal length to another (e.g., long shot to medium shot) should be substantial, and not merely a slight change (Reisz & Millar, 1968).
4. Reverse angle shots (see Sect. III.A.1) are difficult to comprehend (Reisz & Millar, 1968; Vorkapich, 1972).

The first prescription is a cognitive matter, as we discuss in Section III.B.2.b. The remaining prescriptions are mechanical ones, and are regarded not as inviolable prohibitions but as precautions about what will, without special care, be jarring or incomprehensible to the viewer.

We believe most bad cuts involve an initial misidentification of one object with another or an apparent movement between views that differs from what the camera faced. Other factors may of course also be involved. For example, the work by Shepard and his colleagues (Cooper & Shepard, 1973, 1976; Shepard & Judd, 1976; Shepard & Metzler, 1971), showing that "mental rotation" takes time, suggests that (other things being equal) the time taken to grasp a cut between two views will be

[40] There are good reasons to believe that the information about the locations of objects, their parts, and their contexts are treated hierarchically, and are given different priorities even within a single glance (Bayliss & Driver, 1993; Watt, 1988) as well as when the information is being gained from separate displays (Hochberg, 1992; Kahneman and Treisman, 1992).

proportional to the angle that the camera has moved. The false-identity factor is stronger and faster, however, and more directly demonstrated.

a. False Identity and Apparent Movement across Cuts

As noted when discussing stroboscopic movement in Section I.A.2.b.iii (Fig. 7), apparent motion will occur between objects that fall near each other in successive views even if they are quite different objects. Especially if that apparent motion connects noncorresponding objects, momentary disorientation will result (cf. Vorkapich, 1972). For the first moment after the cut in Figure 17A, instead of perceiving a change in camera angle and an invariant layout, as we would expect from the optical kinesthesis explanation of how one perceives overlapping views, apparent motion occurs between the man in the first shot and the woman in the second, and vice versa.

Misleading motions or momentarily mistaken identities may also underlie other kinds of bad cuts. With a change in focal length that is too small, apparent movement, outward or inward, will occur between the contours of the two successive views of the same object when cutting between medium shot and close-up, or medium shot and long shot (as diagrammed in Fig. 16D,E). Corrective prescriptions follow easily from this analysis, and are familiar in the manuals.

Where the main object in the second view is not that in the first, the objects should lie far apart from each other *on the screen* (Fig. 18C) and the cut direction indicated in some other way (see p. 261). Filmmakers often avoid the "jump" inherent in such a discontinuity by so framing the new object (or the same object in a new scene) that its contours remain as undisturbed by the cut as possible, as in Figure 18D. This usually mitigates the jump, but also usually decreases the cut's impact (p. 272) and increases the time needed to grasp that a change has occurred. It would seem better to follow the principle in Figure 18C, so that the viewer at least has correct information about the direction, if not the extent, of the camera movement (extent between objects is probably not well coded and preserved, anyway. Other "rules" are noted in Section III.B.4.a.

In any case, where action to prevent misleading motion or mistaken identity cannot be taken (for reasons of composition, economy, lost opportunity, etc.), the filmmaker should allow a longer dwell time on each shot (see Section I.A.2.b.iii). But this remedy works only if the viewer can recognize the relationship between the two shots. Consider cuts between two highly detailed or cluttered shots. With dense and cluttered shots, one should expect that local short-range apparent motions occurring between unrelated contours will more or less cancel each other, overall, as when D_{max} is exceeded in the random dots of Figure 4A, with displacements greater than d in Figure 5Bi–iii (Hochberg, Brooks, & Roule, 1977).[41]

[41] The cut in Figure 18 brings contours from Figure 18Bi near noncorresponding contours in Figure 18Bii, resulting in whatever local motions occur by chance; something like this must contribute as well to the actual size of D_{max} in the phenomena discussed in Sect. I.A.2.b.i.

Misleading motion between shots has then been prevented, but even the absence of such misleading motion will not ensure that the shots' overlapping spatial relationship is correctly perceived. Such shots generally share no salient landmarks (landmarks are features in some part of the field of view that are readily visible even when the gaze is directed at another region).[42] That lack can be remedied by adding a landmark (e.g., by removing some part of the clutter, as in Fig. 5Biv–vi or by shaped chiaroscuro) because the correct direction of displacement is then perceived even over much larger distances (Hochberg et al., 1977). But it returns us to the problem of misleading apparent motion between these lower spatial-frequency shapes.

The implications these warnings and prescriptions hold for the filmmaker are not huge, although they might help codify otherwise unrelated rules of thumb. The implications for the computer-generated motion picture or slide shows may be greater, if the underlying principles can be built into the software. The implications for perception psychologists (particularly for those who hope for some transformational explanation of the integration of successive views) are, in contrast, probably quite serious.

b. The Limited Utility of Full Transformational Information: Cuts with and without Landmarks and the Importance of Schemas and Cues

At the beginning of Section II.B.2, we asked whether the perception of represented space by means of cuts between overlapping shots is accounted for through the shared stimulus information about the layout revealed by the transformation between views, and if so whether the process could be treated as though it were direct (i.e., without bringing mental representations and/or inferences into the discussion). Some of the stimulus information is indeed rapidly responded to: the low-level apparent motion between corresponding objects or features that occurs with brief views and small displacements. However, it offers no account of the integration of overlapping discontinuous cuts. At greater displacements the very same stimulus-based low-level motion, if it occurs, may be incorrect (as in Fig. 7) or self-canceling (as in Fig. 5Bi–iii). The information as such is present in the stimulus transformation, but by itself it does not account for perception of space through overlapping cuts. What is more, even adding landmarks will not automatically help, as we see next.

i. The limits of the glance: Landmarks and information versus cues and construal. Even when stimulus information is presented on the screen that specifies the relationship between successive shots, that information may well be unusable by the viewer, at least in part depending on what other cues the viewer attends. Here is a demonstration of that point (Hochberg, 1992; Hochberg & Brooks, in press), which has been used repeatedly with large audiences. Figures 19A–D are overlap-

[42] See Hochberg and Gellman, 1977, for experimental measure and demonstration of use; for review of effective search, see Treisman, 1986.

FIGURE 19 Layout in the mind's eye: not "stimulus information," but effective and less effective cues. Successive views (A and D, ca. 1 s; B and C, ca. 0.5 s) presented in continuous cycle (A–D, D–A, . . .). In this sequence, depicted gaze direction is opposed to the information provided by landmarks, and by the line of text; gaze direction prevails, in that A is predominantly reported to appear to be at the right and D at the left of the overall space.

ping shots taken from the layout shown in 19E; the two geometrical landmarks (as well as the "landmark" formed by the familiar sentence) specify the relative locations of the three cartoon characters. Shown at about one view per second in a repeating sequence, (A,B,C,D,D,C,B, . . .) and asked whether the man in the fedora is left or right of the man in the sweater, significant majorities of the viewers say he is to the left.[43] The text itself goes unread or is unreadable when the sequence is so construed.

[43] The procedure lends itself to varying parameters of landmark salience, eye direction, and instructed

Filmmakers routinely use actors' gaze direction to indicate the loci of successive shots. In the demonstration just described, which has incidentally tested that usage, gaze direction alone has overcome the overlap between views that should, by transformational theory, specify the rigidly translating layout. The dots that convey gaze are far smaller than the shared landmarks, but they dominate the construal, and it is the construal, not "stimulus information" measurable by vector calculus, that has determined how the successive (overlapping) views are combined. Viewers who have seen the establishing shot frequently report that if they concentrate on trying to see the actors as looking away from each other, or on seeing a stable line of the nursery rhyme, they can overcome this illusion. Surely, we expect that larger and more salient landmarks would also overcome the gaze and provide the correct construal. But we also assume at this point that if the invariant layout underlying the translation provided by overlapping views does indeed contribute to the comprehension of cuts in motion pictures (and perhaps of successive glances when one makes saccadic eye movements), it is *only one factor among others that contribute to perceptions of the space beyond the screen.* And the viewer's construal may be among the more important factors that determine the perceptual consequence of a succession of views.

If, as seems to be the case, the transformational theory (or any theory that draws *only on the measurable attributes* of the stimulus display) cannot by itself give an account of the perception of discontinuous but overlapping cuts, what alternative is there?

Our answer (for which there has been much precedent: Helmholtz, 1909/1924; Miller, Galanter, Pribram, 1960; Mill, 1865/1965; Neisser, 1967) is that we perceive by formulating and testing expectations (or schematic maps; Hochberg, 1968, 1970) of what one would see if one looked at this place or that, at this time or that. Remember that the saccades that provide successive views of the world are *elective,* the results of a motivated perceptual inquiry, not a neutral raster scan of a physical optic array. *The glances bring answers to questions.* One can ignore that basic fact in talking about experimental tasks in which the questions are posed by the experimenter and the glances are left to the subject. Such conditions are essential for much perceptual research. But that is precisely not what happens at the movies (nor even in reading by free eye movements).

This vague proposal (or some less mentalistic-sounding phrasing of it) may seem the more worth developing when considering that *one of the two or three big facts of motion pictures is that most cuts contain little or no overlap,* and that viewers' expectations greatly affect what information is in fact integrated across such cuts.

---

construal, and such research is in progress; meanwhile, the demonstration by itself is quite effective, even in a videotape version that has a moiré pattern in its shading, and will be sent to anyone for the cost of tape and mailing. The (virtual) camera has centered the "actors," who are the largest objects, within each view to avoid any short-range motion; each shot lasts 1 s and sequences are shown for 3 cycles.

3. Nonoverlapping Successive Views

Most cuts do not bridge overlapping shots. Montage, which many film theorists (Colpi, 1966; Eisenstein, 1942; Godard, 1966; Pudovkin, 1958; Spottiswoode, 1933/1962) regard as the most characteristically cinematic device (with cautions from Bazin, 1967, 1972), provides sequences of views that are not related to each other by overlap or by common background. Only some mental representation can relate them. Consider the sequential partial views of a cross in Figure 15C, which are like those in Figure 15B save that each view is static. (The subject receives the entire sequence, although only a subset is shown here.) Given only views of the corners (1,2, . . .), the viewer sees either an erratic set of clock hands or successive views through a stationary aperture of the corners of an equally erratic square. But if the same sequence begins with the long shot of the cross as in View i of Figure 15D, it is perceived almost as it was with continuous movement in Figure 15B (although the viewer must make more of an effort to follow the figure in its course) and the viewer reports the shortcut across the right arm of the cross (Hochberg, 1968, 1986).

Given an initial long shot and subsequent views that are not too brief (≥ ca. 250 ms), the sequence is comprehensible;[44] without the long shot (or equivalent identification), or if shots are too brief (≤ 100 ms), the sequences are incomprehensible.

Only a mental representation of the cross relates the views. But until we can become more specific about the characteristics of such maps or schemas, and about how they are used, this proposal remains vacuous.

Noting the time scale of the phenomenon in Figure 15D, the assumption that visual schemas must serve (among their other uses) to guide and store the encoded input from successive glances (Hochberg, 1970) seems a reasonable start. As indicated by different tasks, viewers can retrieve and identify content information about entirely unrelated pictures when these are shown successively as fast or faster than the normal glance rate, (i.e., ca. 4/s). Tasks include keypress for a precued picture (Potter, 1975), or asking the viewer to identify each picture that is *unrelated* to some original category (Intraub, 1981). Features from one view are found to intrude into another at rates comparable to those that were too fast for the task in Figure 15D (Intraub, 1985, 1989), and it is interesting to note that viewers cease their eye movements when pictures change too rapidly (Potter & Levy, 1969). Another significant finding in such research on sequences of independent pictures—a finding that filmmakers should probably note—is that viewers encode normal and closeup pictures as being more wide angle than they are, presumably because of the expectancies each view entails (Intraub, Bender, & Mangels, 1992; Intraub & Richardson,

[44] Various measures of comprehensibility have been tried. One, most satisfying as a demonstration, is that, as noted, the viewer can say what part was skipped, converting the "what" of each corner to its "where"; or, using a method more amenable to psychophysical measure, the viewer can tell whether two sequences (that may differ somewhere in the middle), are different (Hochberg, 1986)

1989). Surveying the facts about timing in these studies and in eye movement research, and the phenomena just mentioned, Intraub (1992) argued that such procedures are indeed tapping the guidance function of visual schemas, as well as the actions of a visual buffer that stores some small number of views.

Although the entire scene being filmed may have been presented over the course of the camera movements, or even been presented in longshot, and though some information has clearly been retained (see Section II.B.2.), much is also lost. We do not yet know how to predict what the spatial representation will be, but it is certainly not the layout itself as specified by the information in the overall optic array. This is as true of the mental representations of layout in the real world as it is in the movies, as Kevin Lynch (1960) showed so dramatically in *The Image of the City,* nor is the sequence of events that is registered by the viewer a predictable copy of the events that were presented (see footnote 48). Applied studies on *situation awareness* in the real world (see Endsley, 1990) suggest that an overall knowledge of layouts, locations and status depend on the viewer's goals and intentions as well as on the available stimulus information.

What we lack here is an explicit model for achieving a single schema or mental representation of a scene or event from successive partial views,[45] a model that need not use *all* of the information in each view (which can be subject to differential attention and construal, cf. Fig. 19) and that is not limited to the stimulus information actually given (because of completion that characterizes the schema, Fig. 15). A model for response to a stimulus configuration (or what it specifies) is not needed, but a model that deals with expectations of sequential input is. In short, what is needed is a theoretical framework so that what the pictures describe—the *narrative* aspect of pictorial sequences—might be discussed in terms of visual, rather than merely verbal, content.

With or without such a model, no discussion of motion pictures (or of perception quite generally) can ignore or comfortably defer the discussion of narrative content: As seen in the previous sections, *the viewer's attention is potentially a factor at all levels of perception, and cannot in general be predicted within the framework within which those levels have thus far been discussed.* In the final part of this section, we sample what has been said and done, mostly not by perception psychologists, that might contribute to a study of the role of visual narrative (very broadly defined) in construing and combining the information in motion picture sequences.

4. The Perceptual and Nonperceptual Levels in Motion-Picture Comprehension

a. Fast Stimulus Factors, Slow Stimulus Factors, and "the Language of Film"

It is a basic fact of motion perception, not by any learned convention, that apparent motion occurs between nearby objects in successive views lasting 33–350 ms. That

[45] Biederman (1985) proposed a scheme, which seems potentially very powerful, for achieving very rapid recognition of categories of nameable objects. Although it does not do so presently, it might serve as the starting point for carrying over the semantic expectations from one picture to the next.

same fact provides apparent motion between noncorresponding objects should they chance to be neighbors in successive views, as they may well do at cuts between overlapping and nonoverlapping shots. Only at relatively long viewing times (about 350–500 ms per view) do objects' identities and recognizable shapes determine the direction of apparent motion of object or viewpoint. Although perhaps they may then do so partly by virtue of the mathematical identity of the stimulus invariant undergoing transformation between views (Sect. II.B.2), that has not as yet been demonstrated. Moreover, the views must also offer some landmarks that the viewer can identify from one to the next.

And always, *with any cut the viewer must make a very fast early "decision" as to whether it opens a different scene or event;* lighting levels, sound track, and so forth, are obviously factors here. If the new shot is within the same scene, it may be left unlocalized with respect to the previous shot (perhaps to be resolved later); if it is to be assigned to some location relative to the previous shot, the viewer needs some cue[46] as to where. Here are three cues:

1. When one shot shows a person looking offscreen to right or left, the next shot is usually seen as the region toward which the person was looking (cf. Fig. 19). For it to do otherwise leaves the previous shot with no point or resolution, unless some other is provided.

2. A leftward pan is a strong cue that the next view is to the left of the previous one, *even if the motion is not completed,* and therefore the next view does not actually come onscreen through the leftward motion. Such motion-as-cue-to-location (shown in Fig. 15B and shown too with dancers in limbo in Fig. 20A) would seem to be a slow factor (i.e., one taking time to digest), though we know of no relevant research. It is our observation that if shots are connected by paths much more complicated than a simple translation, their relative locations are not readily perceived. (See Hochberg, 1986, p. 45, for discussion; cf. Losey's *The Servant,* and *The Go-Between,* for motion picture examples; note that Landau and Spelke, 1988, have shown that infants can keep track of objects they are trying to retrieve while they themselves have been moved, but only if they have been moved on simple paths.)

3. Closely related to the previous two cues is that of *maintaining screen direction:* Given an actor or salient object moving offscreen, we expect (but are not certain) that he, she, or it will have moved in a fairly direct line into the next view. In the experimental research by Frith and Robson (1975), visualized in Fig. 20B, children shown either a sequence like i or one like ii, could reconstruct the event (from drawings of the shots in each sequence) better with i than ii. A low-level factor

[46] We use *cue* as Woodworth (1938) did, doing so whenever the currently fashionable "informative variables of stimulation" (which really have meaning only when they refer to the optical expansion pattern and its close relatives) would sound misleadingly specific.

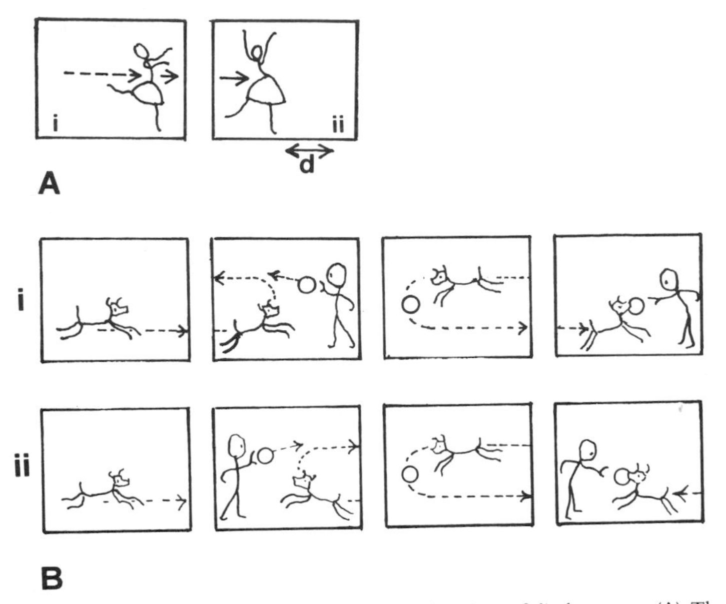

FIGURE 20 Continuity of perceived motion versus direction of displacement. (A) The dancer exiting at right in i is taken as continuing along the same path, as she enters shot ii, even though she is substantially displaced leftward (d) from her last screen position. (B) Visualizations of a short four-shot film designed to show that screen direction is important for directional continuity and narrative comprehensibility in young children (Frith & Robson, 1975).

might contribute here,[47] but as with axis crossing (Fig. 17A), sequence ii has a more general difficulty: If the represented event is continuous, then the camera must be construed as having crossed the layout in which the event occurs. In simple visual narratives (e.g., two main characters in one location), Kraft (1986, 1987, 1991) found that both directional continuity, and the presence of an establishing shot, enhanced subjects' memory of depicted activities.

4. An expectation about the contents of the scene, which may have been presented in a long shot; which may be given in general knowledge (as that the World Trade Center is left of the Empire State Building when viewed from Brooklyn); or which may rest in the memory of a layout built up over several shots.

The speed with which such shots can be comprehended is probably quite variable. And even where one can be sure of the information that the film potentially provides, one cannot usually be sure of what the viewer actually encodes.[48]

[47] Namely, facilitation between directions of apparent motion (Ramachandran & Anstis, 1986; see Sect. I.A.ii).

[48] As Stefan Scharf pointed out during a lecture analyzing Fellini's *Nights of Cabiria,* and was noted in

So far, we have mentioned misleading apparent motion, directional continuity, direction of gaze or heading, and establishing shot, as factors that can relate non-overlapping views to each other within a larger space. Many hints on what to do and what not to do may be elaborated from these alone. There are many instruction manuals for filmmakers and film students. These largely list and illustrate rules most of which we believe can be derived from the few principles described here and earlier (see Hochberg, 1986; Hochberg & Brooks, 1978a; Kraft, 1991). Such illustrations when addressed to film scholars are almost always given as examples of the "language of film," or "film grammar," terms we feel are too loaded with misleading connotations. The prescriptions do not seem like arbitrary rules that we are free to set up as we wish, and the elements of film—the pictures themselves—are not, like phonemes and words, open to be assigned referential meanings at will (see Metz, 1974).[49] The most serious attempt to treat these prescriptions as grammar was made recently by J. Carroll (1980), who tried to treat film within Chomsky's generative linguistics, identifying widely known errors of film editing (e.g., axis-crossing, Fig. 17) as constructions that putatively elicit judgments of ungrammaticality or unfilmicity. Such judgments, made about sequences never before seen, presumably show the viewer to have an underlying generative grammar. But those errors are bad cuts because of misleading apparent motion, and because continuity (or simplicity) is violated.

In short, we believe that most of the visual rules for achieving visually comprehensible cuts derive not from an agreed-on film grammar, but from the prospect of unwanted apparent motion, from the shortcuts characteristic of inference and implicature, and from the economies of encoding and memory that are at work in the normal perception of the world. But the visual comprehension of cuts, and of filmed events more generally, does not depend only on visual stimulus input, and it is clear that nonvisual structures, which may indeed be subject to linguistic or languagelike constraints, are important in how both individual cuts and much longer visual constructions are perceived and comprehended.

Hochberg (1986), a truck seen over one actor's shoulder to be approaching the camera has, after a brief cut away, appeared to have vanished without trace, explanation, or noticeability. On the other hand, one often believes that one has seen something that has not in fact been presented (e.g., an elided punch in a barroom brawl). Basic research on what the viewer accepts as really having been shown earlier, as a function of elapsed time (work that has been done with still pictures and words; see Johnson, Kounios, & Reed, 1994) has not yet been undertaken with movie shots in meaningful sequence. The latter would seem to be a rewarding medium for exploring visual narrative in particular and event perception more generally.

[49] The word *beggar* is an arbitrary array of graphemes or phonemes that must be assigned its meaning, whereas a specific beggar in a specific film—say, the one in *Sullivan's Travels,* who was carefully chosen for his resemblances, and whose later recognizability after a minor earlier appearance is important to the film—is not a coded token.

b. Nonvisual Input and Context

Thus, if the viewer were given verbal information about the cross and the location of the first corner in Figure 15D, the consequence should be essentially the same as having been shown the long shot.[50] Similarly, knowledge gained from an earlier viewing, even though originally visual, must after some time has passed no longer count as visual stimulation. There is a point, in short, at which our perception and comprehension of views being presented on a motion picture (or, for that matter, of glances at our environment) must draw on nonperceptual ("narrative-like") processes. (Perhaps the major difference between the two is that in the case of motion pictures another person—the filmmaker—must make a guess at, and provide for, those nonperceptual processes, as well as providing the moving pictures themselves.)

It seems unlikely that we can ever expect a truly scientific understanding that is both general and explicit enough to predict the consequences of any shot in its context, selected at random from some real, narrative film—viewer's histories and narrative structures are both too unconstrained. But perhaps we can eventually achieve an understanding of the processes by which such narrative, nonvisual input affects the perceptual response to the visual event, and of what that reveals about perception more generally. We consider writings about narrative in film with the latter goal in mind.

C. Noniconic Effects of Sequence: Narrative Syntax, Affect, and Momentum

1. The Directing of Ideation and the Scope of Narrative Inference in Motion Pictures

a. Shots as Ideational Elements

We noted previously that cuts to closeups were often intended to direct the viewers' thoughts to the closeups' subjects. This suggests research on the time course of thinking and problem solving that remains virtually unexplored but that was implied by ideologies of montage. The two classical Russian theorists, Eisenstein (1942, 1949) and Pudovkin (1958), offered conflicting[51] and tentative descriptions about how film sequences affect viewers' thought processes. To Pudovkin, ideas are to be assembled by the sequential presentation of their elements; to Eisenstein new ideas emerge automatically from the conflict of the views presented.

[50] Time would probably be needed to mobilize an image of the cross (about 500 ms; cf. Posner, Boises, Eichelman, & Taylor, 1969), or to activate a feature list (Pylyshyn, 1973) or a schematic map (Hochberg, 1968).

[51] In practice, it is not clear how much these filmmakers really differed in their practice in this regard: To advance a narrative, Eisenstein had to follow its structure, and study of Pudovkin's work shows his heavy reliance on a dialectic of ideas, if only to provide comment by contrast.

Most "research" to this point pursued with motion pictures in mind has used only cuts between a few shots (for a review, see Isenhour, 1975). The most famous "experiment" was probably one by Kuleshov and Pudovkin (Pudovkin, 1958), in which similar closeups of an actor, posed to express no feeling at all, were juxtaposed with a bowl of soup, a dead woman in a coffin, and a child playing. Audience judgments of the actor's expression differed appropriately with context. In an armchair experiment, Pudovkin also pointed out that the meaning of a shot depended on the order of the sequence: a three-shot sequence of a man smiling, a pistol pointing, and the man looking frightened will surely says different things about the man than does the reverse order. Subsequent demonstrations that judgments (using a modified semantic differential) about an actor's expression can be varied by changing both the preceding and following context and order corroborate Pudovkin's assertion that the order of a sequence affects its meaning (Foley, 1966); as Worth (1968) put it, Foley's experiments (and Pudovkin's armchair experiment) showed that AB $\neq$ BA, where A and B are individual shots. This does not seem surprising. We consider more principled approaches to this issue shortly in Section b, below.

Despite Eisenstein's assertion (1949) that two pieces of film of any kind, when placed in juxtaposition, inevitably combine into a new concept or quality, there is no reason to believe that without a specific effort at construal by the viewer anything other than a meaningless flight of visual fragments, relieved by an occasional meaning that chunks the montage into a memorable unit, will be perceived.[52] Gregory (1961), assuming with Osgood, Suci, and Tennenbaum (1957) that the meaning of two signs (shots, in this case) will affect each other only when they are coupled by some "assertion," proposed that associative cues (such as tools on the wall that identify two shots as both occurring in a workshop, or the knowledge about familiar activities that permits us to recognize the shot of a man swinging his arm backward as connected to the next shot's closeup of the hammerhead on the nail) are needed to provide that coupling or assertion.

We should be able to discover the ideas a given cut elicits by using priming procedures and related measures (see for example Dosher & Corbett, 1982; McKoon & Ratcliff, 1989; Ratcliff & McKoon, 1988), so the issue is not beyond

[52] As in verbal communication, in order for a viewer to respond to such sequences, they must be seen as being connected in some way. There is, of course, the inescapable fact that any viewer who thinks about it at all knows that the images that are presented were assembled as they have been by a purposive filmmaker who intends to communicate something to him (cf. Metz, 1974, p. 47). Even viewers who do not think about it (any more than they think about the fact that a soap opera or a comic book has been assembled for some purpose and is not merely a window on the world) have learned that things will be represented and events will occur if they attend to the changing images on the screen, paying attention to their sequences by attempting to fit them into some schematic map or schematic event sequence (Hochberg, 1968, 1970); but one should not assume that such construals must rest on much more than the present sequence and a few preceding shots (Hochberg & Brooks, in press; see McKoon & Ratcliff, 1992).

scientific inquiry, and the rapidity and depth with which motion pictures can convey socially important information might profit from such research.

But shots are not so clear-cut in content, and only if the filmmaker bounds each idea by starting and stopping the camera could such inquiry be pursued. The prior question (similar to the one that linguists face in various ways when seeking speech boundaries) is what marks the boundaries of the narrative structure and its units?

b. Narrative Structure: Perceived, Expected, and Remembered

If an event is recorded without elisions, event time, narrative time, and screen time all agree. But films normally abound in elisions, both within and between events.[53] Because of omissions (and inserts), Bordwell (1985), with writings on verbal stories in mind, distinguished three time scales of motion picture structure: the presumed duration of the event being narrated (the *fabula*), the durations of those portions actually shown (the *syuzhet*), and screen time. Omissions leave gaps. To bridge any gap, the viewer must provide the structure of the sequence, and must expect (or at least accept) the shot after the cut as the outcome of the shot before the cut.

i. The evidence for structural knowledge within and between shots. We obviously know the approximate structure of some physical events: the trajectories of thrown balls, colliding objects, and so on (Michotte, 1946/1963); speed of animal limb and hide movements (which spoils the use of an iguana, say, to portray a dinosaur: Fry & Fourzon, 1977). It must certainly be true that in dealing with events in the world, and therefore with moving pictures of it, people can anticipate more or less the trajectories of motion imposed by physical kinetics. As we noted earlier, some psychologists summarize such anticipations as a form of internalized physics, which in principle would allow the filmmaker to use physics to gauge what viewers expect of the time course of an object intermittently glimpsed in successive views.

Things are not so simple. For example, Freyd (1983, 1987) argues that viewers accept overestimates (overshoots) of where an object should have gotten to, thereby manifesting a *representational momentum,* which of course requires some internalized representation of mass, even though none may be indicated in the pictures concerned (e.g., simple geometrical shapes). Against this, Cooper and Munger (1993), varying factors that should affect the mass and friction portrayed, found no evidence of a momentum or attributed mass, and did find small but significant

[53] Films that record performance events (sports films, dance films, etc.) typically have few elisions; at least one story film approximates a single long take, notably, Hitchcock's *Rope* (the film was eight takes long, with the camera fixed on the same painting at the end of one take and the beginning of the next; Oldham, 1992, p. 346); free of obvious cuts, its event boundaries are marked in other ways. It is fairly common for films to maintain a very long take where a camera movement constitutes an event in itself, while revealing other events and their boundaries: In Scorsese's *Goodfellas,* an uninterrupted (ca. 90 s) steadicam dolly, through the backdoor and kitchens into a nightclub, changes direction only to keep the protagonists near center. Such shots, as do numerous lengthy crane shots in films, from Griffith's *Intolerance* to Forman's *Hair,* present examples of both nested unelided events and the astonishing phenomena of motion parsing discussed in Section I.A.2.c (see Figs. 9–11).

deviations (undershoots) from the expected course. The rules or heuristics which mediate viewers' expectations about physical motions may in their outcome often approximate those motions, but are almost certainly not equivalent to an internalized physics, no matter what else they may be. Moreover, as we have noted elsewhere (Hochberg & Brooks, 1989, in press), in using elisions and repetitive expansions, filmmakers treat off-screen (intermittently viewed) motions, and the times they occupy, as malleable quantities.

In any case, a great many of the events that we encounter in the world, and perhaps most of the events that we perceive through the medium of moving pictures and movie cutting, are driven not by physical kinematics but by biological and social factors. But what other knowledge of event structures can we expect, and in what forms? Consider the animate, purposive, social events portrayed by the dot-protagonists in the Heider-Simmel film of 1944 (see Fig. 12): What can those abstract paths taken by such abstract shapes share with human actors?

We consider that to be a very important question, one that probably can be (but has not yet been) answered. Note that the motions are obviously punctuated or segmented. So are human actions: watching videotapes, viewers agree closely about the *breakpoints,* with more predictable goals yielding larger and fewer segments (Newtson, 1973; Newtson & Rindner, 1979; Wilder, 1978a,b). Lichtenstein and Brewer (1980) had one set of viewers identify the actor's purposes; when other viewers were asked to recall all that the actor did, actions that were part of what Lichtenstein and Brewer called major *plan schemas* (hierarchical "in-order-to" relationships), that were higher in the hierarchy and consistent with the underlying schema, were recalled better than others. Actions drawn from the lower levels of the hierarchical plan structure are forgotten first (Brewer & Dupree, 1983), leaving the more abstract information about plans and goals in memory.

Moreover, in memory experiments with story text, readers (or listeners) often appear to use even larger mental representations—story structures or schemas (Bartlett, 1932)—within which to encode and store more local action segments, inferring connections between such story parts as goals and outcomes (e.g., Mandler, 1978; Mandler & Johnson, 1977). Most research on story structure has used text, but experiments by Cowen (1988) and by Kraft (1991), reordering shot sequences in short narrative films, show that departures from linearity of the story structure reduces comprehension and recall, much as they do in texts. There is therefore an area for discourse between those who study narrative structure as pursued in writings on film and literature, and students of experimental cognitive psychology (see Bordwell, 1985, 1989a; Carroll, 1988).[54]

[54] An area that seems to us potentially more useful than discussions about film grammar or psychoanalytic meaning can be that film theorists have drawn so heavily on what they viewed as the science of psychology that some attention to contemporary cognitive psychology may be a corrective. For discussions of what needs correction, see N. Carroll, 1984 (about media specificity, a dominant theme since Arnheim, 1932/1933) and 1988 (for an analysis of some of the more extravagant psychoanalytic

And returning to the Heider–Simmel film, the sequence of geometrical motions of three simple shapes is too long and complex to remember as a lengthy movement in space, but is easily encoded and recalled when construed as the story—a structure of purposive acts—which it was designed to narrate. That is, as in stories quite generally, the behaviors are recalled in hierarchically nested action units (Bower, Black, & Turner, 1979), filled out with inferred or reconstructed information.

It seems plausible, therefore, that the course of an event bounded by breakpoints is encoded within a plan schema: given its breakpoints and identifying features, the intervening action is redundant. Newtson and his colleagues have shown just that,

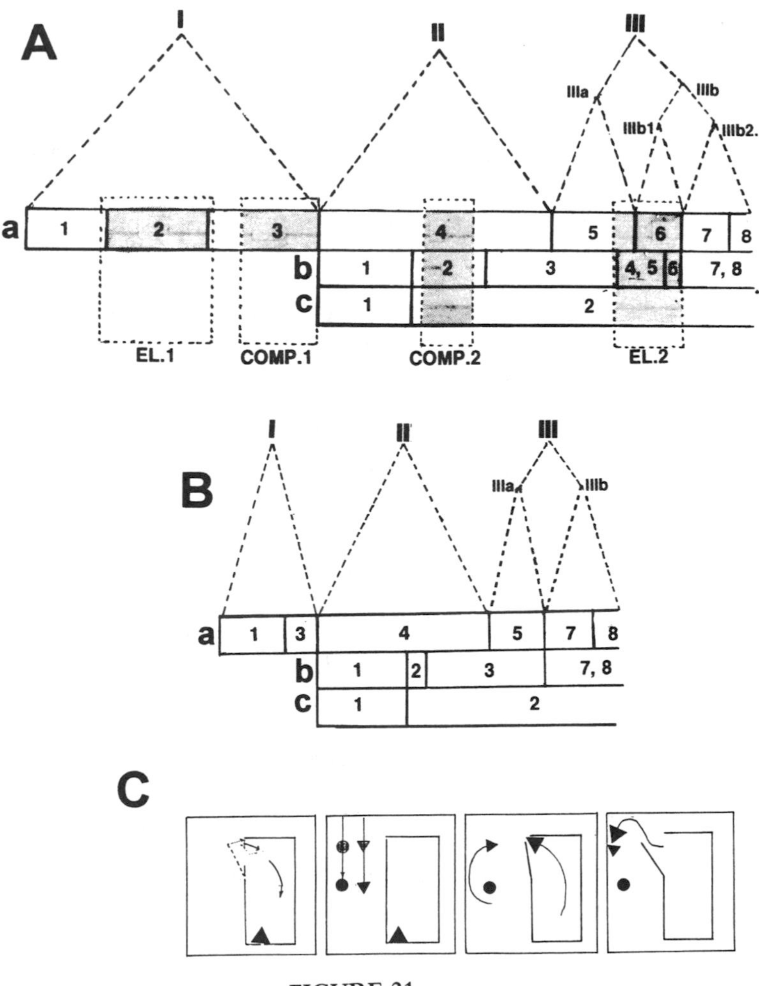

FIGURE 21

using stills taken from videotapes for which subjects had earlier identified breakpoints (Newtson & Enqvist, 1976), and have shown that activity levels are lower at the breakpoints (Newtson, Hairfield, Bloomingdale, & Cutino, 1987). In fact, using an automatic computer-generated activity index of lowest activity (the least pixel-by-pixel change between successive frames) to chose frames from video sequences of an actor using American Sign Language, Parish, Sperling, and Landy (1990) found that sequences constructed only of those lowest activity frames were significantly more intelligible than frames sampled at a constant interval.

An idea of the redundancy, and of omissions that reduce screen time by *compression* or *elision,* may be gained in Figure 21. In a less restricted narrative film, some number of frames might be cut from within an action with little loss because they are not used in recognizing it, compressing the action;[55] quite differently,

assertions about the psychology of film); for introductions of cognitive issues, see Bordwell, 1985, 1989a,b, and Brooks, 1984, 1989, and for papers pro and con, see the whole issue of *Iris* edited by D. Andrew (1989).

[55] As one can usually recognize forms from their vertices alone (or from their distinctive features as a subset), but cannot recognize them readily when those are omitted (see Attneave, 1954; Biederman, 1985; Hochberg, 1980), so with elisions in film: *La Jetée,* by Marker, is a 20-minute fiction film composed (all but a few seconds) of brief stills that so clearly communicate plan schemas and a story

FIGURE 21 Withdrawn from the stream: Elision and compression in the cutting of structured events. In representing a sequence of events, most narrative motion pictures elide events not needed to comprehend the sequence, and also compress individual events by retaining only those segments needed by the narrative (e.g., retaining only those segments of an action that specify its purpose or consequence). (For attempts to identify such segments objectively, see p. 269.) To illustrate this, at (A) we analyze the intact visual narrative segment summarized in Figure 12, altering a notation method developed by de Wied (see p. 276). The actions of the large triangle, small triangle, and circle are represented in these diagrams at rows a, b, and c, respectively. The numbers below give the number of frames elapsed (each of 42 ms duration). Actions are identified by panels (i–vi) in Figure 12 and listed separately for each actor (a–c). (A suggested grouping of events into a hierarchical narrative structure is indicated by the Roman numerals I–III, etc.; these might be named I: Opening, II: Convergence, III: Conflict, etc.)

Actions in Fig. 12, panels i–v

(a) i: 1–3 (1–46) looks out, closes door; 4 (46–76) withdraws to wait; iii: 5 (132–154) returns to door; iv: 6 (155–169) looks out at b; v: 7 (170–178) exits; (179–185) attacks b;

(b) ii: 1 (76–98) enters the scene, 2 (77–112) reconnoiters; iii: 3 (117–151) approaches door, 4 (151–154) pause; iv: 5 (155–163) approaches door, 6 (163–167) negative gesture, 7 (169–170) retreat; v: 8 (170–178) pauses;

(c) ii: 1(76–98) enters the scene, 2(98–312) stands still

Two possible elisions and two possible compressions are shown at EL.1,2 and COMP.1,2, respectively.

(B,C) The sequence after cutting, with narrative structure and movement summary at B and C, respectively.

they might be harmlessly deleted because an active process of inference from the ongoing story structure fills in between the breakpoints; or both. Similarly, with respect to the omission of entire actions, or elisions, these two accounts—stimulus redundancy and inference from story structure—are quite different conceptions of the cognitive processes involved, but their relative roles in narrative text and film have only lately received attention.

ii. Limited scale of on-line inferences. As little as 500 ms is enough to render unrecoverable any briefly presented visual stimulus that has not been encoded within some otherwise recoverable (familiar, named, regeneratable) cognitive structure (Sperling, 1960). With the continually changing views of a moving-picture film, it takes repeated close scrutiny to identify and remember the film itself, and to consider how and why it was made as it was. For a particularly instructive (and visually sensitive) example of such close scrutiny, see Scharff, 1982. Such analysis is not the rule. Most experimental studies of narration, and all humanistic writings about literature and film, use only recall memory to demonstrate the importance of the story structures. That does not tell us what happens during the reading itself, and in the case of motion pictures, it is the experience on first viewing—the *on-line* processing, or perception—that may be more important (Hochberg & Brooks, in press) to the psychologist if not to the critic, the film student, or the cultist.

Tan and van den Boom (1992) obtained continuous measurement of viewer interest during a film screening (using a slide apparatus) and found the data obtained in this manner to be more precise for specific narrative segments than were the responses gotten from a postscreening questionnaire, although the two measures were significantly correlated. Their results "indicated that it is well possible to describe the core of the affect structure of a narrative film by testing a causal model in which situational determinants and two types of related affective response have been included" (p. 75).

In studies of reading, on-line measures suggest much less inference from overall story structure than do memory-based studies. Thus, using a form of priming measure, Dosher and Corbett (1982) found that readers did not, in fact, appear to have "spoon" in mind when reading "Mary stirred her coffee" (or make similar inferences in other examples), unless they had been asked explicitly to guess at the instruments used. With related implications, but using reading rate as their measure, Glanzer, Fischer, and Dorfman (1984) showed that after an interruption, readers were normally brought up to speed by being given the preceding lines of text (essentially the contents of short-term memory) rather than a story outline. Reviewing much of their own and others' recent work, McKoon and Ratcliff (1992) concluded that while reading normally only such inferences are made

---

structure that the absence of motion is largely unnoticed; by way of contrast, in *Neighbors,* by McLaren, elision of the break points between the actors' steps leaves a continuous undifferentiated levitated glide (see *pixillation,* Hochberg, 1986, Fig. 36e) within which the individual actions are undiscernible.

automatically as are needed to provide *local*[56] coherence to statements encountered in the text, and those inferences are based either on explicit statements or on general knowledge (i.e., not *deduced* from the story itself).

Invitations to inference are fundamental in speech acts (Grice, 1968; Searle, 1969), namely, the ubiquitous kinds of communication that require not just the grasp of syntactic conventions but also shared knowledge about the life and social reality (Schmidt, 1976). In film, familiarity with the medium, the genre, and indeed the individual passage, as well as familiarity with life, should provide the local inferences routinely needed to bridge cuts and elisions, without reference to the story structure. But knowing what the viewer expects locally also gives the filmmaker the opportunity to *invite inferences* about the story: To the engaged audience, emphases or violations of expectations that are unexplained by their context *but are apparently deliberate* probably invite inference. For example, a shot that is much longer than its neighbors requests an explanation, as does an otherwise-unexplained visual emphasis to some detail, such as the glass of milk in Hitchcock's *Suspicion* (Truffaut, 1967, p. 103), with a concealed lamp that increases its luminance and leads the viewer to think of the possibilities (poison, murderous intentions, etc.).[57]

In sum, the notion that shots are ideational units is an ideological assertion and not a cognitive truth. A great deal of elision between and within represented events characterize almost all motion pictures, but viewers bring enough structure to their viewing (in terms of plan schemas and hierarchical story structures) to bridge the ellipses. The completion processes actually involved may include distinctive feature recognition, local inferences, or inferences from the more abstract or distant story structures, but no research has been done with motion pictures to separate these processes. Judging from the studies of verbal narrative that have examined on-line processing as distinguished from retrospective memory studies (the former would seem to be the more relevant to film), inferences are minimal and local unless some challenge is posed. Cutting may indeed be used to challenge the viewer intellectually, but that is not *inherent* in the nature of narrative elision. With effort, cutting may be avoided over much or all of the motion picture, and made seamless where it is used (see Sect. III.B.2.a). As we see in the remaining two sections, however, cuts

[56] Evidence for local as opposed to more distant, story-based effects will probably be easy to find in narrative films. Cowen's finding (1988) that those viewers who received narrative film sequences out of linear order, and thereby received negative information at the end of the film, tended afterward toward more negative impressions of the protagonist than viewers who received that same information earlier, *even when both sets of viewers showed that they had achieved the same story structure*, is reminiscent of the effectiveness of local depth cues even when these contradict the overall spatial information (Hochberg, 1968; Hochberg & Peterson, 1987).

[57] Local inferences can surely be supplied from immediate content and genre, without need for particular story structure (for a sample of what a dictionary of motifs might look like, cf. Durgnat, 1971, pp. 229–235.). Even if there were evidence of a systematic film grammar (J. Carroll, 1980; Pryluck, 1968; Pryluck & Snow, 1967; Worth, 1968), of which we are unconvinced, all these other sources of structured expectations will surely make it difficult to identify.

are also used for purposes that are only remotely related to narration or, for that matter, to representation, although those purposes do affect and are affected by the processes of perception.

2. Affective and Connotative Effects of Cutting

Whether or not Eisenstein and Pudovkin really disagreed, at a working level, about the cognitive effects of cutting, they did agree (as do most writers) about the emotional or affective effects of cutting. Because each cut produces a momentary state of arousal, it was thought, the pace and rhythm of the cutting will presumably be reflected in the physiological state of the viewer. Two approaches have been distinguished by Leeuwen (1985): the "expressive" approach (quoting from Martin, 1962), which holds that slow rhythm imparts yearning, powerlessness, monotony (as in *L'Avventura*), whereas a fast, nervous rhythm imparts anger, feverish activity (as in the montage of indignant faces in *Potemkin*); and the "psychological" approach (quoting from Chartier, 1946), in which film rhythm is attained by cutting each shot at the moment attention begins to wane.

The idea that cutting rate in itself has expressive consequences, and can be plugged in when those effects are needed, would give a specific role in filmmaking to tempo as such. But despite heavy film usage, it has not been established that tempo has such stable effects. Any word, gesture, sound or camera movement, Leeuwen (1985) observed, will be given increased salience by placing it at a rhythmic beat, which implies an interaction in which the meaning is selected by the rhythm but determined by what the rhythm selects.[58] Indeed, Reisz and Millar (1968), in their classic text on editing, caution that cutting rate interacts with the motion picture content, for which we now have some laboratory research as well as filmmakers' writings. There are indeed reports that films are judged as more active at higher cutting rates (Gregory, 1961; Malpass, Dolan, & Coles, 1976), and Penn (1971) found that increased cutting rate did yield an overall increase of judged activity and potency (measuring the nonnarrative experimental film's connotative meaning by a semantic differential, and its meaningfulness by Noble's, 1952, measure). But Penn did not find the same effects of cutting rate with his three different subject matters,[59] and more recently, Heft and Blondal (1987), coming closer to

[58] By cutting rhythmically just before each beat of the music and of the dance move associated with it (as Brooks, 1987/1988, notes), or cutting just before each rhythmic beat of the hammer, and so on, the filmmaker both assures the visibility of the action to be emphasized and provides a synergy of rhythms, whereas cutting between beats would presumably preserve visibility but mute the rhythm. Given the multiple levels of rhythms in motion pictures, Leeuwen's assertion invites what should be fruitful research.

[59] Penn constructed films with stationary or moving objects (pairs of autos, people, or rectangles); cutting rates were either constant or accelerating and decelerating. Viewers rated concepts drawn from each film (e.g., white car, black car, the pair, the scene as a whole) on a semantic differential, and used Noble's (1952) measure of meaningfulness. Motion and content interacted significantly in their effects on the potency and activity scales in the films using cars and people, and a scattering of other specific comparisons were significant.

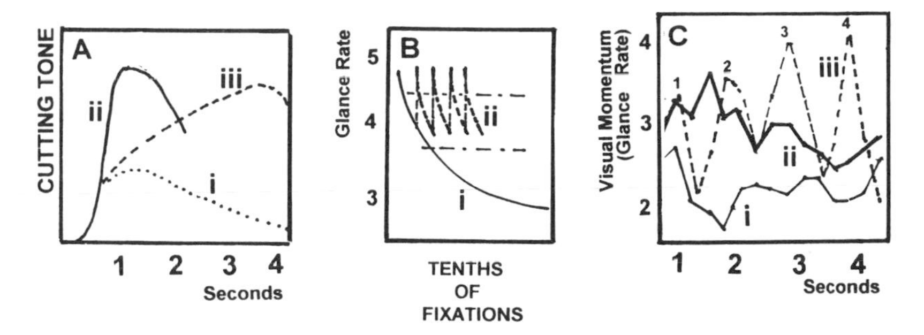

FIGURE 22 Affective cutting tone, glance rate, and visual momentum. (A) Spottiswoode's introspective description of how *affective tone* first rises and then falls after a cut to i, a simple shot, ii, a more interesting shot, and iii a more complex or difficult shot (Spottiswoode, 1933/1962). (B) Eye movements are made at a higher rate when a new picture is first presented, as shown at i, falling with time (Antes, 1974). (The abscissa here scales tenths of the total number of fixations.) If the picture were changed after the rate drops, the curve for five such cuts would presumably look something like ii, raising the average rates at which looking is sustained. (C) Affective cutting tone is presumably reflected in the impetus to keep looking—what we have called *visual momentum* (Hochberg & Brooks, 1978b). Antes's data (in B) suggests glance rate as one measure of visual momentum. In C, glance rates were measured after each cut between simple 4-s shots (i), more complex 4-s shots (ii), and simple 1-s shots, like the four cuts shown in curve (iii). Curves i and ii show that more eye movements are made after a cut and that more are made with the complex shots, and curve iii is much as we expected from the hypothetical curve at Bii, so it may be that glance rate in fact offers a measure of momentum. Other measures and consequences are needed. Work by d'Ydewalle and Vanderbeeken (1990) suggested that viewers' reaction times, when they are asked to detect cuts, may provide another usable measure.

narrative motion pictures by using two brief films (ca. 2.5 min each, of an instructor addressing his class, angry in one film and lighthearted in the other), found different effects of cutting rate in the two films: with the lighthearted instructor, faster cutting increased positive ratings of the speaker, as measured on an emotion state inventory (Mehrabian & Russell, 1974), whereas in the other film judgments were more negative at faster rates.

As in all such laboratory research with films, one must be concerned whether the results are specific to the situation and stimuli used, but it seems to us very likely that further research will corroborate the finding that the effects of cutting rate, varies with the content of what is being cut. That is, one cannot hope for a list of rates and corresponding emotions. What seems needed now are specific theoretical models about what measurable effects one should then expect with cutting rate as an independent variable.

Spottiswoode, in 1933, offered an excellent first pass at a theoretical model, based only on his introspections about montage. Figure 22A, a simplified version of his description, graphs how *affective tone* (his estimate of the quantity of emotional effect) varies over time after each cut. The time function depends on two independent factors, cutting rate and the new shot's content: Curve i shows the time course of affect after a simple and striking shot has been presented; curve ii shows the effect of a more interesting, beautiful, or meaningful shot; and iii is more complex or difficult to grasp. Although the model appears to be both potentially valuable and *potentially* specific, it has received no direct research. In the course of studies of glance frequency that were guided loosely by Spottiswoode's introspections, research discussed in the next section, Hochberg and Brooks (1978a, b) failed to find an interaction of complexity and cutting rate[60] on measures of looking rate or on pleasingness ratings, but those measures do not necessarily equate to affective tone; furthermore, none of the stimuli were involving at the level to be expected in real cinema. More recently, Heft and Blondal (1987), getting viewers' ratings of pleasure and dominance, did find an interaction between film content and cutting rate, but complexity as such was not one of their independent variables, and the measures on which they found an interaction are once again not necessarily what Spottiswoode meant by cutting tone. The model would still seem to deserve formalization and research.

3. Visual Momentum, Suspense, and the Maintenance of Perceptual Inquiry

Gaining visual information is the result of elective skilled purposeful actions. When first confronted by a still picture or scene, a viewer whose eye movements are being recorded takes several glances at a declining rate (Antes, 1974), as sketched in the curve i of Fig. 22B. The initial glances are directed toward parts of the field that are likely to be informative, or that subjects rate as being prominent (Antes, 1974; Brooks, 1961; Hochberg & Brooks, 1962; Loftus, 1976; Mackworth & Morandi, 1967; Pollack & Spence, 1968).

How might we elicit a high rate of looking for a longer period? For one thing, we can give the viewer more to look at. Dense text can keep the viewer busy for a much longer period than a simple picture, and Berlyne (as part of a long tradition relating complexity and novelty to esthetic value—cf. Berlyne, 1971; Birkhoff, 1933; Rashevsky, 1960; Vitz, 1966; etc.) showed that viewers look longer at more novel and more complex stimulus patterns (Berlyne, 1958; also Faw & Nunally, 1967). Or one could change to a new picture before the viewer has quite finished his initial rapid exploratory search. To Faulstich (1989), engaged in quantitative analysis of the measurable attributes of individual narrative films, the frequency of cuts (or takes) defines what he calls the *tension* of each segment, and reveals the dramatic structure of the montage. In any case, pictorial change is, quite tautologically, what distinguishes motion pictures from stills, and abrupt change has generally been thought to command visual attention (see Jonides & Yantis, 1988; Yantis & Jonides, 1984), although perhaps not inescapably so (see Folk, Remington, & Johnston, 1993; Yantis, 1993).

[60] Spottiswoode quite explicitly posits that cutting rate and complexity have interactive effects on affective tone: although increased rate should increase tone in simple shots, more complex shots should not reach their peak at those higher rates.

We expect that the impetus to maintain visual exploration or *visual momentum* (Brooks & Hochberg, 1976; Hochberg & Brooks, 1978b), will depend on two independent stimulus factors, cutting rate and the substantive content. With Spottiswoode (1962), we assume each abrupt view change to be at least minimally arousing, or "surprising." Keeping in mind the way in which eye movements increase at each picture change (Fig. 22B), one can take glance frequency as *one* measure of visual momentum, with reservations about what happens at the extremes. (Although clearly related to perceptual inquiry and shifts in attention, eye movements are not the only route to such shifts, nor does the sequence of glances necessarily reveal the sequence of perceptual questions. Attention can shift from one place to another much faster than eye fixation can be changed (Hoffman, 1975), and although experimental data strongly suggest that viewers first attend to some place in a display before moving their eye to it (Hoffman & Subramaniam, 1995; Shepherd, Findlay & Hockey, 1986), we know that viewers can also attend changing pictures while keeping their gaze fixed (see Figure 15). Indeed, as noted in Sect. III.B.3, Potter and Levy, 1969, found that eye movements stopped when pictures were changed more than 4 or 5 per second). Curve i in Figure 22B charts the glance frequency expected with a single view. Replacing each view before the curve falls too far should keep the average glance rate (ii) higher. With cutting rate constant, content will presumably affect momentum too; for example, in a picture that has more centers of attention momentum should fall more slowly, as in curve iii in Figure 22A. Using abstract film sequences made up of views having one, two, or four centers of attention, and presented at rates from 1 view/s to 1 view/4 s while glances were recorded, these simple expectations were very approximately fulfilled (Brooks & Hochberg, 1976; Hochberg & Brooks, 1978) (see Fig. 22C).[61]

Something like a measure of visual momentum seems worthwhile, particularly in the pacing of instructional or interactive material, where the task itself motivates the viewer to keep searching the changing displays.[62] But it is still far too primitive to apply either to an understanding of motion pictures in their recreational and cultural uses, or to the cognitive psychology of looking that is engaged by such uses: Given how important what the viewer seeks is to what is perceived, at all levels, it is naive to expect that the value of a shot within a montage of shots can be considered independent of its narrative function.

Testing their ideas about the filmic equivalent of grammatical clauses, J. Carroll and Bever (1976) measured the apparent brightness of flashes inserted as probes at

[61] Visual momentum is not, however, simply another name for eye-movement frequency. A sequence of shots with higher momentum should also keep the viewer looking at it longer, which in fact seems to be the case when two sequences are projected next to each other and the viewer is relatively free to look at either or neither (Hochberg & Brooks, 1978b). (Again, one should remember that with fast sequences, eye movements are curtailed, Potter & Levy, 1969, even though viewers are striving to snatch the fleeting glimpses.)

[62] For adaptation to the human factors problem of cognitive coupling of human to computer, see Woods, 1984; for display design, see Andre, Wickens, Moorman, and Boschell, 1991, and Aretz, 1991.

different points during short narrative films. The probes were rated as less bright just before structural boundaries, brighter at the boundaries themselves, and generally less bright during sections that require more cognitive processing. D'Ydewaller and Vanderbeeken (1990) found that viewers were slower to detect the occurrence of cuts in a motion picture that retained the narrative order of its segments, and presumably therefore engaged the viewer's attention to what was happening, than in the same segments out of order. This last procedure sounds promising, but it has not been used as yet to answer questions about motion picture usage.

In a recent dissertation that should serve as an excellent starting point for future work on timing in motion picture narration, de Wied (1991) undertook a sophisticated analysis of the relationship between story time, screen time, and suspense (untying story time and screen time by elision and compression, *not* by rate of cutting; see footnote 27). The stimulus sequence, 9 minutes from Leone's *Once upon a Time in the West,* was edited to vary the time (objective and subjective) that viewers spent in anticipating the harmful event; the main dependent variables were an on-line measure of suspense, which viewers indicated by their pressure on a bulb, and ratings after the sequences ended.

Surely suspense is one form of sustained interest, one route to visual momentum (although it would be good to have a bridge between this phenomenal variable and some perceptual or behavioral consequence, like the effort the viewer will expend to keep the narrative flowing, or to keep channels from switching). De Wied's (1991) main hypothesis was that rated suspense would first increase and then decrease as the time spent in anticipation increased, which was confirmed. Several subsidiary hypotheses were also tested, notably that the effect of time on suspense was mediated by subjective durations that depend on viewers' temporal expectancies (in turn influenced by the pace of narrative structure), not clearly confirmed in these experiments.

Although her findings may actually be taken into account by some filmmakers planning suspense scenes, what seems most important to us in her research is that it is guided by a model that uses cognitive concepts like plan schemas, breakpoints, and so on, to predict quite different consequences (e.g., judged suspense). In short, these concepts now do more than account for how subjects rate some social event, or how they remember it, which were their original defining operations: the concepts have been given explanatory consequences in a very different context. If various properties of motion pictures—suspense, interest, persuasiveness—are predicted by models of cognition that draw on such concepts of mental representation, those concepts are then subject to test and revision in ways that seem less subject to the prejudgments of the experimenter and the limitations of the laboratory.

The motion picture is not an ecologically isolated psychological domain: There is, as we have argued repeatedly in this chapter, little reason to consider the perception of motion pictures as a set of conventions learned specifically for watching film and TV; but even if such perception were so learned, a process of learning that is so quick and widespread must be regarded as a much different and more important ability than is involved in the laborious acquisition of reading, arithmetic, and interpersonal responsibilities.

IV. SUMMARY AND CONCLUSIONS

We have reviewed present explanations of how motion is perceived in motion pictures (sequential static pictures), and then surveyed problems, important to perceptual psychologists, that are advantageously studied with motion pictures and that are solved for their own purposes by filmmakers. Most of these problems arise because filmmakers rely very heavily on changes in the image on the screen that do not result from the viewer's own perceptuomotor actions. The comprehensibility of those changes therefore cannot rest on the viewer's efferent record or proprioceptions, with consequences for cognitive theory quite generally.

Two general explanations are considered. The first, which can be made quite explicit, is that viewers respond (directly or indirectly) to the visual information about the invariant scene that is mathematically specifiable as undergoing the transformations on the screen; there is now much research and analysis to this point. The second, much less explicit notion is that the viewer's attention to a sequence of sensory patterns (such as a succession of views) amounts to the formulation and testing of schematic maps or schematic events, which are structures of perceptuomotor expectations, or mental representations. The bulk of motion picture usage that we review, from precautions needed to avoid bad cuts (most of which result from proximity-based failure to respond to the invariant under transformation, or to misleading expectations), through the effects of the moving camera, to the comprehension of animate movements and nonoverlapping cuts in montage, and of plan schemas and story structures in filmic elision and compression, is better described in terms of schematic events and other mental representations characterized by at least two levels of structure (i.e., local and global).

The counterpart of trying to base our explanations solely on physical stimulus descriptions and to avoid any mentalisms, is to make the mental representations themselves so consistent with a real or virtual physical world, and the perceptual inferences based upon them so mandatory, that we have returned to a disguised physicalism. In fact, much film usage suggests that viewer's inferences in the on-line viewing of film are predominantly local and minimal, as has now also been strongly suggested by on-line analyses of how viewers read or hear verbal narrative. In both cases the job of discovering how the inquiring mind actually works is at least as important as explicating the possible structures of an ideal mental representation. And to do that we need to study the inquiring mind at what it does *by choice.*

Cinematic practice, we argue, taps abilities that we employ in the real world to guide our purposeful perceptual inquiries that themselves guide our larger actions. Film editors say, in fact, that good, rapidly comprehended cuts are those that provide the viewer with the answer to the visual question that he or she would normally be free to answer. With motion pictures, we can intervene in and thereby study the course of active looking. For the filmmaker (and that includes the writer of animation-narrative software), such perceptual research promises to provide a compact set of principles (sometimes counterintuitive, and sometimes difficult to disentangle in practice) to replace a larger number of ad hoc rules and tricks, and to guide the construction of structures that are both suspenseful or engaging, and comprehensible. They cannot, of course, substitute for the filmmaker's creativity, experience, or imagination.

For the psychologist, motion pictures are particularly suited to the study of the dynamic and purposeful aspects of perception. Beyond the pure perceptual aspects of cinema, to which we have limited ourselves in this chapter, motion pictures offer a visual approach, increasingly easy to employ, to the nature of narration and discourse, and to the study of social psychology in its perceptual and cognitive aspects, that is almost impossible to achieve by other means.

References

Adelson, E. H. (1991). Mechanisms for motion perception. *Optics & Photonics News,* Aug., 24–30.

Adelson, E. H., & Bergen, J. (1985). Spatiotemporal energy models for the perception of motion. *Journal of the Optical Society of America, A.2,* 284–299.

Allen, W. H., & Cooney, S. M. (1964). Nonlinearity in filmic presentation. *Audio-Visual Communication Review, 12,* 164–176.

Ames, A. (1951). Visual perception and the rotating trapezoidal window. *Psychological monographs, 65:* whole no. 324.

Anderson, G. J., & Braunstein, M. L. (1985). Induced self-motion in central vision. *Journal of Experimental Psychology: Human Perception and Performance, 11,* 122–132.

Anderson, J., & Anderson, B. (1993). The myth of persistence of vision revisited. *Journal of Film and Video, 45,* 3–12.

Andre, A. D., Wickens, C. D.., Moorman, L., & Boschelli, M. M. (1991). Display formatting techniques for improving situation awareness in the aircraft cockpit. *International Journal of Aviation Psychology, 1,* 205–218.

Andrew, D. (Ed.). (1989). Cinema and cognitive psychology: Issue no. 9 of *Iris.*

Andrew, J. D. (1976). *The major film theories.* New York: Oxford University Press.

Anstis, S. M. (1980). The perception of apparent movement. *Philosophical transactions of the Royal Society of London, 290B,* 153–168.

Antes, J. R. (1974). The time course of picture viewing. *Journal of Experimental Psychology, 103,* 62–70.

Aretz, A. J. (1991). The design of electronic map displays. *Human factors, 33,* 85–101.

Arnheim, R. (1932/3) *Film.* (L. M. Sieveking & I. F. D. Morrow, Trans.). London: Faber and Faber. (Originally work published 1932)

Arnheim, R. (1960). *Film as art,* Berkeley, California: Univ. of California Press. (A revised edition of the 1932 work.)

Attneave, F. (1954). Some informational aspects of visual perception. *Psychological Review, 61,* 183–193.

Attneave, F. (1974). Apparent movement and the what-where connection. *Psychologia, 17,* 108–120.

Attneave, F., & Block, G. (1973). Apparent movement in tridimensional space. *Perception & Psychophysics, 13,* 301–307.

Aubert, H. (1886). Die Bewegungsempfindung [Kinaesthesia]. *Archive fur die Gesamte Physiologie, 39,* 347–370.

Baker, C. L., & Braddick, O. J. (1982). The basis of area and dot number effects in random dot motion perception. *Vision Research, 22,* 1253–1260.

Barfield, W., & Weghorst, S. (1993). The sense of presence within virtual environments: A conceptual framework. In Om G. Salvendy & M. Smith (Eds.) *Human-computer interaction: Software and hardware interfaces* (Vol. B, pp. 699–704). Amsterdam: Elsevier.

Bartlett, F. C. (1932). *Remembering.* Cambridge, UK: Cambridge University.

Bayliss, G. C., & Driver, J. (1993). Visual attention and objects: Evidence for hierarchical coding of location. *Journal of Experimental Psychology: Human Perception and Performance, 19,* 451–470.

Bazin, A. (1967). *What is cinema?* (Vol. 1). Berkeley: University of California Press.

Bazin, A. (1972). *What is cinema?* (Vol. 2). Berkeley: University of California Press.

Beer, J. M. A. (1993). Perceiving scene layout through an aperture during visually simulated self-motion. *Journal of Experimental Psychology: Human Perception and Performance. 19,* 1066–1081.

Begelman, D. A. (1991). Virtual realities and virtual mistakes: A comment on Tart. *Dissociation Progress in the Dissociative Disorder, 4,* 214–215.

Berlyne, D. E. (1958). The influence of complexity and novelty in visual figures on orienting responses. *Journal of Experimental Psychology, 55,* 289–296.

Berlyne, D. E. (1971). *Aesthetics and psychobiology.* New York: Appleton.

Biederman, I. (1985). Human image understanding: Recent research and a theory. *Computer Vision, Graphics and Image Processing, 32,* 29–73.

Birkhoff, G. D. (1933). *Aesthetic measure.* Cambridge, MA: Harvard University Press.

Bordwell, D. (1985). *Narration in the fiction film.* Madison: University of Wisconsin Press.

Bordwell, D. (1989a). A case for cognitivism. *Iris, 9* (Spring), 11–40.

Bordwell, D. (1989b). *Making meaning: Inference and rhetoric in the interpretation of cinema.* Madison: University of Wisconsin Press.

Bordwell, D. (1992). Foreword. In D. Gomery, *Shared pleasures* (pp. ix–xxii). Madison: University of Wisconsin Press.

Boring, E. G. (1942). *Sensation and perception in the history of experimental psychology.* New York: Appleton-Century-Crofts.

Bower, G., Black, J., & Turner, T. (1979). Scripts in memory for text. *Cognitive Psychology, 11,* 177–220.

Braddick, O. J. (1974). A short range process in apparent movement. *Vision Research, 14,* 519–528.

Braddick, O. J. (1980). Low-level and high-level processes in apparent motion. *Philosophical Transactions of the Royal Society of London, 290B,* 137–151.

Brandt, T., Dichgans, J., & Koening, E. (1973). Differential effects of central versus peripheral vision on egocentric and exocentric motion perception. *Experimental Brain Research, 16,* 476–491.

Braunstein, M. L. (1962). Depth perception in rotating dot patterns: Effects of numerosity and perspective. *Journal of Experimental Psychology, 64,* 415–420.

Braunstein, M. L. (1968). Motion and texture as sources of slant information. *Journal of Experimental Psychology, 78,* 247–253.

Braunstein, M. L., Hoffman, D. D., & Pollack, F. E. (1990). Discriminating rigid from nonrigid motion: Minimum points and views. *Perception & Psychophysics, 47,* 205–214.

Breitmeyer, B. G., & Ganz, L. (1976). Implications of sustained and transient channels for theories of visual pattern making, saccadic suppression, and information processing. *Psychological Review, 83,* 1–36.

Breitmeyer, B. G., & Julesz, B. (1975). The role of on and off transients in determining the psychophysical spatial frequency response. *Vision Research, 15,* 411–415.

Breitmeyer, B. G., Love, R., & Wepman, B. (1974). Contour suppression during stroboscopic motion and metacontrast. *Vision Research, 14,* 1451–1455.

Brewer, W. F., & Dupree, D. A. (1983). Use of plan schemata in the recall and recognition of goal-directed actions. *Journal of Experimental Psychology: Learning, Memory and Cognition, 9,* 117–129.

Bridgeman, B., van der Heijden, A. H. M., & Velichkovsky, B. M. (1994). A theory of visual stability across saccadic eye movements. *Behavioral and Brain Sciences, 17,* 247–292.

Brooks, V. (1961). *An exploratory comparison of some measures of attention.* Unpublished master's thesis, Cornell University, Ithaca, NY.

Brooks, V. (1977). *The perceptual factors involved in filming dance: I. The effect of frame area on perceived velocity* (unpublished research report). New York: Columbia University.

Brooks, V. (1984). Why dance films do not look right: A study in the nature of the documentary of movement as visual communication. *Journal of Visual Communication, 10(2),* 44–66.

Brooks, V. (1984/1985). Film, perception and cognitive psychology. *Millenium Film Journal, 14/15,* 105–126.

Brooks, V. (1987/1988). Conventions in the documentary recording of dance: Research needs. *Dance Research Journal, 19(2),* 15–26.

Brooks, V. (1989). Restoring the meaning in cinematic movement: What is the text in a dance film? *Iris, 9* (Spring) 69–103.

Brooks, V., & Hochberg, J. (1976). Control of active looking by motion picture cutting rate. *Proceedings of the Eastern Psychological Association,* 49, (Abstract).

Brosgole, L., & Whalen, P. M. (1967). The effect of meaning on the allocation of visually induced movement. *Perception & Psychophysics, 2,* 275–277.

Brown, R. H., & Conklin, J. E. (1954). The lower threshold of visible movement as a function of exposure time. *American Journal of Psychology, 67,* 104–110.

Brunswik, E. (1956). *Perception and the representative design of psychological experiments* (2nd ed.). Berkeley: University of California Press.

Burch, N. (1973). *Theory of film practice.* New York: Praeger.

Burt, P., & Sperling, G. (1981). Time distance and feature trade-offs in visual apparent motion. *Psychological Review, 88,* 171–195.

Calvert, E. S. (1950). Visual aids for landing in bad visibility, with particular reference to the transition from instrument to visual flight. *Transactions of the Illuminating Engineering Society, London, 15,* 183–219.

Carroll, J. M. (1980). *Toward a structural psychology of cinema.* The Hague: Mouton.

Carroll, J., & Bever, T. G. (1976). Segmentation in cinema perception, *Science, 191,* 1053–1055.

Carroll, N. (1984). Film, video and photography. *Millenium Film Journal, 14/15,* 127–153.

Carroll, N. (1988). *Mystifying movies: Fads and fallacies of contemporary film theory.* New York: Columbia University Press.

Cavanagh, P., Arguin, M., & von Grunau, M. (1989). Interattribute apparent motion. *Vision Research, 29,* 1197–1204.

Cavanagh, P., & Mather, G. (1989). Motion: The long and short of it. *Spatial Vision, 4,* 103–129.

Chartier, J.-P. (1946). *Art et réalité au cinéma* [Art and reality in film]. *Bulletin de L'IDHEC,* 4.

Chubb, C., & Sperling, G. (1989). Two motion perception mechanisms revealed by distance-driven reversal of apparent motion. *Proceedings of the National Academy of Sciences, 86,* 2985–2989.

Cohene, L. S., & Bechtoldt, H. P. (1975). Visual recognition of dot-pattern bigrams: An extension and replication. *American Journal of Psychology, 88,* 187–199.

Colpi, H. (1966). Debasement of the art of montage. *Cahiers du Cinema in English, 3,* 44–45.

Comalli, P. E., Jr., Werner, H., & Wapner, S. (1957). Studies in physiognomic perception: III. Effect of directional dynamics and meaning-induced sets on autokinetic motions. *Journal of Psychology, 43,* 289–299.

Cook, O. (1963). *Movement in two dimensions.* London: Hutchinson.

Cooper, L. A. (1976). Demonstration of a mental analog of an external rotation. *Perception & Psychophysics, 19,* 296–302.

Cooper, L. A., & Munger, M. P. (1993). Extrapolating and remembering positions along cognitive trajectories: Uses and limitations of analogies to physical motion. In N. Eilan, W. Brewer, & R. McCarthy (Eds.), *Spatial representation* (pp. 112–131). London: Blackwell.

Cooper, L. A., & Shepard, R. N. (1973). Chronometric studies in the rotation of mental images. In W. G. Chase (Ed.), *Visual information processing* (pp. 75–176). New York: Academic Press.

Cooper, L. A., & Shepard, R. N. (1976). *Transformations of objects in space (Tech. Rep. No. 59).* San Diego: University of California, Center for Human Information Processing.

Corbin, H. H. (1929). The perception of grouping and apparent movement. *British Journal of Psychology, 19,* 268–305.

Cowen, P. S. (1988). Manipulating montage: Effects on film comprehension, recall, person perception, and aesthetic responses. *Empirical Studies of the Arts, 6,* 97–115.

Crowell, J. A., & Banks, M. S. (1993). Perceiving heading with different retinal regions and types of optic flow. *Perception & Psychophysics, 53,* 325–337.

Cutting, J. E. (1986). *Perception with an eye for motion.* Cambridge, MA: MIT Press.

Cutting, J. E. (1987). Rigidity in cinema seen from the front row, side aisle. *Journal of Experimental Psychology: Human Perception and Performance, 13,* 323–334.

Cutting, J. E., & Kozlowski, L. T. (1977). Recognizing friends by their walk: Gait perception without familiarity cues. *Bulletin of the Psychonomic Society, 9,* 353–356.

Cutting, J. E., & Proffitt, D. R. (1982). The minimum principle and the perception of absolute, common, and relative motions. *Cognitive Psychology, 14,* 211–246.

Cutting, J. E., Proffitt, D. R., & Kozlowski, L. T. (1978). A biomechanical invariant for gait perception. *Journal of Experimental Psychology: Human Perception and Performance, 4,* 357–372.

Cutting, J. E., Springer, K., Braren, P. A., & Johnson, S. H. (1992). Wayfinding on foot from information in retinal, not optical, flow. *Journal of Experimental Psychology: General, 121,* 41–72.

Dawson, M. R. (1987). Moving contexts do affect the perceived direction of apparent motion in motion competition displays. *Vision Research, 27,* 799–809.

Dawson, M. R. (1990). Apparent motion and element connectedness. *Spatial Vision, 4,* 241–251.

Dawson, M. R. (1991). The how and why of what went where in apparent motion: Modeling solutions to the motion correspondence problem. *Psychological Review, 98,* 569–603.

DeLorme, A., & Martin, C. (1986). Roles of retinal periphery and depth periphery in linear vection and visual control of standing in humans. *Canadian Journal of Psychology, 40,* 176–187.

DeLucia, P. R. (1991). Pictorial and motion-based information for depth perception. *Journal of Experimental Psychology: Human Perception and Performance, 17,* 738–748.

De Silva, H. R. (1929). An experimental investigation of the determinants of apparent visual movement. *American Journal of Psychology, 37,* 569–501.

de Wied, M. A. (1991). *The role of time structures in the experience of film suspense and duration: A study on the effects of anticipation time upon suspense and temporal variations on duration experience and suspense.* Unpublished doctoral dissertation, University of Amsterdam.

Dimmick, F. L., & Scahill, H. G. (1925). Visual perception of movement, *American Journal of Psychology, 36,* 412–417.

Dodge, R. (1907). An experimental study of visual fixation. *Psychological Review Monograph Supplement, 8* (Whole No. 35).

Dosher, B. A., & Corbett, A. T. (1982). Instrument inferences and verb schemata. *Memory & Cognition, 10,* 531–539.

Dosher, B. A., Landy, M. S., & Sperling, G. (1989). Kinetic depth effect and optic flow: 1. 3D shape from Fourier motion. *Vision Research, 29,* 1789–183.

Duncker, K. (1929). Uber induzerte Bewegung [On induced motion]. *Psycholigishe Forschung, 12,* 180–259.

Durgnat, R. (1971). *Films and feelings,* Cambridge, MA: MIT Press.

Durlach, N. I., & Mavor, A. S. (Eds.) (1994). *Virtual reality: Scientific and technological challenges.* Washington, DC: NAS Press.

D'Ydewalle, G., & Vanderbeeken, M. (1990). Perceptual and cognitive processing of editing rules in film. In R. Groner, G. d'Ydewalle, & R. Parham (Eds.), *From eye to mind: Information acquisition in perception, search and reading* (pp. 129–139). Amsterdam: North-Holland.

Eisenstein, S. M. (1942). *Film sense.* New York: Harcourt.

Eisenstein, S. M. (1949). *Film form.* New York: Harcourt.

Endsley, M. R. (1990). Predictive utility of an objective measure of situation awareness. *Proceedings of the Human Factors Society 34th Annual Meeting* (pp. 41–45). Santa Monica, CA: Human Factors Society.

Eriksen, C. W., & Collins, J. F. (1967). Some temporal characteristics of visual pattern perception. *Journal of Experimental Psychology, 74,* 476–484.

Faulstich, W. (1989). Film aesthetics and new methods of film analysis, *Empirical Studies of the Arts, 7,* 170–190.

Farber, J., & McConkie, A. (1977). Linkages between apparent depth and motion in linear flow fields. *Bulletin of the Psychonomic Society, 10,* 250 (Abstract).

Faw, T. T., & Nunnally, J. C. (1967). The effects on eye movements of complexity, novelty and affective tone. *Perception and Psychophysics, 2,* 263–267.

Festinger, L., & Easton, A. M. (1974). Inferences about the efferent system based on a perceptual illusion produced by eye movements. *Psychological Review, 81(1),* 44–58.

Foley, J. M. (1966). *The bilateral effect of film context.* Unpublished master's thesis, University of Iowa, Iowa City, IA.

Folk, C., Remington, R. W., & Johnston, J. C. (1993). Contingent attentinal capture: A reply to Yantis (1993), *Journal of Experimental Psychology: Human Perception and Performance, 19,* 682–685.

Freyd, J. J. (1987). Dynamic mental representations. *Psychological Review, 94,* 427–438.

Freyd, J. J. (1983). The mental representation of movement when static stimuli are viewed. *Perception & Psychophysics, 33,* 575–581.

Frith, U., & Robson, J. E. (1975). Perceiving the language of film. *Perception, 4,* 97–103.

Fry, R., & Fourzon, P. (1977). *The saga of special effects.* Englewood-Cliffs, NJ: Prentice-Hall.

Gibson, J. J. (1947). *Pictures as substitutes for visual realities.* (Army Airforces Aviation Psychology Program Research Reports, No. 7.) In Motion picture testing and research. Washington, D.C.: U.S. Government Printing Office.

Gibson, J. J. (1950). *The perception of the visual world.* Boston: Houghton Mifflin.

Gibson, J. J. (1954). The visual perception of objective motion and subjective movement. *Psychological Review, 61,* 304–314.

Gibson, J. J. (1957). Optical motions and transformations as stimuli for visual perception. *Psychological Review, 64,* 288–295.

Gibson, J. J. (1966). *The senses considered as perceptual systems.* Boston: Houghton Mifflin.

Gibson, J. J. (1979). *The ecological approach to visual perception.* Boston: Houghton-Mifflin.

Gillam, B. (1972). Perceived common rotary motion of ambiguous stimuli as a criterion for perceptual grouping. *Perception & Psychophysics, 11,* 99–101.

Gillam, B. (1979). Even possible figures can look impossible! *Perception, 8,* 229–232.

Glanzer, M., Fischer, B., & Dorfman, D. (1984). Short-term storage in reading. *Journal of Verbal Learning and Verbal Behavior, 23,* 467–486.

Godard, J. L. (1966). Montage, mon beau souci [Editing, my great concern]. *Cahiers du Cinema in English, 3,* 45–46.

Gogel, W. C. (1977). The metric of visual space. In W. Epstein (Ed.), *Stability and constancy in visual perception.* New York: Wiley.

Gogel, W. C. (1990). A theory of phenomenal geometry and its applications. *Perception & Psychophysics, 48,* 105–123.

Goldstein, B. (1975). The perception of multiple images. *AV Communication Review, Spring, 23 (1),* 34–68.

Gomery, D. (1992). *Shared pleasures.* Madison: University of Wisconsin Press.

Green, B. F., Jr. (1961). Figure coherence in the kinetic depth effect. *Journal of Experimental Psychology, 62,* 272–282.

Green, M., & Odom, J. V. (1986). Corresponding matching in apparent motion: Evidence for three-dimensional spatial representation. *Science, 233,* 1427–1429.

Greenberg, A., & Strickland, L. (1973). "Apparent behavior" revisited. *Perceptual and Motor Skills, 36,* 227–233.

Gregory, J. R. (1961). *Some psychological aspects of motion picture montage.* Unpublished doctoral dissertation, University of Illinois, Urbana, IL.

Gregory, R. L. (1964). Human perception. *British Medical Bulletin, 20,* 21–26.

Gregory, R. L. (1970). *The intelligent eye.* London, Weidenfeld.

Gregory, R. L. (1980). Perceptions as hypotheses. In H. C. Longuet-Higgins & N. S. Sutherland (Eds.), *The psychology of vision* (pp. 137–149). London: The Royal Society.

Grice, H. P. (1968). Utterer's meaning, sentence meaning and word meaning. *Foundations of language, 4,* 225–242.

Harmon, L. D., & Julesz, B. (1973). Masking in visual recognition: Effects of two-dimensional filtered noise. *Science, 180,* 1194–1197.

Hayhoe, M., Lachter, J., & Feldman, J. (1991). Integration of form across saccadic eye movements. *Perception, 20,* 393–402.

Heft, H., & Blondal, R. (1987). The influence of cutting rate on the evaluation of affective content of film. *Empirical Studies of the Arts, 5(1),* 1–14.

Heider, F., & Simmel, M. (1944). An experimental study of apparent behavior. *American Journal of Psychology, 57,* 243–259.

Held, R. (1968). Dissociation of functions by deprivation and rearrangement. *Psychologische Forschung, 31,* 338–348.

Held, R., Dichgans, J., & Bauer, J. (1975). Characteristics of moving visual scenes influencing spatial orientation. *Vision Research, 15,* 357–365.

Held, R., & Durlach, N. (1991). Telepresence, spotlight on: The concept of telepresence. *Presence, 1,* 109–112.

Helmholtz, H. L. F. von. (1909/1924). *Treatise on physiological optics.* Vol. III (Trans. from the 3rd German ed., 1909–1911, J. P. C. Southall, Ed. and Trans.). Rochester, NY: Optical Society of America, 1924–1925.

Hinton, G. E., & Parsons, L. M. (1981). Frames of reference and mental imagery. In J. Long & A. Baddeley (Eds.), *Attention and performance IX* (pp. 261–277). Hillsdale, NJ: Erlbaum.

Hippel, W. von, Jonides, J., Hilton, J. L., & Narayan, S. (1993). Inhibitory effect of schematic processing on perceptual encoding. *Journal of Personality and Social Psychology, 64,* 921–935.

Hochberg, J. (1964). *Perception.* Englewood Cliffs, NJ: Prentice-Hall.

Hochberg, J. (1968). In the mind's eye. In R. N. Haber (Ed.), *Contemporary theory and research in visual perception* (pp. 304–331). New York: Holt, Rinehart & Winston.

Hochberg, J. (1970). Attention, organization and consciousness. In D. I. Mostofsky (Ed.), *Attention: Contemporary theory and analysis.* New York: Appleton-Century-Crofts.

Hochberg, J. (1971). Perception: I. Color and shape (Ch. 12). II. Perception: space and movement (Ch. 13). In J. A. Kling & L. A. Riggs (Eds.), *Woodworth and Schlosberg's experimental psychology* (pp. 395–550). New York: Holt, Rinehart & Winston.

Hochberg, J. (1974b). Higher order stimuli and inter-response coupling in the perception of the visual world. In R. B. MacLeod & H. L. Pick (Eds.), *Perception: Essays in honor of James J. Gibson.* Ithaca, NY: Cornell University Press.

Hochberg, J. (1980). Pictorial functions and perceptual structures. In M. A. Hagen (Ed.), *The perception of pictures* (Vol. 2, pp. 47–93). New York: Academic Press.

Hochberg, J. (1982). How big is a stimulus? In J. Beck (Ed.), *Organization and representation in perception* (pp. 191–217). Hillsdale, NJ: Erlbaum.

Hochberg, J. (1984, August). *Visual worlds in collision: Invariants and premises, theories vs. facts.* Presidential address presented at the 92nd Annual Convention of the American Psychological Association, Toronto, Ontario, Canada.

Hochberg, J. (1986). Representation of motion and space in video and cinematic displays. In K. Boff, J. Thomas, & L. Kaufman (Eds.), *Handbook of perception and human performance* (Vol. 1, pp. 1–64). New York: Wiley.

Hochberg, J. (1987). Machines should not see as people do, but must know how people see. *Computer Vision, Graphics, and Image Processing, 37,* 221–237.

Hochberg, J. (1988). Visual perception. In R. Atkinson, R. Herrnstein, G. Lindzey, & D. Luce (Eds.), *Stevens' Handbook of Experimental Psychology* (Vol. I, pp. 295–375). New York: Wiley.

Hochberg, J. (1989). The perception of moving images. *Iris, 5,* 41–68.

Hochberg, J. (1992). Paths of perceptual consequence (Abstract). *Proceedings of the Psychonomic Society, 36,* 471.

Hochberg, J. (1994). Perceptual theory and visual cognition. In S. Ballestreros (Ed.) *Cognitive Approaches to Human Perception* (pp. 269–289). Hillsdale, NJ: Erlbaum.

Hochberg, J., & Beer, J. (1990). Alternative movement organizations: Findings and premises for modeling (Abstract). *Proceedings of the Psychonomic Society, 28,* 503.

Hochberg, J., & Brooks, V. (1962). The prediction of visual attention to designs and paintings. *American Psychologist, 17,* 437. (Abstract)

Hochberg, J., & Brooks, V. (1973). *The perception of television displays.* New York: The Experimental Television Laboratory of the Education Broadcasting System.

Hochberg, J., & Brooks, V. (1974). The integration of successive cinematic views of simple scenes. *Bulletin of the Psychonomic Society, 4,* 263. (Abstract)

Hochberg, J., & Brooks, V. (1978a). The perception of motion pictures. In E. C. Carterette & M. Friedman (Eds.), *Handbook of perception* (Vol. 10, pp. 259–304). New York: Academic Press.

Hochberg, J., & Brooks, V. (1978b). Film cutting and visual momentum. In J. W. Senders, D. F. Fisher, & R. A. Monty (Eds.), *Eye-movements and the higher psychological functions.* Hillsdale, NJ: Erlbaum.

Hochberg, J., & Brooks, V. (1989). Perception of still and moving pictures. In E. Barnouw (Ed.), *International encyclopedia of communications* (Vol. 3, pp. 255–262). New York: Oxford University Press.

Hochberg, J., & Brooks, V. (in press). Movies in the mind's eye. In D. Bordwell & N. Carroll (Eds.) *Post-theory: Reconstructing film studies.* Madison, WI: University of Wisconsin Press.

Hochberg, J., Brooks, V., & Roule, P. (1977). Movies of mazes and wallpaper. *Proceedings of the Eastern Psychological Association, April,* 179. (Abstract)

Hochberg, J., & Fallon, P. (1976). Perceptual analysis of moving patterns. *Science, 194,* 1081–1083.

Hochberg, J., Fallon, P., & Brooks, V. (1977). Motion organization in "stop action" sequences. *Scandanavian Journal of Psychology, 18,* 187–191.

Hochberg, J., & Gellman, L. (1977). Feature saliency, "mental rotation" times and the integration of successive views. *Memory & Cognition, 5,* 23–26.

Hochberg, J., & Peterson, M. A. (1987). Piecemeal organization and cognitive components in object perception: Perceptually coupled responses to moving objects. *Journal of Experimental Psychology: General, 116,* 370–380.

Hochberg, J., & Peterson, M. A. (1993). Mental representations of occluded objects: Sequential disclosure and intentional construal. *Italian Journal of Psychology, xx,* 805–820.

Hochberg, J., & Smith, O. W. (1960). Landing strip markings and the "expansion pattern": I. Program, preliminary analysis, and apparatus. *Perceptual and Motor Skills, 5,* 81–92.

Hoffman, J. E. (1975). Hierarchical stages in the processing of visual information. *Perception & Psychophysics, 18,* 348–354.

Hoffman, J. E., & Subramaniam, B. (1995). The role of visual attention in saccadic eye movements. *Perception & Psychophysics, 57,* 787–795.

Hogben, J., & Di Lollo, V. (1972). Effects of duration of masking stimulus and dark interval on the detection of a test disk. *Journal of Experimental Psychology, 95,* 245–250.

Holst, E. von (1954). Relations between the central nervous system and the peripheral organs. *British Journal of Animal Behavior, 2,* 89–94.

Howard, I. P., & Heckmann, T. (1989). Circular vection as a function of the relative sizes, distances, and positions of two competing visual displays. *Perception, 18,* 657–665.

Hunter, I. W., Tilemachos, D. D., Lafontaine, S. R., Charette, P. G., Jones, L. A., Sager, M. A., Mallinson, G. D., & Hunter, P. J. (1993). A teleoperated microsurgical robot and associated virtual environment for eye surgery. *Presence, 2,* 265–280.

Ingle, D., Schneider, G., Trevarthen, C., & Held, R. (1967). Locating and identifying: Two modes of visual processing: A symposium. *Psychologische Forschung, 32,* 44–62, 299–348.

Intraub, H. (1981). Rapid conceptual identification of sequentially presented pictures. *Journal of Experimental Psychology: Human Perception and Performance, 7,* 604–410.

Intraub, H. (1985). Visual dissociation: An illusory conjunction of pictures and forms. *Journal of Experimental Psychology: Human Perception and Performance, 11,* 431–442.

Intraub, H. (1989). Illusory conjunction of forms, objects and scenes during rapid serial visual search. *Journal of Experimental Psychology: Learning, Memory, and Cognition, 15,* 98–109.

Intraub, H. (1992). Contextual factors in scene perception. In E. Chekaluk & K. R. Llewellyn (Eds.), *The role of eye movements in perceptual processes* (pp. 45–72). New York: Elsevier.

Intraub, H., Bender, R., & Mangels, J. (1992). Looking at pictures but remembering scenes. *Journal of Experimental Psychology: Learning, Memory and Cognition, 18,* 180–191.

Intraub, H., & Richardson, M. (1989). Wide-angle memories of close-up scenes. *Journal of Experimental Psychology: Learning, Memory and Cognition, 12(2),* 179–187.

Irwin, D. E. (1991). Information integration across saccadic eye movements. *Cognitive Psychology, 23,* 420–456.

Isenhour, J. P. (1975). The effects of context and order in film editing. *Audio-Visual Communications Review, 23 (1),* 69–80.

Ittelson, W. W. H. (1952). *The Ames demonstrations in perception.* Princeton, NJ: Princeton University Press.

Jastrow, J. (1899). The mind's eye. *Popular Science Monthly, 54,* 299–312.

Jensen, G. D. (1960). Effect of past experience upon induced movement. *Perceptual and Motor Skills, 11,* 281–288.

Johnson, M. K., Kounios, J., & Reed, J. A. (1994). Time-course studies of reality monitoring and recognition. *Journal of Experimental Psychology: Learning, Memory & Cognition, 20,* 1409–1419.

Johansson, G. (1950). *Configurations in event perception.* Uppsala, Sweden: Almqvist & Wiksell.

Johansson, G. (1973). Visual perception of biological motion and a model for its analysis. *Perception & Psychophysics, 14,* 201–211.

Johansson, G. (1974). *Spatio temporal differentiation and integration in visual motion perception.* (Report No. 160). Department of Psychology, Upsala University, Sweden.

Johansson, G. (1977). Studies on visual perception of locomotion. *Perception, 6,* 365–376.

Johansson, G. (1982). Visual space perception through motion. In A. H. Wertheim, W. A. Wagenaar, & H. W. Leibowitz (Eds.), *Tutorials on motion perception.* New York: Plenum.

Johansson, G., & Jansson, G. (1968). Perceived rotary motion from changes in a straight line. *Perception & Psychophysics, 4,* 165–170.

Jonides, J., & Yantis, S. (1988). Uniqueness of abrupt visual onset in capturing attention. *Perception & Psychophysics, 43,* 345–354.

Johnston, I. R. (1972). *Visual judgments in locomotion.* Unpublished doctoral dissertation, University of Melbourne.

Julesz, B. (1971). *Foundations of cyclopean perception.* Chicago: University of Chicago Press.

Kaiser, M. K., & Phatak, A. V. (1993). Things that go bump in the light: On the optical specification of contact severity. *Journal of Experimental Psychology: Human Perception and Performance, 19,* 194–202.

Kahneman, D., & Treisman, A. (1993). The reviewing of object files: Object-specific integration of information. *Cognitive Psychology, 24,* 175–21.

Kaplan, G. (1969). Kinetic disruption of optical texture: The perception of depth at an edge. *Perception & Psychophysics, 6,* 193–198.

Kaufman, L. (1974). *Sight and mind.* New York: Oxford University Press.

Kaufman, L., Cyrulnick, I., Kaplowitz, J., Melnick, G., & Stof, D. (1971). The complementarity of apparent and real motion. *Psychologische Forschung, 34,* 343–348.

Keesey, U. T. (1972). Flicker and pattern detection: A comparison of thresholds. *Journal of the Optical Society of America, 62,* 446–448.

Kinchla, R. A. (1971). Visual movement perception: A comparison of absolute and relative movement discrimination. *Perception & Psychophysics, 9,* 165–171.

Koenderink, J. J. (1986). Optic flow. *Vision Research, 26,* 161–180.

Koenderink, J. J., & van Doorn, A. J. (1976). Local structure of movement parallax of the plane. *Journal of the Optical Society of America, 66,* 717–723.

Koenderink, J. J., & van Doorn, A. J. (1981). Exterospecific component of the motion parallax field. *Journal of the Optical Society of America, 71,* 953–957.

Koffka, K. (1935). *Principles of Gestalt psychology.* New York: Harcourt.

Kolers, P. A. (1972). *Aspects of motion perception.* Oxford: Pergamon.

Kolers, P. A., & Pomerantz, J. R. (1971). Figural changes in apparent motion. *Journal of Experimental Psychology, 87,* 99–108.

Kolers, P. A., & von Grunau, M. (1976). Shape and color in apparent motion. *Vision Research, 16,* 329–335.

Kopfermann, H. (1930). Psychologische Untersuchungen uber die Wirkung Zweidimensionalar Darstellungen korperlicher Gebilde [Psychological research on the efficacy of two-dimension representations of solid figures]. *Psychologische Forschung, 13,* 293–364.

Korte, A. (1915). Kinematoskopische Untersuchungen [Cinematoscopic studies]. *Zeitschrift fur Psychologie, 72,* 193–296.

Kraft, R. N. (1986). The role of cutting in the evaluation and retention of film. *Journal of Experimental Psychology: Learning, Memory and Cognition, 121(1),* 155–163.

Kraft, R. N. (1987). The influence of camera angle on comprehension and retention of pictorial events. *Memory & Cognition, 25,* 291–307.

Kraft, R. N. (1991). Light and mind: Understanding the structure of film. In R. R. Hoffman & D. S. Palermo (Eds.), *Cognition and the symbolic processes: Applied and ecological perspectives* (pp. 351–370). Hillsdale, NJ: Lawrence Erlbaum.

Kraft, R. N., & Green, J. S. (1989). Distance perception as a function of photographic area of view. *Perception & Psychophysics, 45(5),* 459–466.

Krumhansl, C. L. (1984). Independent processing of visual form and motion. *Perception, 13,* 535–546.

Landau, B., & Jackendoff, R. (1993). "What" and "where" in spatial language and spatial cognition. *Behavioral and Brain Sciences, 16,* 217–265.

Landau, B., & Spelke, E. (1988). Geometric complexity and object search in infancy. *Developmental Psychology, 24,* 512–521.

Lappin, J. S., & Fuqua, M. (1982). Nonlinear recruitment in the visual detection of moving patterns. *Investigative Optalmology Visual Science Supplement, 22,* 123.

Layer, H. A. (1974). The aesthetic basis for media design. *Audio-Visual Communications Review, 22,* 328–330.

Lee, D. N. (1976). A theory of visual control of braking based on information about time-to-collision. *Perception, 5,* 437–459.

Lee, D. N. (1980). The optic flow field: The foundation of vision. *Philosophical Transactions of the Royal Society of London, B, 290,* 169–179.

Leeuwen, T. van (1985). Rhythmic structure of the film text. In T. A. van Dijk (Ed.), *Discourse and communication* (pp. 216–232). Berlin, New York: De Gruyter.

Leibowitz, H. W. (1955). The relation between the rate threshold for the perception of movement and luminance for various durations of exposure. *Journal of Experimental Psychology, 49,* 209–214.

Lichtenstein, E. H., & Brewer, W. F. (1980). Memory for goal directed events. *Cognitive Psychology, 12,* 412–445.

Livingstone, M., & Hubel, D. H. (1988). Segregation of form, color, movement and depth: Anatomy, physiology, and perception. *Science, 240,* 740–749.

Loftus, G. R. (1976). A framework for a theory of picture recognition. In R. A. Monty & J. W. Senders (Eds.), *Eye movements and psychological processes.* Hillsdale, NJ: Erlbaum.

Longuet-Higgens, H. C., & Prasdny, K. (1980). The interpretation of a moving retinal image. *Proceedings of the Royal Society of London, B, 223,* 165–175.

Lynch, K. (1960). *The image of the city.* Cambridge, MA: MIT Press.

Mack, A. (1986). Perceptual aspects of motion in the frontal plane. In K. Boff, J. Thomas, & L. Kaufman (Eds.), *Handbook of perception and human performance* (Vol. 1, Ch. 17, pp. 1–38). New York: Wiley.

Mack, A., & Herman, E. (1973). Position constancy during pursuit eye movement: An investigation of the Filehne illusion. *Quarterly Journal of Experimental Psychology, 25,* 71–84.

MacKay, D. M. (1980). Discussion. *Philosophical Transactions of the Royal Society of London, 290B,* 167–168.

Mackworth, N., & Morandi, A. J. (1967). The gaze selects informative details within pictures. *Perception & Psychophysics, 2,* 547–552.

Malpass, R. S., Dolan, J. A., & Coles, M. (1976). *Effects of film content and technique on observer's arousal.* Mimeograph communication, SUNY College of Arts and Science, Plattsburgh, New York.

Mandler, J. M. (1978). A code in the node: The use of a story schema in retrieval. *Discourse Processes, 2,* 14–35.

Mandler, J. M., & Johnson, N. S. (1977). Remembrance of things parsed: story structure and recall. *Cognitive Psychology, 9,* 111–151.

Martin, M. (1962). *Le langage cinématographique* [The language of cinema]. Paris: Editions du Cerf.

Massad, C. M., Hubbard, M., & Newtson, D. (1979). Selective perception of events. *Journal of Experimental Social Psychology, 15,* 513–532.

Matin, E. (1975). The two-transient (masking) paradigm. *Psychology Review, 82,* 451–461.

Matin, L. (1986). Visual localization and eye movements. In K. R. Boff, L. Kaufman, & J. P. Thomas (Eds.), *Handbook of perception and human performance: Vol. 1. Sensory processes and perception* (pp. 20-1–20–45). New York: Wiley.

Maunsell, J. H. R., & Gibson, J. R. (1992). Visual response latencies in striate cortex of the macaque monkey. *Journal of Neurophysiology, 68,* 1332–1344.

McBeath, M. K., Morikawa, K., & Kaiser, M. K. (1992). Perceptual bias for forward-facing motion. *Psychological Science, 3,* 363–367.

McKoon, G., & Ratcliff, R. (1989). Inferences about textually defined categories. *Journal of Experimental Psychology: Learning, Memory, and Cognition, 15,* 1134–1136.

McKoon, G., & Ratcliff, R. (1992). Inference during reading. *Psychological Review, 99,* 440–466.

McLuhan, M. (1964). *Understanding media: The extensions of man.*

McLuhan, M. (1969). *Counterblast.* New York: Harcourt.

McVey, G. F. (1970). Television: Some viewer-display considerations. *Audio-Visual Communication Review, 18,* 277–290.

Mehrabian, A., & Russell, J. A. (1974). The basic emotional impact of environments. *Perceptual and Motor Skills, 38,* 283–301.

Metz, C. (1974). *Film language: A semiotics of the cinema* (M. Taylor, Trans.). New York: Oxford University Press.

Metzger, W. (1953). *Gezetse des Sehens* [Laws of Vision] Frankfurt am Main: Waldemar Kramer.

Michotte, A. (1963). *The perception of causality.* (T. & E. Miles, Trans.) London: Methuen. (Original work published in 1946)

Mill, J. S. (1965). An examination of Sir William Hamilton's philosophy. In R. J. Herrnstein & E. G. Boring (Eds.), *A source book in the history of psychology* (pp. 363–377). Cambridge, MA: Harvard University Press (Original work published in 1865)

Miller, G. A., Galanter, E., & Pribram, K. (1960). *Plans and the structure of behavior.* New York: Holt, Rinehart & Winston; 2nd Ed., New York: Adams-Bannister-Cox.

Minter, P. C., Albert, F. A., & Powers, R. D. (1961). Does method influence film learning? *Audio-Visual Communications Review, 9(4),* 195–200.

Mishkin, M., Ungerleider, L. G., Macko, K. (1983). Object vision and spatial vision: Two cortical pathways. *Trends in Neurosciences, 68,* 414–417.

Munsterberg, H. (1970). *The film: A psychological study.* New York: Dover. (Original work published in 1916)

Mutch, K., Smith, I. M., & Yonas, A. (1983). The effect of two-dimensional and three-dimensional distance on apparent motion. *Perception, 12,* 305–312.

Muybridge, E. (1882). The attitudes of animals in motion. Paper read before the Society of Arts, London, April 4, 1882. *The Scientific American Supplement, 14(343),* 5469–5470.

Nakayama, K., & Silverman, G. H. (1984). Temporal and spatial characteristics of the upper displacement limit for motion in random dots. *Vision Research, 24,* 293–299.

Navon, D. (1976). Irrelevance of figural identity for resolving ambiguities in apparent motion. *Journal of Experimental Psychology: Human Perception and Performance, 2,* 130–138.

Neisser, U. (1967). *Cognitive Psychology.* New York: Appleton-Century-Crofts.

Neuhaus, W. (1930). Experimentelle Untersuchung der Scheinbewegung [Experimental study of apparent motion]. *Archiv fur gesamte psychologie, 75,* 315–458.

Newtson, D. (1973). Attribution and the unit of perception of ongoing behavior. *Journal of Personality and Social Psychology, 28,* 28–38.

Newtson, D.., & Enqvuist, G. (1976). The perceptual organization of ongoing behavior. *Journal of Experimental Social Psychology, 12,* 436–450.

Newtson, D., Hairfield, J., Bloomingdale, J., & Cutino, S. (1987). The structure of action and interaction. *Social Cognition, 5,* 191–237.

Newtson, D., & Rindner, R. (1979). Variation in behavior perception and ability attribution. *Journal of Personality and Social Psychology,·37,* 1847–1858.

Noble, C. E. (1952). An analysis of meaning. *Psychological Review, 59,* 421–430.

Norman, J. F., & Todd, J. T. (1993). The perceptual analysis of structure from motion for rotating objects undergoing affine stretching transformations. *Perception & Psychophysics, 53,* 279–291.

O'Regan, J. K., & Levy-Schoen, A. (1983). Integrating visual information from successive fixations: Does trans-saccadic fusion exist? *Vision Research, 23,* 765–768.

Oldham, G. (1992). *First cut: Conversations with film editors.* Berkeley: University of California Press

Orlansky, J. (1940). The effect of similarity and difference in form on apparent visual movement. *Archives of Psychology, 246,* 85.

Osgood, C., Suci, G. J., & Tannenbaum, P.H. (1957). *The measurement of meaning.* Urbana: University of Illinois Press.

Palmer, J. (1986). Mechanisms of displacement discrimination with and without perceived movement. *Journal of Experimental Psychology: Human Perception and Performance, 12,* 411–421.

Parish, D. H., Sperling, G., & Landy, M. S. (1990). Intelligent temporal subsampling of American Sign Language using event boundaries. *Journal of Experimental Psychology: Human Perception and Performance, 16,* 282–294.

Parish, D. H., & Sperling, G. (1991). Object spatial frequencies, retinal spatial frequencies, noise and the efficiency of letter discrimination. *Vision Research, 31,* 1399–1415.

Penn, R. (1971). Effects of motion and cutting rate in motion pictures. *Audio-Visual Communication Review, 19,* 29–50.

Penrose, L., & Penrose, R. (1958). Impossible objects: A special type of visual illusion. *British Journal of Psychology, 49,* 31–33.

Perrin, D. G. (1969). A theory of multiple-image communication. *Audio-Visual Communication Review, 17,* 368–382.

Petersik, J. T. (1975). Two types of stroboscopic movement in the same display. *Bulletin of the Psychonomic Society, 6,* 435 (Abstract).

Petersik, J. T. (1989). The two-process distinction in apparent motion. *Psychological Bulletin, 106,* 107–127.

Peterson, M. A., & Hochberg, M. A. (1983). The opposed-set measurement procedure: The role of local cues and intention in form perception. *Journal of Experimental Psychology: Human Perception and Performance, 9,* 183–193.

Pollack, I., & Spence, D. (1968). Subjective pictorial information and visual search. *Perception & Psychophysics, 3(1-B),* 41–44.

Pollatsek, A., Rayner, K., & Henderson, J. M. (1990). Role of spatial location in integration of pictorial information across saccades. *Journal of Experimental Psychology: Human Perception and Performance, 16,* 199–210.

Posner, M. I., Boies, S. J., Eichelman, W. H., & Taylor, L. (1969). Retention of visual and name codes of single letters. *Journal of Experimental Psychology, 7,* 1–16.

Post, R. B. (1988). Circular vection is independent of stimulus eccentricity. *Perception, 17,* 737–744.

Potter, M. C. (1975). Short-term conceptual memory for pictures. *Journal of Experimental Psychology: Human Learning and Memory, 2,* 509–522.

Potter, M. C., & Levy, E. I. (1969). Recognition memory for a rapid sequence of pictures. *Journal of Experimental Psychology, 81,* 10–15.

Prasdny, K. (1980). Egomotion and relative depth map from optical flow. *Biological Cybernetics, 36,* 87–102.

Prasdny, K. (1986). What variables control (long-range) apparent motion? *Perception, 15,* 37–40.

Pratt, G. (1973). *Spellbound in darkness.* Greenwich, CT: New York Graphics Society.

Previc, F. H., Kenyon, R. V., Boer, E. R., & Johnson, B. (1993). The effects of background visual roll stimulation on postural and manual control and self-motion perception. *Perception & Psychophysics, 54,* 93–107.

Proffitt, D. R., & Cutting, J. (1980). Perceiving the centroid of curvilinearly bounded rolling shapes. *Perception & Psychophysics, 28,* 484–487.

Proffitt, D. R., Cutting, J. E., & Stier, D. M. (1979). Perception of wheel-generated motions. *Journal of Experimental Psychology: Human Perception and Performance, 5,* 289–302.

Proffitt, D. R., Gilden, D. L., Kaiser, M. K., & Whelan, S. W. (1988). The effect of configural orientation on perceived trajectory in apparent motion. *Perception & Psychophysics, 43,* 46–474.

Pryluck, C. (1968). Structural analysis of motion pictures as a symbol system. *Audio-Visual Communication Review, 16,* 372–402.

Pryluck, C., & Snow, R. E. (1967). Toward a psycholinguistics of cinema. *Audio-Visual Communication Review, 15,* 54–75.

Pudovkin, V. I. (1958). *Film technique and film acting.* London: Vision Press.

Pylyshyn, Z. W. (1973). What the mind's eye tells the mind's brain: A critique of mental imagery. *Psychological Bulletin, 30,* 1–24.

Pylyshyn, Z. W. (1989). The role of location indexes in spatial perception: A sketch of the FINST spatial-index model. *Cognition, 32,* 65–97.

Ramachandran, V. S., & Anstis, S. M. (1986). The perception of apparent motion. *Scientific American, 254(6),* 102–109.

Rashevsky, N. (1960). *Mathematical biophysics (Vol. 2).* New York: Dover.

Ratcliff, R., & McKoon, G. (1988). A retrieval theory of priming in memory. *Psychological Review, 95,* 385–408.

Regan, D., & Beverly, K. I. (1982). How do we avoid confounding the direction we are looking and the direction we are moving? *Science, 215,* 194–196.

Reisz, K., & Millar, G. (1968). *The technique of film editing.* New York: Hastings House.

Restle, F. (1979). Coding theory of motion configurations. *Psychological Review, 86,* 1–24.

Richards, R. (1992). *A director's method for film and television.* Boston: Focal Press.

Rock, I. (1977). In defense of unconscious inference. In W. Epstein (Ed.), *Stability and constancy in visual perception* (pp. 321–373) New York: Wiley.

Rock, I. (1983). *The logic of perception.* Cambridge, MA: MIT Press.

Rock, I., & Ebenholtz, S. (1962). Stroboscopic movement based on change of phenomenal rather than retinal location. *American Journal of Psychology, 75,* 193–207.

Salomon, G. (1972). Can we affect cognitive skills through visual media? An hypothesis and initial findings. *Audio-Visual Communication Review, 20,* 401–422.

Saucer, R. T. (1954). Processes of motion perception. *Science, 120,* 806–807.

Schmidt, C. F. (1976). Understanding human action: Recognizing the plans and motives of other persons. In J. S. Carroll & J. W. Payne (Eds.), *Cognition and social behavior* (pp. 47–67). Hillsdale, NJ: Lawrence Erlbaum.

Searle, J. R. (1969). *Speech acts: An essay in the philosophy of language.* New York: Cambridge University Press.

Sekuler, R., & Ganz, L. (1963). Aftereffect of seen motion with a stabilized retinal image. *Science, 139,* 419–420.

Sgro, R. J. (1969). Beta motion thresholds. *Journal of Experimental Psychology, 66,* 281–285.

Sharff, S. (1982). *The elements of cinema: Toward a theory of cinesthetic impact.* New York: Columbia University Press.

Shepard, R. N. (1981). Psychophysical complementarity. In M. Kubovy & J. Pomerantz (Eds.), *Perceptual organization* (pp. 279–341). Hillsdale, NJ: Erlbaum.

Shepard, R. N. (1984). Ecological constraints on internal representations: Resonant kinematics of perceiving, imaging, thinking and dreaming. *Psychological Review, 91,* 417–447.

Shepard, R. N., & Cooper, L. A. (1982). *Mental images and their transformations.* Cambridge, MA: M.I.T. Press.

Shepard, R. N., & Judd, S. A. (1976). Perceptual illusion of rotation of three-dimensional objects. *Science, 171,* 701–703.

Shepard, R. N., & Metzler, J. (1971). Mental rotation of three-dimensional objects. *Science, 171,* 701–703.

Shepherd, M., Findlay, J. M., & Hockey, R. J. (1986). The relationship between eye movements and spatial attention. *Quarterly Journal of Experimental Psychology, 38A,* 475–491.

Shor, R. (1957). Effect of pre-information upon human characteristics attributed to animated geometric figures. *Journal of Abnormal and Social Psychology, 54,* 124–126.

Sigman, E., & Rock, I. (1974). Stroboscopic movement based on perceptual intelligence. *Perception, 3,* 9–28.

Smith, W. M., & Gulick, W. L. (1956). Visual contour and movement perception. *Science, 12,* 316–317.

Smith, W. M., & Gulick, W. L. (1962). A statistical theory of dynamic contour perception. *Psychological Review, 69,* 91–108.

Sperling, G. (1960). The information available in brief visual presentations. *Psychological Monographs, 74,* 1–29.

Sperling, G. (1976). Movement perception in computer-driven visual displays. *Behav. Res. Instrum. 8,* 144–151.

Sperling, G. (1989). Three stages and two systems of visual processing. *Spatial Vision, 4,* 183–207.

Sperling, G. (1990). Comparison of perception in the moving and stationary eye. In E. Kowler (Ed.), *Eye movements and their role in visual and cognitive processes* (pp. 307–351). Amsterdam: Elsevier Biomedical Press.

Spottiswoode, R. (1962). *A grammar of the film.* Berkeley: University of California Press. (Original work published in 1933)

Stoper, A. E. (1973). Apparent motion of stimuli presented stroboscopically during pursuit movement of the eye. *Perception and Psychophysics, 13,* 201–211.

Stratton, G. M. (1911). The psychology of change: How is the perception of movement related to that of succession? *Psychological Review, 18,* 262–293.

Tan, E. S. H., & van den Boom, I. J. M. (1992). The psychological affect structure of narrative film. In E. F. Nardocchio (Ed.), *Reader response to literature* (pp. 57–94). Berlin & New York: Mouton de Gruyter.

Tart, C. T. (1991). On the uses of computer-generated realities: A response to Begelman. *Dissociation Progress in the Dissociative Disorders, 4,* 216–217.

Telford, L., & Frost, B. J. (1993). Factors influencing the onset and magnitude of linear vection. *Perception & Psychophysics, 53,* 682–292.

Ternus, J. (1938). The problem of phenomenal identity. In W. D. Ellis (Trans. & Ed.), *A sourcebook of Gestalt psychology* (pp. 149–160). London: Routledge & Kegan Paul.

Thompson, W. B., Mutch, K. B., & Berzius, H. A. (1985). Dynamic occlusion analysis in optical flow fields. *I.E.E.E. Transactions on Pattern Analysis and Machine Intelligence, 7,* 374–383.

Todd, J. T. (1982). Visual information about rigid and nonrigid motion: A geometric analysis. *Journal of Experimental Psychology: Human Perception & Performance, 8,* 238–252.

Todd, J. T. (1985). Perception of structure from motion: Is projective correspondence of moving elements a necessary condition? *Journal of Experimental Psychology: Human Perception & Performance, 11,* 689–710.

Todd, J. (1990). Models of static form perception. In J. I. Elkind, S. K. Card, J. Hochberg, & B. M. Huey (Eds.), *Human performance models for computer-aided engineering* (pp. 75–88). New York: Academic Press.

Todd, J. T., & Norman, J. F. (1991). The visual perception of smoothly curved surfaces from minimal apparent motion sequences. *Perception & Psychophysics, 50,* 509–523.

Treisman, A. (1986). Properties, parts, and objects. In K. Boff, J. Thomas, & L. Kaufman (Eds.), *Handbook of perception and human performance* (Vol. 2, pp. 1–70). New York: Wiley.

Trevarthen, C. (1968). Two mechanisms of vision in primates. *Psychologische Forschung, 31,* 299–337.

Truffaut, F. (1967). *Hitchcock.* New York: Simon & Schuster.

Ullman, S. (1978). Two dimensionality of the correspondence process in apparent motion. *Perception, 7,* 683–693.

Ullman, S. (1979). *The interpretation of visual motion.* Cambridge, MA: MIT Press.

van Santen, J. P. H., & Sperling, G. (1985). Elaborated Reichardt detectors. *Journal of the Optical Society of America, A.2,* 300–321.

VanderMeer, M. A. (1954). Color versus black and white in instructional films. *Audio-Visual Communication Review, 2,* 121–134.

Victor, J. D., & Conte, M. M. (1990). Motion mechanisms have only limited access to form information. *Vision Research, 30,* 289–301.

Vitz, P. C. (1966). Preference for different amounts of visual complexity. *Behavioral Science, 11,* 105–114.

Vorkapich, S. A. (1972). A fresh look at the dynamics of film-making. *American Cinematographer, 53,* 182–195.

Wallach, H. (1959). The perception of motion. *Scientific American, 201,* 56–60.

Wallach, H., & O'Connell, D. N. (1953). The kinetic depth effect. *Journal of Experimental Psychology, 45,* 205–217.

Warren, R. (1976). The perception of egomotion. *Journal of Experimental Psychology Human Perception and Performance, 2,* 448–456.

Warren, R. (1990). Preliminary questions for the study of egomotion. In R. Warren & A. H. Wertheim (Eds.), *Perception & control of self-motion* (pp. 3–32). Hillsdale, NJ: Erlbaum.

Warren, W. H., & Hannon, D. J. (1988). Direction of self-motion is perceived from optical flow. *Nature, 336,* 162–163.

Warren, W. H., & Kurtz, K. J. (1992). The role of central and peripheral vision in perceiving the direction of self-motion. *Perception & Psychophysics, 51,* 443–454.

Warren, W. H., Mestre, D., Blackwell, A., & Morris, M. (1991). Perception of circular heading from optical flow. *Journal of Experimental Psychology: Human Perception and Performance, 17,* 28–43.

Watson, A. B., & Ahumada, A. J., Jr. (1985). Model of human visual-motion sensing. *Journal of the Optical Society of America, A,1,* 322–342.

Watt, R. J. (1988). *Visual processing: Computational psychophysical, and cognitive research.* Hillsdale, NJ: Erlbaum.

Wertheimer, M. (1912). Experimentelle Studien uber das Sehen von Bewegung [Experimental studies on seeing motion]. *Zeitschrift fur Psychologie, 61,* 161–265.

Wilder, D. A. (1978a). Effect of predictability on units of perception and attribution. *Personality and Social Psychology Bulletin, 4,* 281–284.

Wilder, D. A. (1978b). Predictability of behaviors, goal and unit perception. *Personality and Social Psychology Bulletin, 4,* 604–607.

Williamson, S. J., Kaufman, L., & Brenner, D. (1977). Biomagnetism. In B. Schwartz & S. Foner (Eds.), *Superconductor applications: Squids and machines.* New York: Plenum.

Wohlgemuth, A. (1911). On the aftereffect of seen movement. *British Journal of Psychology Monographs, 1,* 1–117.

Woodworth, R. S. (1938). *Experimental psychology.* New York: Holt.

Woods, D. D. (1984). Visual momentum: A concept to improve the cognitive coupling of person and computer. *International Journal of Man-Machine Studies, 21,* 229–244.

Worth, S. (1968). Cognitive aspects of sequence in visual communication. *Audio-Visual Communication Review, 16,* 121–1.

Yantis, S. (1993). Stimulus-driven attentional capture. *Current Directions in Psychological Science, 2,* 156–161.

Yantis, S., & Jonides, J. (1984). Abrupt visual onsets and selective attention: Evidence from visual search. *Journal of Experimental Psychology: Human Perception and Performance, 10,* 601–621.

Yonas, A., Sen, M., & Knill, D. C. (1993). Reconceptualizing accretion and deletion of texture. In D. Valenti & J. Pittenger (Eds.), *Studies in Perception and Action II* (288–292). Hillsdale, NJ: Lawrence Erlbaum.

Zettl, H. (1973). *Sight-sound-motion: Applied media aesthetics.* Belmont, CA: Wadsworth.

CHAPTER 7

The Art of the Puzzler*

Arthur Schulman

It is the exceptional phenomenon which is likely to explain the usual one.

Jacques Hadamard (1945)

I wish to investigate plausible reasoning in the manner of the naturalist: I collect observations, state conclusions, and emphasize points in which my observations seem to support my conclusions. I cannot tell the true story [of] how the . . . discovery did happen. . . . Yet I shall try to make up a likely story.

Georg Polya (1954)

I. WORDPLAY AND THE COGNITIVE PSYCHOLOGIST

It is easy to compartmentalize one's life, to try to separate the frivolous from the serious, one's hobbies from one's research, and so for a long time I did just that, rarely thinking about cognitive theory when I was making up crossword puzzles, and rarely thinking about the mind of the puzzler when I was doing research on memory or preparing courses on imagery and invention. But gradually I came to realize that it was wrong to act as if the worlds of the puzzler and the cognitive psychologist were fundamentally different, and wrong to thing that puzzlers, unlike scientists, "merely" play. Finally I became convinced that the kinds of things that word puzzlers do, if properly described and understood, could shed light on aspects of creativity and the nature of discovery.

The word *puzzle* has been applied, at times indiscriminately, to many different tasks. The reader will find little discussion here of jigsaw puzzles, and none at all of chess puzzles or mathematical puzzles, even when what is required to make or solve them are some of the same skills the word puzzler uses. Puzzles like Mastermind and the Tower of Hanoi, which may be solved by deductive reasoning alone, will not concern us at all. Instead we will concentrate on word puzzles—crosswords, anagrams, cryptograms, and the like—and on those who make and solve them. The word puzzler is an inductive, not a deductive creature, whose intimate knowl-

* Dedicated to the memory of Elizabeth Schulman

Cognitive Ecology

edge of letters and words enables him to identify and capitalize on hidden constraints. Like any expert, he can assess *promise:* for the puzzler this means the fruitfulness of paths toward the creation or solution of puzzles. It is the word puzzler's world, then—an esoteric one, but one more penetrable by outsiders than, say, that of the nuclear physicist—that we shall explore here. What the puzzler *does* with his esoteric knowledge is, in my view, what all of us must do with *ours* if we are to make real discoveries of our own.

II. THE METHODS OF SCIENTIFIC PSYCHOLOGY

> Though there is a certain amount of empirical material that can be brought to bear upon [the means by which musicians] imagine the music that they play, . . . much of what follows is introspective . . . in nature; but the reader is invited to validate, or refute, what is said by checking it against his own experience.
>
> *Nicholas Cook (1990)*

More than one hundred years ago, James (1890/1983) described psychology as the "science of mental life," though he never supposed that our science could be sustained solely by the fruits of experimental research. It is ironic that generations of James's intellectual heirs defined themselves as "experimental" psychologists, over the years becoming seekers of questions that might be answered through experiment rather than questioners seeking paths most likely to lead to their answers. Although the cognitive revolutionaries of the 1960s rejected as sterile the mindless psychology of their times, few of them seemed to suspect that what was in need of change had itself been brought about by exclusive reliance on experiment.

Those who did, like Simon, became so enamored of their computer-simulation approach to cognition that they failed to see *its* limitations. Their "successes" have come (not by accident) with problem-solving tasks that call for programmable skills: heuristic search through a "problem space" and a methodical plan of attack. In a typical task, the subject knows what confronts him and knows what a solution will look like; all he needs to do is narrow the gap between where he is and where he wants to go. But many of our real problems, and certainly most of our puzzles, are not like that at all. "I do not seek, I find," said Picasso; and it is hard to see how Simon's methods could simulate paths toward discovery like those of, say, Gutenberg, or of Watson and Crick. If one knows what one must prove, one need only verify it. Machines are better verifiers than people, but people remain better problem finders and pattern recognizers than machines. Simon's breathtaking assertions (Simon, 1993) that "how people solve problems is no great mystery," and that we already know how we "learn language and make scientific discoveries" (p. 151) seem more sanguine, even, than the hopes of early behaviorists for what their methods would ultimately explain.

Very little of what I will have to say about the puzzler's world depends on experimentally produced evidence. Most of it is a product of my own experience and of my association with America's best puzzlers, nearly all of whom are members of the National Puzzlers' League (NPL).[1] The puzzler's expertise, as I hope to show, derives from intimate knowledge of his materials—letters and words, primarily—as well as from disciplined play. I am convinced that all expertise springs from these sources and can be properly understood only by appreciating just what the expert knows and how he plays.

III. THE ORIGINS OF WORDPLAY

Wordplay must be as old as language itself. The sounds of any language comprise a set small enough so that words built from them inevitably share, and are heard to share, prosodic features. "The pattern is there, and will make itself felt" (Wood, 1936, p. 11). Words that have nothing to do with one another come to be associated because they rhyme or because they have common patterns of stress. Poetry and mnemonics arise naturally from these discoveries about the structural properties of words, as chance juxtapositions suggest what might be produced on purpose and through play. English *lends* itself to iambic pentameter, allowing poets to explore and exploit its possibilities. Slips of the tongue, especially those rare ones that result in messages not only unintended but amusing, stimulate the deliberate generation of puns and spoonerisms.

IV. WRITTEN LANGUAGE, CONCEALMENT, AND REVELATION

> The labyrinth is made so that whoever enters it will stray and get lost. But the labyrinth also poses the visitor a challenge: that he reconstruct the plan of it and dissolve its power. If he succeeds, he will have destroyed the labyrinth; for one who has passed through it, no labyrinth exists.
>
> *Hans Magnus Enzensberger (quoted in Calvino, 1986)*

Poetry and song and riddles and puns all are possible with spoken language alone. Written language, of course, multiplies the opportunities for verbal play; the alphabet may be the greatest human invention. The art of the modern puzzler may be better understood against the background of earlier attempts at concealment and revelation.[2] Although this is not the place for a systematic historical review of wordplay, the examples that follow may suggest some of the fascination that written

[1] The NPL is a century-old organization devoted to sophisticated forms of wordplay. It publishes a monthly journal (*The Enigma*) of as many as 100 puzzles in verse, as well as anagrams, cryptograms, and other word puzzles. All puzzles are designed by NPL members to entertain and mystify their fellows. The NPL's annual convention features even more elaborate and challenging puzzles, including some puzzle-within-puzzle games that are nothing short of brilliant in conception and design. One of these is outlined in Appendix B.

[2] In *After Babel,* George Steiner (1975) suggests that the vast number of natural languages, past and present, is not merely the result of random proliferation from a few *Ursprache,* but rather derives from the need to conceal at least some of one's thoughts from too-easy penetration by outsiders. Some of the linguistic structures that have evolved to meet this need must therefore pose formidable, if not insuperable, problems for the translator.

language has always had for its users, and may help the nonpuzzler to enter the puzzler's cognitive universe.

In any written language, the symbols themselves may inspire wonder and invite contemplation. According to Jewish legend, the letters of the Hebrew alphabet "have been around since before the creation of the world and are mysteriously linked with the creative process itself" (Kushner, 1975, p. 5). A cabalistic manual by Abulafia (see Scholem, 1941) tells the reader how these holy letters should be mentally reordered so as to lead the apt student to the proper contemplation of the name of God.[3] Exercises in wordplay—not just letter transposals and anagrams but acrostics, ciphers, word squares, and many other types—have long histories. Biblical passages, especially in the Psalms, may conceal an acrostic message—one carried by the initial letters of successive words—that is at least as important as the apparent meaning of the verse and which, of course, is likely to be lost in translation. There are also a few *enciphered* words in the Old Testament: the last letter of the Hebrew alphabet is used to replace the first, the next-to-last to replace the second, and so on, so as to transform *Babel* into *Sheshach*. When medieval monks realized what was going on here, they began to develop more sophisticated encipherment schemes, and modern cryptography was born. The biblical substitution cipher was never meant to conceal or to deceive, however; it arose merely "from the predilection of scribes for amusing themselves with word and alphabet games" (Kahn, 1967, p. 78).

Wordplay has often been incorporated in pattern poetry, whose message is conveyed by the spatial arrangement of its characters as well as by the words themselves. A familiar example is "Mouse's Tale" from *Alice in Wonderland* (Carroll, 1960), but others of the genre date back at least to the Hellenistic Greek and occur in the works of modern writers as diverse as Apollinaire and the Dadaists. A particularly sophisticated seventeenth-century Portuguese composition (Hatherly, 1986), illustrated in Figure 1, incorporates several types of wordplay and must be puzzled out before it can be properly read. The title provides reading instructions for this "labyrinth sonnet." It tells the reader that each circle is a verse; that each verse is composed of two anagrams; and that each of these anagrams uses precisely those fifteen letters found on the globe's periphery. Before a verse can be read it must be deciphered, using the substitution-cipher key provided: 1 = D, 2 = O, and so on. The completed 14-line verse is in sonnet form; the first letters of each line constitute an acrostic, spelling out the name (D)OM SANCHO MANOEL. This

[3] Anagrams—the rearrangement of the letters of a word or phrase to produce another of similar meaning—have held their fascination since antiquity. Louis XIII, we are told by the Britannica, actually employed a court anagrammatist. Many words may be transposed into others (e.g., REGION into IGNORE), but the transposal seldom preserves the original meaning, and is not a proper anagram. The longer the phrase, of course, the easier it becomes to make its letters into an entirely new set of words; this makes possible such puzzles as double-crostics. Galileo, wishing to record a discovery of his while at the same time concealing it, transposed "*Cynthiae figuras aemulator Mater Amorum*" into the innocuous "*Haec immatura a me jam frustra leguntur—oy*". In English, this would be like transforming "Venus has phases like the moon" into "Male evokes passion? Hush, then!"

A O SENHOR CONDE DE VILLAFLOR.

Labyrintho: Egnima: Soneto: Encomiastico, Acrostico, Anagrammatico: em vinte, e oito Anagramas rigurosos. He cada circulo hum verso, cada verso dous Anagramas. Compoemse as letras pellos numeros, e os numeros pellas letras, da periferia deste Orbe.

FIGURE 1 A Portuguese "labyrinth sonnet" of the seventeenth century.

astonishing example of pattern poetry also conceals a riddle that one must find and answer, but I am unable to indicate either the riddle or its solution.

Word squares, in which the same words read across and down, are at least two thousand years old.[4] Figure 2 shows a Latin word square supposed by many, in medieval times, to have magical properties; note that AREPO and SATOR are

[4] Moschion's third-century tribute to Osiris, inscribed on an alabaster slab in a 39 × 39 grid, has its letters arranged in such a way as to permit that tribute to be read in many directions, all emanating from the central square. Because of its labyrinthine character it somewhat resembles modern "word search" puzzles.

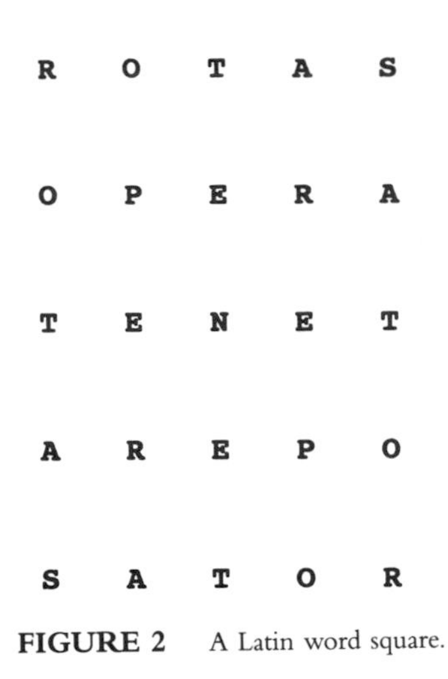

FIGURE 2 A Latin word square.

reversals of OPERA and ROTAS, and that TENET is a palindrome. In a "double word square," *different* words go across and down, as in a crossword puzzle. This principle underlies the "crossword hymn to Mut" (H. M. Stewart, 1971), one of a handful of surviving inscribed stelae from ancient Egypt that may be seen as crossword constructions, even though they were not meant to be "solved." Hieroglyphs, rather than letters, fill the cells of the stele's enormous grid. (Rebuses, one of the mainstays of the modern-day puzzler, derive ultimately from the ideas that gave rise to hieroglyphs.) The grid's hieroglyphs have been entered in such a way as to allow different meaningful readings going across (from right to left) and down. This was an achievement that the maker could only have *suspected* was possible, for only by inserting particular hieroglyphs in the grid can one appreciate the constraints their placement imposes. Similar constraints govern the words that today's crossword makers may place in *their* grids, and may even play a key role in literary invention. In *The Castle of Crossed Destinies,* Italo Calvino (1977) determined what his characters did and what happened to them by reading, once across and once down, from a rectangular array of tarot cards.

V. WORD PUZZLES AND PERCEPTUAL PUZZLES

Puzzlers are drawn to puzzles precisely because they puzzle, and because they enjoy the process by which patternlessness yields to the discovery of constraint, and bafflement ends in revelation. The puzzle maker makes familiar things hard, but not impossible, to discover, by making their appearance strange. The solver, like the reader of a good metaphor, must somehow find familiarity in the strangeness.

Certain perceptual puzzles pose analogous problems. Consider those fragmentary pictures, initially undecipherable, that eventually disclose their coherence. Probably the best-known example of these, a mutilated image of a dalmatian, is shown in Figure 3. When the naive perceiver is first shown such a picture, he is told that there is something seeable in it, but is not told what is there to be seen. If the observer trusts the experimenter (the perceptual puzzle setter) he will proceed to try to find the picture's hidden meaning. How this is done is far from clear, because the mutilation has destroyed the integrity of those diagnostic parts that now must presumably be imagined if redintegration is to take place. And yet all of us, as expert perceivers, must surely count on our discovery heuristics—intuitions that help us find hard-to-see evidence—whenever we suspect that meaning lurks within the noise.

As perceivers, all of us have learned to see things that we had never seen before; the unknown and unnoticed becomes familiar and more easily discovered when it is there. Neither the perceiver nor the puzzler may be able to characterize their discovery processes, even if it is clear that what both must do is cast about for indications that might be diagnostic. We know intuitively what we must do, but we are unable to characterize the mental operations our intuitions reflect. And yet we continue to recognize *undegraded* patterns, and so always expect our evidence-gathering heuristics, whatever they may be, to continue to produce perceptual

FIGURE 3 Finding the mutilated dalmatian is a puzzle.

revelations. With mutilated pictures like that of the dalmatian, our usual data-gathering heuristics prove inadequate to the task of finding what is there. Thus we must hope for revelation from improbable sources, from parts we would not ordinarily expect to play a role in the discovery process. Puzzlers know that if they are to solve *their* puzzles they will have to identify sometimes unsuspected constraints—hidden but diagnostic evidence—that point the way to discovery.

The puzzler's ability to take apart and put together again, and so to see the "same thing" in different ways, is not an ability that develops overnight. The puzzler, like anyone else, is first and foremost a perceiver, and making sense of the world ordinarily calls for an automaticity of function to which forever-shifting perspectives would be fatal. Under normal conditions, veridical perception is the rule: our sampling heuristics yield evidence that implicates the only source that might have given rise to it. The world of the puzzler, however, requires shifting perspectives. The habits that serve so well in everyday sense-making do not work—by definition—in a puzzle. The puzzler thus must be prepared to reject one conventional hypothesis after another so that he may finally turn up the evidence upon which solution depends.

VI. INTUITION AND DISCOVERY

> The processes of poetry and art, says Gombrich, are analogous to those of a play on words. It is the childish pleasure of the combinatorial game that leads the painter to try out arrangements of lines and colors, the poet to experiment with juxtapositions of words. At a certain moment things click into place, and one of the combinations obtained—through the combinatorial mechanism itself, independently of any search for meaning . . . —becomes charged with an unexpected meaning . . . which the conscious mind would not have arrived at deliberately.
>
> *Italo Calvino (1986)*

> The mechanic should sit down among levers, screws, wedges, wheels etc. like a poet among the letters of the alphabet, considering them as the exhibition of his thoughts; in which a new arrangement transmits a new idea to the world.
>
> *Robert Fulton (quoted by C. O. Philip, 1985)*

> Be patient, and reshuffle the cards.
>
> *Old Spanish proverb*

The puzzler's discoveries—finding a new anagram, filling a crossword grid, or seeing through the disguise of a cryptogram—typically arise, like most others we regard as interesting, in a sudden, revelatory manner. A systematic search through a problem space may be the first refuge of a simulation program, but it is the last resort of the expert: no puzzler will be methodical if he can help it. Intuition is what marks the expert, and intuition is acquired only with the sorts of experience that permit intimate knowledge to develop. This is true for the great improvisers of music and sport and for all experts—birdwatchers, chick sexers, wine tasters, masters of the games of go and of chess—who must be able to perceive things that novices cannot in order to assess a situation and its promise. It is surely true for much of creative mathematics, perhaps especially in cases of mathematical genius. As a youth, the brilliant Indian mathematician Ramanujan (see Kanigel, 1991) spent countless hours studying numbers, contemplating them—*playing* with them—in the process coming to know them so intimately that when some recurred on a later day, in a later thought, they could give rise to extraordinary revelations.

Nobody knows, of course, precisely what Ramanujan had come to know about numbers, or what he did when he "thought" about them. And nobody knows precisely what any word puzzler knows that enables him to solve an eight-letter anagram at sight, to penetrate a cryptogram's disguise almost as quickly, or to construct a crossword in which rows of long words going across will be layered in a way that makes possible columns of equally long intersecting words going down. But even though we cannot specify precisely what it is that experts know that permits their expertise to be manifested, we can, I think, describe the sorts of knowledge upon which it depends.

VII. INTIMACY AND THE DEVELOPMENT OF EXPERTISE

The development of expertise within a realm depends on intimate knowledge of its materials. Such intimacy requires time and appropriate experience; it cannot be *transmitted* from teacher to student, but rather must be acquired through the right sorts of play and practice. The best teachers know what a student must do to develop expertise, but there is no escaping the need for play. Real understanding of the nature of expertise within any realm can only arise through an appreciation of its developmental course.

The mathematician Mark Kac (quoted by Kanigel, 1991) has said that an "ordinary" genius is one whose mind works like yours and mine, only much better or faster; but that the mind of a genius like Ramanujan seems to operate on its own, perhaps unique, principles, impossible even for "ordinary" geniuses to imagine. To agree with Kac is to despair of ever understanding creativity, something that few cognitive psychologists are prepared to do. And yet it seems clear that expert–novice distinctions reflect differences in kind, and not merely of degree. What experts know, if not how experts deal with their knowledge, allows them to see and do things that novices cannot. Much of what experts know are things that novices can scarcely imagine: meaningful patterns that only opportunities for perceptual learning can reveal. Whenever expertise hinges on sensitivity to subtle patterns, novices will not be able to see what experts see, and so will not be able to act as experts do. With time, experts not only get to see things they did not see before, but acquire "superredundant" knowledge of their materials: they come to know, as second nature, a host of properties and their implications. When things known this well turn up again, often identified from fragmentary evidence of no use to the novice, they may give rise to revelations. Apparently unsought solutions and understandings come to mind, presumably from affinity-based rules of mental combination that we only dimly understand. Kac's views notwithstanding, geniuses, like ordinary experts, may owe their discoveries more to the richness of what they have come to know—entities and the relations among them—than to privileged or unique ways of using their knowledge. And the very activity of play may increase what we know of our playthings, thereby making possible new discoveries with continued play.

VIII. WHAT THE PUZZLER KNOWS AND DOES

Like all experts, puzzlers know their materials and how profitably to play with them. They know where to look to turn up potentially useful evidence, and what to do to find it when they get there; they know how to camouflage and how to see through camouflage; they know how to synthesize wholes from parts that a novice would find impoverished, and they know how to play by rules, to modify those rules, and to appreciate when the rules have been changed. Scientists must be able to do many of these things, too, as everyone must in everyday perception. Thus the puzzlers' skills, though applied to the specialized materials they know so well, are much more general than one might suspect.

A. Diagnostic Parts and Redintegration

Puzzlers know extraordinary things about the structure of words, and this knowledge is the basis of their redintegrative skills: they can tell what a word or phrase must be from structural cues that are unique to them. Hangman, the "Wheel of Fortune" game, poses little challenge to them, and they would find words easily for virtually all of the fragments supplied as cues in implicit memory experiments. (I suspect they would outperform the typical subject in a tachistoscopic word-recognition experiment as well). They can redintegrate a familiar phrase from its initial letters alone, A. M. T. H. G. F. yielding "absence makes the heart grow fonder," and O. O. S., O. O. M. yielding "out of sight, out of mind." Even the *last* letters of familiar phrases may be enough to synthesize them: K. E. U. P. can only be "look before you leap," and some puzzlers will see that. (The solution process may be redintegrative at both word and phrase levels: in this example, U may produce "you" which in turn may be enough to bring the entire phrase to mind.) All of us, puzzlers or not, may redintegrate phrases from stress patterns or word lengths alone: the sentence fragment "__ __ __ ___ __ __" can only be "to be or not to be." Some book titles and movies titles are similarly constrained by their word lengths, as the reader is invited to verify. Puzzlers appreciate constraints like these because of their intimacy with words, and so can tell what a word must be from much less evidence than a novice would require.

Many puzzlers are expert cryptographers. To solve a substitution cipher, the sort of cryptogram one finds in our newspapers, puzzlers rely on telltale patterns; they do *not,* if they can help it, count the frequencies of letters in the enciphered text. Sensitivity to the patterns of English words enables experienced puzzlers to see through a cryptogram's disguise and to read individual words as if they were not enciphered at all. This is because XURX is very likely THAT; because AGWANG is almost surely either PEOPLE or PROPER; and because FLEEFY *must* be LITTLE. Such diagnostic patterns, one should note, are recognized even when the *particular* encipherment, such as FLEEFY, has never been encountered before. And because of the redundancy of messages of even modest length, the solver knows that *his* decipherment, if it makes sense, is the only one possible; no alternative hypothesis, no different way of looking at the cryptic message, could lead to new meaning.

The following example illustrates how puzzlers use a word fragment as a cue that may lead to the discovery of the word that contains it. Each of the trigrams below occurs within only one familiar English word, at least according to the dictionary used to determine their uniqueness.

HBI AKF ANW IPW IWO RIJ RYD CHC SYG UDU DPR

That means that each trigram, like the fieldmarks learned by birdwatchers, is perfectly diagnostic: no further data are needed to identify the word. But these trigrams are not integral parts of the words that contain them, and so most of them can play no redintegrative role, even for most puzzlers, until they are suitably transformed. For this task, the puzzler's transformational heuristics are themselves redintegrative in nature. In the case of HBI, for example, the puzzler calls to mind morpheme candidates that end in -H (north, high, etc.) and that begin with BI- in the expectation that one of these productions will lead to revelation of the answer. In the end the target word—for HBI it is *archbishop*—is discovered just as a perceptual pattern is recognized, and with the same apparent speed.[5] If phenomenal "pop-out" ever occurs, it surely happens here.

Another redintegrative game that puzzlers play is based on the fact that the order of consonants in any known word—TR, say, in TEAR—is likely to be found in others—say, in OUTER and in TRUE. Players are challenged to find as many of these words as they can. To do so, they may insert any number of vowels (but no additional consonants) within the given consonantal sequence. From LSN, for example, one might rediscover such familiar words as LOOSEN, LESION, ELISION, LIAISON, and LOUISIANA. In a variation of this game, two words linked by "consonantcy" must be discovered; one is always the name of a country (say), while a definition is provided for the other. The definition *schedule,* for example, might suggest AGENDA, whose GND redintegrates UGANDA; if *schedule* makes one think of PLAN, however, no country name can come to mind, the redintegra-

[5] The unique solutions for the remaining trigrams are, in order, *breakfast, meanwhile, shipwreck, handiwork, marijuana, everyday, witchcraft, easygoing, fraudulent,* and *soundproof.*

tive failure telling the puzzler that it is another of *schedule's* synonyms that he must be after.

B. Combinatorial Play: Anagrams and Transposals

Games like Scrabble, its ancestor Anagrams, and many others reward the ability to discover words that an available set of letters might give rise to. Such discovery skills depend on intimate knowledge of frequently occurring—and not-so-frequently occurring—letter clusters. If the letters belonging to a promising cluster become available, the cluster is likely to be redintegrated no matter how the individual letters are arrayed. Frequent play, it is true, may result in the puzzler's knowing what can be made from particular sets of letters in various permutations—finding SMOTHER, e.g., the puzzler may recall that THERMOS is its transposal—but the cuing word must always be rediscovered through the combinatorial redintegrative process.

Given a set of letters—either an arbitrary one or one that is known to yield a meaningful word or phrase—it is seldom clear, a priori, whether it can be transposed into anything else that is familiar. Usually it cannot. True anagrams—rearrangements that carry more or less the same meaning as the word or phrase one begins with, as MOONSTARER for ASTRONOMER and BENEATH CHOPIN for THE PIANO BENCH, are genuine discoveries; the puzzler who finds them cannot know they are there until his heuristics of play turn them up. The discovery process is not mysterious, however. ASTRONOMER, after all, suggests MOON, whose letters are obviously present in the available pool, so that STARER emerges easily from the rest. As a rule, then, the would-be anagrammatist finds a word within the source phrase that relates to it, and goes on to see what, if anything, can be made of the letters that remain. The task is straightforward and usually frustrating, but finds such as SNUB I USE FOR NOSY ONE for MIND YOUR OWN BUSINESS can motivate the quest for new ones.

Psychologists have studied anagram solution—really transposal solution, to make the puzzler's distinction—for decades, but their research has taught us little about how such "anagram" problems are attacked. Some of the variables that affect likelihood and latency of solution are known—word frequency and word length, for example—but how a solution emerges is not clear. According to one review (Richardson & Johnson, 1980), the subject "selects an orthographically regular sequence of letters from the set given and uses this as a probe to interrogate the lexicon" (p. 247). Just how the selection is made, and just what the interrogation might entail, are not spelled out, and so these remain vague conjectures about the anagram solution process. Subjects in anagram experiments, moreover, are by and large novices at the tasks they are set. Their performance reflects their naïveté, with solutions ten times as long as those of expert puzzlers, when they occur at all (Schulman, Jonides, & Cohen, 1990). Order-of-magnitude effects like these suggest that the expert is not merely a faster novice but rather one whose manner of solution is different. I suspect that everyone solves an anagram, when we do so at all, by a process more like redintegration than interrogation, but the expert's vaster knowledge of word structure enables him to build more letter clusters with redintegrative promise. The cluster-assembly process depends on mutual affinities that for the expert have become second nature: not only will Q "attract" U, but many other letters and letter pairs will be drawn to one another. Hofstadter (1983) argues that all pattern identification, and not merely anagram solution by experts, in the end works in this bottom-up fashion. ("Looking for" may affect evidence gathering, but not the redintegrative use of the evidence itself.) I think he is right, but convincing evidence of any kind will be hard to muster. For the moment, Hofstadter tries to persuade by an example (see Gleick, 1983). He asks the reader to try to transpose the nonsense string LOONDERK into a familiar word. What comes to mind are wordlike "solutions" like KLONDORE and KNOODLER, while impossible and unpronounceable strings like RKONLDEO never seem to be entertained. It turns out, after all, that there is no solution word lurking behind LOONDERK; it is simply impossible to permute its letters into an eight-letter English word. But the plausible word parts one creates along the way suggest that Hofstadter's scenario for sense-making is essentially correct: discovery takes place through the synthesis of diagnostic parts, obviating the need for a systematic search.

Many games that puzzlers play require the recombinatorial skills of the anagrammatist. Some of these are more demanding, and therefore of greater cognitive interest, than simple transposals. In Double Jumbles, an offshoot of the newspaper puzzle, the letters of a related pair of words are first pooled and then alphabetized. From the alphabetical listing the puzzler must determine the two answer words, whose relation to each other he may, or may not, know. For example, AABDEGOPRST presents the letters of BOGART, the actor, and SPADE, a character that Bogart played. Knowing the relation between the words, it is possible to reach solution from either of two directions: in a bottom-up fashion from the letters, or through an initially top-down process in which one thinks of actor-role candidates whose letters might be found in the alphabetical list. Ignorance of the linking relation does not necessarily defeat the puzzler, because the letter set itself may be so constraining as to make the intended solution, and only that solution, discoverable. From ACIJJKLL, for example, one may derive only JACK and JILL, and from ABCEHIKLTW only BLACK and WHITE. One might suppose that AAAEEEHILMRSSTVY would provide insufficient constraint, but one would be wrong.[6]

[6] The solution words are HIMALAYAS and EVEREST. A good anagrammatist might seize upon the most constraining letters in this set and, having done so, discover either of these two solution words. The number of relations into which a word might easily enter is not large, and so EVEREST would suggest HIMALAYAS, or HIMALAYAS would suggest EVEREST, whose letters would immediately be found in the unused set. The whole discovery process can take much less time than it takes to describe it, sometimes merely a second or two.

C. Puzzles in Verse

An exhaustive description of the puzzler's recreations, even if it could be accomplished, would not serve my purposes here. A representative sample of puzzles currently in vogue may be found in Appendix A. All of these are puzzles in verse—*flats* in the jargon of the National Puzzlers' League. Such puzzles have a long history, take many different forms, and involve many types of wordplay. Rebuses and charades are probably the most common; the former go back thousands of years, the latter at least hundreds. Flats of the rebus type offer the solver two paths to solution: through the "rubric," (i.e., the rebus itself, which if properly "read" gives the answer); and through the verse, in which the answer is embedded and indicated by a place marker. The puzzle maker looks for words and phrases that lend themselves to rubrics that are hard to decipher, and tries to embed them in verses in a way that makes their identity hard to ascertain. The puzzle solver, as always, tries to see through the maker's disguises.[7]

D. The Crossword Maker and the Crossword Solver

Since crosswords as we know them first appeared, solvers have had to determine the words in the diagram by using (1) the clues and (2) the constraints of word length and of the letters already entered. In a rare discussion of crosswords by a psychologist, Nickerson (1977) explores such cues' effectiveness in the general context of prevailing thinking about memory. In this section I will consider what the crossword maker does, and how he does it. Then I will consider the solver's task, especially in "cryptic" crosswords in which the clues, unlike those in conventional American crosswords, pose real challenges.

1. The Constructor: Finding Words That Fit

Writers and painters know more or less what they intend to accomplish when they set themselves a particular task, but they also know that what they finally create may look very different from what they had in mind at the start. Creators examine, evaluate, and edit what they do as they do it; the discoveries made along the way may disappoint, please, or—very often—surprise. That is because creators cannot fully anticipate what their habits of mind and artistic skills will produce. They must play with their materials if they are to discover what can be made of them.

So it is with the puzzle maker as well. A crossword constructor who sets out to fill a grid with words cannot know what the final product will look like. Indeed, if the grid-filling task is severely constraining, the puzzle maker may not be able to accomplish it at all: nobody, for example, can create a daily-newspaper sized

[7] Psychologists have used the "cloze" technique to see if a reader can use the surrounding sentence context to infer the identity of deleted words. Writers of puzzles in verse do the same thing, but their purpose is to provide context that prevents, or at least delays, discovery. Because the verse must be seen to make sense once the answer is inserted in its slot, no disguise can be impenetrable.

crossword that has only 20 black squares, and the prospects for creating one would seem bleak if 1-Across had to be RAZZMATAZZ. Whether the constructor's self-imposed challenges can be met is something that only the attempt at creation, only the activity of play, can reveal.[8]

Merl Reagle, one of America's best crossword constructors, once made a puzzle (Reagle, 1982) in which abbreviations for each month of the year were embedded within 12 nine-letter words and phrases. Thus JAN appeared in TROJAN WAR, just above FEB in LIFEBOATS, and JUL appeared in MINT JULEP, just above AUG in BEAU GESTE. The months ran down the rows in calendrical order, making the construction feat even more of a *tour de force.* Reagle could not possibly have known as he embarked on this enterprise that he would be able to complete it. Even when he had discovered the candidate entries (TROJAN WAR, LIFEBOATS, and so on) he could not know that he would be able to fit them into a 17 × 17 grid so that they would mesh well enough to allow discoverable words to run at right angles to them.

Let me give an example from my own experience in crossword making. Some time ago I wondered whether it would be possible to make a crossword in which every entry, across and down, would be a word that was a transposal of another, as SILENT is to LISTEN, TINSEL, INLETS, and ENLIST. It was not obvious that this *would* be possible—most words are not transposable into another—but it soon became clear that the task was not hopeless, and a completed grid emerged in much less time than a good composer might have suspected would be needed.

In another constrictive task, I tried to build a puzzle in which all grid entries would be words whose first letter had been removed (so that IMAGINE, for example, would be entered as MAGINE). Now, crossword makers know, as second nature, how words with particular patterns of vowels and consonants fit together,[9] so that words that cannot fit with those already in place will simply never come to mind. (In the same way, poets can generate words of a given prosodic structure: no trochees intrude when one wants anapests. And we all know that the rhythmic structure of a tango will prevent waltz tunes from coming to mind). Although puzzlers know how *whole* English words mesh, they are not used to thinking about headless words and how *they* might fit together. I suspected that solutions existed,

[8] Until recently, computer-generated crosswords were crude and unimpressive, owing largely to the programmer's odd belief that it was possible to create an expert's product without knowing much of what an expert knows. Eric Albert (Albert, 1992) saw the error in those ways, and proceeded to install a database of more than 250,000 words and phrases that his program could operate on. I cannot tell, exactly, how his complex program works—although its heuristic search methods must be different from a human constructor's more failure-prone redintegrative methods—but Albert's system is a great success, generating filled diagrams that its human rivals would be pleased with, and doing so in reasonable times.

[9] The written forms of natural languages differ in their structure, and so the possibilities for crossword-puzzle interlock vary with those languages. Open patterns with few black squares are easy in Italian, whose words tend to alternate vowels and consonants, but are very difficult in German, where consonantal runs are common.

in the mathematical sense, but I was unsure that I'd be able to find one. Despite formidable problems of keeping track, as well as other atypical cognitive demands, in the end I was able to complete the puzzle.

Other successful exercises in coping with special constraints include (1) completing a grid in which every answer word is purged of one of its letters, as often as it appears: thus THISTLE might be entered as THISTE, omitting its L, or as HISLE, omitting both of its Ts; (2) a puzzle in which all answers must be stripped of their first *and* last letters before being entered in the diagram: JAZZ UP thus would be entered as AZZU; (3) a puzzle in which all letters from the first half of the alphabet (A—M) are deleted from words before grid entry, so that PLASTER OF PARIS appears as PSTROPRS; (4) a puzzle in which every entry represents either of two words from which a single letter has been removed: PLAUE might have been either PLAGUE or PLAQUE. (Just as in the all-transposal puzzle described earlier, it was not obvious a priori that there existed enough words like this in English to support this enterprise. I could anticipate ultimate success only after seeing the results of my early play). This puzzle, it is interesting to note, turned out to be solvable when only the possible deletions (e.g., GQ for PLAUE) were provided as clues; and (5) puzzles whose entries are vowelless so that JIGSAW PUZZLE, for example, is entered as JGSWPZZL. Stacking up such vowelless entries calls for unfamiliar cognitive challenges even for experienced constructors, but the task turns out to be easier than expected and to open up surprising and aesthetically appealing possibilities.

The black squares in a crossword, as we all know, serve to mark the boundaries of adjacent entries. But this "block" terminator is merely a convention, an accepted rule of the constructor's game. We could use a thick line (a "bar"), instead of a block, to separate diagram entries. Doing so would change not only the appearance of the crossword but, more significantly, the ways in which words might be conjoined. Since bar puzzles rarely are used (except in some cryptic crosswords, where the many bars make the grid-filling task easy), their possibilities remain unexplored; even the best constructors cannot know what sorts of interlock might emerge as a result of this altered constraint.

2. Cryptic Crosswords

A crossword constructor must fit words into a diagram and devise clues for them. The word-fitting task may be a challenge, as we have seen, but the clue-writing task, usually an afterthought in American crosswords, need not be. Until fairly recently in American crossword history, in fact, the clues that a solver encountered were rarely puzzling at all. One might not know, or not remember, what a clue was getting at, but there was no need to decipher it.

All of that changed with the development of the "cryptic" crossword, in which the constructor's ingenuity goes largely into clue writing. Most cryptics, in fact, have so many black squares as to make the maker's grid-filling task very easy; only about half of the letters in an across entry also appear in words going down. This makes the solver's task doubly difficult, of course. Not only must he try to decipher the cryptic clues, but only about half the letters he has placed in the diagram can help to decide what the remaining entries must be.

Those who can decipher the clues of a cryptic crossword know that "Beethoven's 5th" probably refers to the letter H and not to the symphony; that "drawback" indicates WARD, the reversal of DRAW; and that words like "confused" and "strange" usually signal a transposal, so that "messy BEDROOM" would clue BOREDOM, and "changing of THE GUARD" would clue DAUGHTER. Clues containing such cryptic instructions require the solver to figure out what a clue is telling him to *do* (i.e., to figure out *how* it means). The solver's task is complicated by the fact that each clue contains cryptic instructions for the answer's assembly as well as a straight definition; it is not always easy to tell where one begins and the other leaves off. In order to decipher such a clue, in any case, the solver must overcome a lifetime's parsing habits. Those habits were developed for getting at the "deep structure" of a text, exactly what the cryptic solver hopes to avoid.[10] Reading a cryptic clue as if it were an ordinary sentence traps one in the deep structure, and so prevents the answer from being found. The solver must dwell on the surface in order to see through the clue writer's deviousness.[11]

Here are some representative cryptic clues, which I hope will convey their flavor:

1. Active women iron some skirts and shirts.

The definition here is "active" women; the cryptic portion of the clue generates FE + MINIS + T'S = FEMINISTS, with the symbol for iron, "skirts," and "shirts" being strung together as a "charade." As in most cryptic clues, mental repunctuation is called for; it's part of the game.

2. Met stranger on train.

Here, "stranger" signals a transposal; ON TRAIN produces RAN INTO, which means "met."

3. Drive to beach resort, picking up a couple of bimbos.

This one is more elaborate. Beach resort is LIDO; a "couple of bimbos" is, literally, BI; LIDO "picks up," or inserts within it, BI, thereby producing LIBIDO, a rather different sort of "drive" than might initially have been suspected.

[10] Proofreaders looking for typos must, like the puzzler, avoid being caught up in a text's meaning. Although the proofreader's job is easier than the puzzler's, both occasionally may find themselves diverted by surface meanings when reading a novel or newspaper: their specialized ways of reading intrude, and must themselves be suppressed.

[11] As one would expect memory for cryptic clues differs for those who do and those who do not know how to decipher them (Schulman, 1983). Experts tend to recall the cryptic indication (i.e., what the clue had told them to do in order to assemble the answer). Novices, on the other hand, usually give a paraphrase of what a clue appears to mean.

4. Language that needs translating, to some extent.
LATIN is "hidden," (i.e., present) "to some extent," in TRANSLATING.

A skilled solver of cryptic crosswords may decipher clues like these and find their answers in no more time than it would take most of us to read "capital of France" and remember "Paris." A clever clue writer, of course, will try to induce plausible misreadings. This is done by understanding how language is ordinarily processed and by inducing those parsings that, although sensible, cannot lead to a clue's answer. Painters' understanding of human perceptual processes enables them to get viewers to see what they want them to see; puzzle makers' understanding of the solver's linguistic habits enables them to play to those habits and to make it hard to escape from them.

IX. ON BEING SURE THERE'S NOTHING THERE

> The sterile combinations do not even present themselves to the mind of the inventor. Never in the field of his consciousness do combinations appear that are not really useful, except some that he rejects but which have to some extent the characteristics of useful combinations.
>
> *Henri Poincare (1904/1956)*

Intimate knowledge makes it easier to appreciate something's presence, as every birdwatcher knows. Samples drawn by nonexperts are less likely to strongly constrain what might be made of them, less likely to be clear in their promise. To be knowledgeable, however, has a second benefit easily overlooked: it permits us to be sure there is *nothing* there. When as expert "draws a blank," another look would be a waste of time. This is so in everyday pattern identification, even if one does not always consider the implications: elephants not noticed are surely elephants that aren't there. Indeed, one need not verify the absence of foul odors, explosions, and other compelling events, for one's metacognitive knowledge of how pattern identification works virtually guarantees that they *cannot* be there.

Although one can rule out the presence of an elephant one failed to notice, one cannot be equally sure about the nonpresence of things one is ill equipped to detect. Experts will know not only what they can make of the evidence at hand, but what they cannot; to assess a situation's promise, one must be able to do both. Word puzzlers, for example, should be able not only to make a word out of a set of letters if it is possible to do so, but should be able to tell when, as with LOONDERK, a set of letters leads nowhere. With myself as the principal subject, I was able to provide experimental evidence to support this claim (Schulman, 1989). The task was one in which I had to try to discover a six-letter word that could be made from three given vowels (say, AEI) and three given consonants (say, NRT). In this example one might find either RETAIN or RETINA. If nothing came to mind in five seconds or so, I would conclude that there was nothing there and go on to the next letter set, of which there were about 1000 in all. About 20% of the sets I worked with could, in fact, be transposed into one or more English words (*Chambers Anagrams,* 1985); my hit rate for these, in the time allotted, was about 95%. The chance that no solution existed, given that I had dismissed the possibility of one, was about 97.5%, or considerably better than the a priori probability of 80%. In this situation, at least, failure to discover a solution was tantamount to demonstrating that none was there to be found.

Not all failures of discovery carry such clear implications, of course, even for the expert. Most absences and deletions are notoriously difficult to detect (Hearst, 1984), even though their detection may be the key to a problem's solution. In a host of games—Twenty Questions, Battleships, word games like Hangman and Jotto, and many more—what has been ruled out may be more diagnostic than what has been positively ascertained. Certainly the knowledge of the cards not yet played in a hand of bridge, or the tiles not yet exposed in a game of Scrabble, are keys to successful play. Puzzlers are more sensitive than most of us to the diagnostic value of certain absences—puzzle solutions often hinge upon seeing their implications—and yet keeping track of absences is never easy. Thinking about absences may, in fact, be an impossibility; only mental *presences* lend themselves to recombination and the operations of thought. The Jotto or Hangman player must mentally transform the spent letters into the still available letters, for only the latter may participate in the redintegrative process of solution. Failure to transform what cannot be the case into what might be the case puts the solution out of immediate reach.

X. DISGUISE AND DECIPHERMENT

> The art of jigsaw puzzling begins with wooden puzzles cut by hand, whose maker undertakes to ask himself all the questions the player will have to solve, and, instead of allowing chance to cover his tracks, aims to replace it with cunning, trickery, and subterfuge. . . . [E]very move the puzzler makes, the puzzle-maker has made before; every piece the puzzler picks up, . . . every combination he tries, . . . each hope and each discouragement have all been designed, calculated, and decided by the other.
>
> *Georges Perec (1987)*

Things are not always what they seem. Camouflage and mimicry are commonplace in the animal and plant worlds—for Nabokov (1980), nature is the "arch-cheat"—and so appearances may deceive. What we see at first in a trompe l'oeil painting is not what we may come to see.[12] And a secret message may be so cleverly disguised,

[12] It is through the penetration of the painting's disguise that one comes to understand it. It cannot delude again, any more than a puzzle solved can continue to puzzle. "It is when we have ceased to be the unwitting targets of a practical joke, and we have decided to reflect upon the experience we have just gone through, that the painting acquires its meaning" (Kubovy, 1986, p. 78).

or so artfully embedded in an innocuous one, that the unprepared observer may remain unaware that there was anything there to be found.

A good puzzler sees through various disguises, some of them ingenious. But he must first realize that there is a disguise to penetrate. Efforts at decipherment can begin only after one appreciates that there is something to decipher, that there may be discoverable but as yet undisclosed meaning lurking beyond a surface. Some hidden meanings may yield to a changed perspective: focusing beyond the picture plane may enable one to see the elusive three-dimensionality of a single-image random-dot stereogram (Grossmann, 1992). Other hidden meanings, however, are not so easily found. The "message" must first be deciphered; to do so, the rules by which structure conveys meaning must be understood. Scholars were unable to read Egyptian hieroglyphs until the implications of the inscriptions on the Rosetta stone permitted their decipherment. Though decipherment requires skills, puzzlers are by no means its only experts. Indeed, children must decipher the speech they hear as infants—we are all puzzlers from birth—while the "reading" of any text presupposes intimate knowledge of its symbolic characters.

What our visual system does not equip us to see cannot be rendered by any artist, but anything seeable can. Still, the "cryptograms of art" (Gombrich, 1960) need to be "deciphered" by the viewer; we must learn to look in new ways, so that we can come to see new things. Unfamiliar styles of painting and music may prove undecipherable and consequently unreadable to many.

Just as the artist must understand human perception, the puzzle maker must understand human cognition. But the puzzle maker's intent is subversive; it is to hide the revelatory path, or to induce the solver to abandon it. Thus the maker must know how solvers think in order to lead them astray, whereas solvers need to identify in the puzzle's structure any telltale clues, not all of which can be obliterated, which betray the maker's design.

Substitution ciphers, as well as most cryptograms of other kinds, are easily seen for what they are; their surfaces virtually proclaim that behind the unintelligible facade lurks a decipherable message. Some hidden messages, however—acrostics are examples—may be embedded in a meaningful surface that looks so "normal" as to call little attention to itself and arouse no suspicion. Other surfaces are not at all what they seem; only by pronouncing "*mots d'heures gousses, rames*" in French, and hearing the result as accented English, can one find the disguised coherence of "Mother Goose rhymes" (Rooten, 1967).

Bronowski (1978) has written that archaeologists can tell how a strange artifact must have been made, and that this in turn may suggest the artifact's function. So it is with puzzlers, too. A cryptic clue, however artfully constructed, is always penetrable if it "plays by the rules." Its very wording suggests how it must have been crafted, and therefore shows the solver what must be done to decipher it. Although any such decipherment attempt must be somewhat tentative, the solver can tell when he is right because the answer flows from both parts (cryptic and definitional) of his reading of the clue.

XI. CONSTRAINT AND CREATIVITY

In the 1960s in France, an extraordinary group of writers and mathematicians began to meet regularly to explore the literary possibilities that might be realized under conditions of severe constraint. They called themselves OuLiPo, which stands (in French) for something like "workshop in potential literature"; their linking of creativity and constraint allies them, at least in spirit, with the puzzle makers I have been discussing. (Mathews, 1976, provides a brief introduction to the OuLiPo; Motte, 1986, a more comprehensive one.) The OuLiPo believe that it is delusory to think that our choices, literary or otherwise, can ever be completely free. "Assume nothing" would be a laughable slogan to them; it is much better, and paradoxically more liberating, to identify the constraints under which we operate. Early attempts to write in sonnet form opened up possibilities that would not otherwise have been explored, and that could not have been anticipated; theories deriving from a handful of axioms have, on occasion, turned out to be surprisingly rich.

The Oulipian lesson is that one cannot tell how confining rules really are until one sees how they play out; sometimes they will close doors, sometimes they will open them. Oulipians "must build the labyrinth from which they propose to escape" (Motte, 1986, p. 22). Some of their chosen constraints may seem frivolous, even pointless, but it would be a mistake to dismiss them before knowing their fruits. A "lipogrammatic" constraint, for example, forbids the use of a particular letter, or letters, in a text. (Charles Hockett once prepared for his linguistics students a set of verses, each of which deliberately excluded phonemes such as stop consonants or sibilants. These lipogrammatic verses were designed to illustrate how a language would sound if it were missing such phonemes.) Georges Perec, one of the OuLiPo's most interesting and productive members, produced an entire novel, *La Disparition* (Perec, 1969), without using the letter *e*. The novel is "*about* the disappearance of [that] letter. It is both the story of what it recounts and the story of the constraint that creates that which is recounted" (Roubaud, cited in Motte, 1986, p. 12). It is unlikely that Perec set out to write an entire novel that would be *e*-less; more likely he began with more modest goals, but discovered as he wrote that a more ambitious work was achievable. People are adaptable creatures. When faced with sudden deprivation, they try to find substitutes for what they have lost, or to find new ways to achieve old results. Lipograms are products of self-denial, but the challenge to adapt is a very human one. As one begins to compose a lipogram, especially one that rules out the use of such a common letter as *e*, words containing the taboo letter come to mind and need to be actively suppressed, and *e*-less paraphrases need to be found to take their place. With practice, however, the lipogrammatic task becomes less strange and so "natural" that even words like *the* and *are* seldom intrude on the writer's thoughts. Perec thus came to inhabit a linguistic universe that had been more or less effectively purged of words containing *e*'s, in the end thinking and writing fluently enough so that they rarely came to mind. *La Disparition,* it may be noted, was a popular success, though it is hard to believe the claim that some readers failed to notice that the work was lipogrammatic.[13]

Raymond Queneau, one of OuLiPo's founders and himself a writer of great inventiveness, is the author of a remarkable work called *Exercises in Style* (Queneau, 1958). Here, the same simple story is retold 99 times—the reader is never bored—the stylistic variations including various poetic and rhetorical forms, character sketches, spoonerisms, permutations of word groups, and much more. The work was stimulated by a performance of Bach's "Art of the Fugue" (whose variations on a rather slight theme proliferate, for Queneau, "almost to infinity") and was composed as a literary analog of Bach's creation.

Other notable Oulipian works[14] include "The Dialect of the Tribe" (Mathews, 1980), a very funny short story about efforts to translate Pagolak, a language whose meaning cannot be separated from its palindromic structure; Calvino's *If on a Winter's Night a Traveler . . .* (Calvino, 1981), which includes an episode suggesting, however playfully, that the meaning of a text is sufficiently constrained by its vocabulary as to be recoverable from a mere listing of word frequencies; and Perec's masterwork *Life a User's Manual* (Perec, 1987), constrained in many ways and on several levels, filled with wordplay of all sorts, and possibly made up almost entirely, according to Perec's translator (Bellos, 1987), of materials drawn so artfully from other works of literature as to conceal their provenance.

XII. CONCLUDING REMARKS

> The distinguishing characteristic of reality is that it is played.
>
> *Jacques Ehrmann (quoted by S. Stewart, 1980)*

> Combinatory play seems to be the essential feature in productive thought.
>
> *Albert Einstein (quoted by Hadamard, 1945)*

> Puzzling to us is an art.
>
> *Archimedes (pen name of a past President of the National Puzzlers' League)*

Puzzlers come to know the structure of words more intimately than most. What such intimacy purchases is the ability to take advantage of constraints that are always there, but that one must come to appreciate. It allows one to find what others might miss, to be confident that what one has not turned up cannot be there, and to assess the promise of a line of attack. Perhaps above all, intimacy with one's

[13] Gilbert Adair has translated Perec's novel into *e*-less English, with the title *A void* (Perec, 1994). It is easier to compose novel-length lipograms in English than in French, but Adair's challenge was to do so while capturing the sense and spirit of Perec's work.

[14] Examples of literary wordplay are hardly confined to the members of the OuLiPo, and may be found in writers as diverse as Lewis Carroll, James Joyce, Vladimir Nabokov, Jorge Luis Borges, and James Thurber.

materials opens up possibilities for discoveries—most of our interesting ones, I think—that can only result from play. Children learn that play provides paths to discovery, and that play is always more fun and more rewarding when one deeply understands the game one is playing. Whenever anyone sets out to make something interesting, one is obliged to play. As Stephen Sondheim has said, "all art—symphonies, architecture, novels—it's all puzzles. The fitting together of notes, the fitting together of words have by their very nature a puzzle aspect. It's the creation of form out of chaos" (Schiff, 1993, p. 76).

References

Albert, E. (1992). Crosswords by computer. *Games, 16*(1), 10–13.

Augarde, T. (1984). *The Oxford guide to word games.* Oxford: Oxford University Press.

Bellos, D. (1987). Literary quotations in Perec's *La Vie Mode d'Emploi. French Studies, 41,* 181–194.

Bronowski, J. (1978). The power of artifacts. In Piero E. Ariotti (Ed.) in collaboration with R. Bronowski, *The visionary eye: Essays in the arts, literature, and science* (pp. 59–74). Cambridge, MA: MIT Press.

Calvino, I. (1977). *The castle of crossed destinies.* London: Secker & Warburg.

Calvino, I. (1981). *If on a winter's night a traveler . . .* New York: Harcourt Brace Jovanovich.

Calvino, I. (1986). Cybernetics and ghosts. In *The uses of literature* (pp. 3–27). San Diego: Harcourt Brace Jovanovich.

Carroll, L. (1960). *The annotated Alice.* (Introduction and Notes by M. Gardner). New York: Clarkson N. Potter.

Chambers anagrams (1985). Edinburgh: W & R Chambers Ltd.

Cook, N. (1990). *Music, imagination, and culture.* Oxford: Clarendon Press.

Gleick, J. (1983, August 21). Exploring the labyrinth of the mind. *New York Times Magazine,* p. 23 ff.

Gombrich, E. (1960). *Art and illusion.* Princeton: Princeton University Press.

Grossman, J. (1992). Look, ma, no glasses! *Games,* 16(2), 12–14.

Hadamard, J. (1945). *The psychology of invention in the mathematical field.* Princeton: Princeton University Press.

Hatherly, A. (1986). Reading paths in Spanish and Portuguese Baroque labyrinths. *Visible Language,* 20, 52–64.

Hearst, E. (1984). Absence as information: Some implications for learning, performance, and representational processes. In H. L. Roitblat, T. G. Bever, & H. S. Terrace (Eds.), *Animal cognition* (pp. 311–332). Hillsdale, NJ: Erlbaum.

Hofstadter, D. R. (1983). The architecture of Jumbo. *Proceedings of the International Machine Learning Workshop.* (pp. 161–170) Monticello IL.

James, W. (1890/1983). *The principles of psychology.* Cambridge, MA: Harvard University Press.

Kahn, D. (1967). *The codebreakers: The story of secret writing.* New York: Macmillan.

Kanigel, R. (1991). *The man who knew infinity.* New York: Maxwell Macmillan International.

Kubovy, M. (1986). *The psychology of perspective and Renaissance art.* Cambridge, UK: Cambridge University Press.

Kushner, L. (1975). *The book of letters.* New York: Harper & Row.

Mathews, H. (1976). Oulipo. *Word Ways, 9*(2), 67–74.

Mathews, H. (1980). The dialect of the tribe. In *Country cooking & other stories* (pp. 39–51). Providence: Burning Deck.

Motte, W. (1986). *Oulipo: A primer of potential literature.* Lincoln: University of Nebraska Press.

Nabokov, V. (1980). *Lectures on literature.* New York: Harcourt Brace Jovanovich.

Nickerson, R. S. (1977). Crossword puzzles and lexical memory. In S. Dornic (Ed.), *Attention and performance, VI* (pp. 699–718). Hillsdale, NJ: Erlbaum.
Perec, G. (1969). *La disparition*. Paris: Denoël.
Perec, G. (1987). *Life a user's manual*. Boston: Godine.
Perec, G. (1994). *A void*. London: Harvill.
Philip, C. O. P. (1985). *Robert Fulton: A biography*. New York: Franklin Watts.
Poincare, H. (1956). Mathematical creation. In J. R. Newman (Ed.), *The world of mathematics* (pp. 2041–2050). New York: Simon and Schuster. (Original work published in 1904)
Polya, G. (1954). *Mathematics and plausible reasoning*. Princeton: Princeton University Press.
Queneau, R. (1958). *Exercises in style*. (Barbara Wright, Trans.). New York: New Directions.
Reagle, M. (1982). Short months. In M. Farrar (Ed.), *Puzzling through 1982*. New York: Simon and Schuster. Unpaginated.
Richardson, J. T. E., & Johnson, P. B. (1980). Models of anagram solution. *Bulletin of the Psychonomic Society, 16*(4), 247–250.
Rooten, van, L. d'A. (1967). *Mots d'heures: gousses, rames*. New York: Grossman.
Schiff, S. (1993). Deconstructing Sondheim. *New Yorker*, 49(3), 76–87.
Scholem, G. (1941). *Major trends in Jewish mysticism*. Jerusalem: Schocken.
Schulman, A. (1983, November). *Cryptic crosswords, mental repunctuation, and memory*. Paper presented at the meeting of the Psychonomic Society, San Diego.
Schulman, A. (1989, November). *On being sure there's nothing there*. Paper presented at the meeting of the Psychonomic Society, Atlanta.
Schulman, A., Jonides, J., & Cohen, D. (1990, November). *Anagram solution by experts*. Paper presented at the meeting of the Psychonomic Society, New Orleans.
Simon, H. (1993). What is an "explanation" of behavior? *Psychological Science, 3,* 150–161.
Steiner, G. (1975). *After Babel: Aspects of language and translation*. London: Oxford University Press.
Stewart, H. M. (1971). A crossword hymn to Mut. *Journal of Egyptian Archaeology, 57,* 87–104.
Stewart, S. (1980). *Nonsense: Aspects of intertextuality in folklore and literature*. Baltimore: Johns Hopkins University Press.
Wood, C. (1936). *The complete rhyming dictionary and poet's craft book*. Garden City: Doubleday.

APPENDIX A: PUZZLES IN VERSE

(Note: these examples are representative, not exhaustive; puzzle types described in the main text are not repeated. The reader interested in a more general survey of word games and word puzzles should consult Augarde (1984), who also provides a selective bibliography. Puzzles like these are termed *penetralia* in *The Enigma,* the monthly publication of the National Puzzlers' League.[15] It is no accident that this word means "the innermost or most private parts of any thing or place, especially of a temple or palace; hidden things or secrets." Readers for whom these puzzles are new should note what is done by the maker to conceal, and what must be done by the solver to reveal, a solution).

1. Rebus: A word or phrase is represented by letters, numbers, or symbols. The word "abalone"—read as "a B alone"—might be represented by:

B

A rebus-type flat may be solved through the verse, by deciphering the cryptic representation (the rebus proper, or the "rubric"), or both.

[15] With whose permission these examples are reproduced.

REBUS (12)
$\dot{T} = T$
My holiday plans have been ruined this year;
I was going to go to the south.
But my ANSWER's bill took all of my savings away;
Why should he, then, look down in the mouth?

Solution: Periodontist (i.e., PERIOD ON T "IS" T). Most rubrics have "readings" that are much harder to arrive at than this.
2. Charade: A longer word is broken down into two or more shorter words. Example: ONE = scar, TWO = city, and TOTAL = scarcity.

CHARADE (10)
My migraine was pounding; I needed some rest.
"There's WHOLE," said my FIRST, "in the medicine chest."
The SECOND on all of the labels looked blurred.
I took something at random and promptly got THIRD.

Solution: FIRST = pa, SECOND = ink, THIRD = iller; WHOLE = painkiller. By convention, only the length of the whole answer word is provided; solvers must discover the length of the words that constitute the charade. The enumeration—'10' in this case—is a redundant clue for solution. An experienced solver can often do without it.
3. Beheadment: A word becomes a new one when its first letter is removed. Example: ONE = factor, TWO = actor.

BEHEADMENT (6, 5)
I wonder if this SHORTER stole,
Though slightly tattered, still is WHOLE.

Solution: WHOLE = usable, SHORTER = sable. Beheadments that change the longer word's pronunciation are harder to solve than those that do not.
4. Deletion: A word becomes a new one when an interior letter is removed. Example: ONE = simile, TWO = smile.

DELETION (5, 4)
One year fat, one year lean.
Never anywhere between.
Diet changes every day;
Either ONE or TWO, I say.

Solution: ONE = feast, TWO = fast.
5. Word Deletion: A word or phrase is deleted from a longer one, leaving a third. For example, TOTAL = performance, ONE = man, TWO = perforce. The length is given of only the longest word or phrase.

WORD DELETION (10)
Since math is not our country's forte,
We'll soon be short of engineers
Unless it's made a high-school sport
With pom-pom girls to lead the cheers:
"Read that journal! Comprehend!
Take that ALL from the minuend!
Integrate from max to min!
OUT that angle! IN, IN, IN!"

Solution: ALL = subtrahend, IN = rah, OUT = subtend. Many word puzzles—this is an example—are more entertaining than they are difficult. The composer of this puzzle in verse was obviously delighted to discover that "subtend" was on the "outskirts" of "subtrahend," another mathematical term, leaving only "rah" in the middle. The idea for the cheerleading verse must have immediately suggested itself.
6. Spoonergram: A phrase becomes another phrase when the initial sounds in the words are swapped. Example: ONE = Morse code, TWO = course mowed.

SPOONERGRAM (6 3, 5 6)
No farce is Wagner's *Ring;* it is a ONE
Of music, drama. Fearful deeds are done.
The hero, Siegfried, battles mighty odds,
And TWO are played on dwarfs and men by gods.

Solution: ONE = tragic mix, TWO = magic tricks. The mine of spoonerisms ("our queer old dean" for "our dear old queen") was long ago very nearly exhausted. Spoonergrams and its variants make it profitable to reenter the mine; the transformation of "Morse code" into "course mowed" is not an amusing one, but it might still be appreciated if the phrases were hard to insert at the place markers within the puzzle's verse.
7. Acrostical Enigma: a relatively hard puzzle to master. Its solution is a word or phrase, divided into chunks of two or more letters each. Each chunk is clued in an unusual way. The chunk plus the first two or more letters of each couplet form a word or phrase (called the *part word*) that is clued somewhere within the couplet itself. A final couplet clues the entire solution.

ACROSTICAL ENIGMA (10)
A. Relaxing on Saturday morning I am;
I smell something burning, but who gives a damn?
B. Gargantuan flames billow smoke in the air;
Some building is blazing, but why should I care?
C. I only want something absorbing to read.
A Harlequin tearjerker: just what I need!

D. Red engines zip by with complete audibility,
Disturbing, with bells, my beloved tranquility.

It's getting quite warm now. What can be the matter?
Let's turn up the fan and ignore all the clatter.

The part words are AFIre, CIgar, ONion, and ADOred, clued respectively by *burning, smoke, tearjerker,* and *beloved;* note the possibilities for polysemous play. The capitalized chunks, strung together, spell the solution word AFICIONADO, clued in the final couplet by *fan.* Note that only the length of the solution word is divulged, the part words' length being left to the solver to discover.

8. Rebade: A hybrid of the rebus and alternade (in which a word or phrase is divided into two or more others by taking alternate letters in order (e.g., *Algeria* divides into *Agra* and *lei*). When the rubric is deciphered and its letters entered in columns in a rectangular, or nearly rectangular, grid, the rows display the words clued by the verse. Making a rebade thus calls for the combined skills of the rebus maker and the crossword constructor. Note that the words in the rebus phrase, as the example shows, may be continued from one column to the next.

REBADE (4, 4, 3)
6
My shoulders THREE; my face grows ashen:
TWO mini—enter ONE! says fashion.

Solution: ONE = midi, TWO = exit, THREE = sag. When these answer words are written as

M I D I
E X I T
S A G

the columns, from left to right, spell out the cryptic reading of the rubric: "ME: SIX, A DIGIT."

9. False Derivative: A word or phrase becomes another when some grammatical change is inappropriately applied. For example, a false plural: ONE = inter, TWO = interim (by analogy with the plurals *seraphim* and *cherubim*).

FALSE FEMININE (6, 8)
To FINAL her ego, he made up his mind
To PRIMAL her up—ah, a clever man, he.
For flattery gets you, not nowhere, you'll find,
But right to the place you're most eager to be.

Solution: PRIMAL = butter, FINAL = buttress (analogous to *waiter* and *waitress*).

APPENDIX B: AN EXEMPLARY PUZZLE-FILLED GAME

In "Where's Gorby", a game devised when Gorbachev's political failure was in question, players were told that the Soviet leader has been abducted, and that they would need to find out where he'd been taken and by whom. Then they were given a list of the states of the former U.S.S.R.—Ukraine, Georgia, Armenia, and the rest—and were told that Gorby was being held hostage in the capital city of one of them. Atlases were supplied for later reference. Finally, a clue naming the kidnappers, but in the disguise of a cyclic keyword cipher, was provided. The keyword's identity, essential to encipherment, could be determined only by solving the sequence of puzzles described below.

1. A word-search puzzle was first. The words to be located were not merely listed, however, but rather were clued by single letters: Q for TIP, V for NECK, X for RAY, and so on. When all of the clued words had been found in the grid, the unused letters spelled out the first of five answer words.

2. Next was a "pattern matching" puzzle, in which the solver tried to discover the attribute shared by each of the nine sets of four words and phrases and then to identify another item (from four possibilities offered) that also shared that attribute. For example, SEA, CROSS, TAPE, and ALERT had to be associated with *red* so that CARPET (rather than GAME, ENVELOPE, or TROUSER) might be chosen as the correct answer; similarly, MERE, CHAT, PAYS, and MAIN had to be seen as words in French, as well as in English, so that LOUD, OVEN, and COWS could be rejected and PAIN selected. In this puzzle a letter happened to mark each alternative to be considered, and the solver had to rearrange the letters of the nine correct answers so that they spelled out the second word necessary to track Gorby down.

3. The third puzzle presented a set of nine word blocks. Each of these was a 5 × 5 grid whose cells contained one or two letters. By choosing a cell from each column, and proceeding from left to right, the solver might find five related words that were hidden in the grid. The nature of the relation varied from block to block, and had to be discovered from the evidence of the letters themselves. For example, one may notice the Q in the second column of the illustrated block; realize that it must be preceded by the ES in the first column; note that the U, I, and RE in columns 3–5 allows the completion of ESQUIRE; and surmise, correctly, that the other words in this grid are also names of magazines. Finally, the "?" in the central square would have to be filled with the single letter needed to complete one of the to-be-discovered words. From the set of nine such missing letters the solver could find the third word that the ultimate solution would require.

P	AC	U	U	LE
ES	L	G	NT	RE
V	E	?	R	IC
AT	O	WO	I	LD
M	Q	O	P	E

4. A fourth puzzle provided cryptic clues for five 5-letter words; these had to be entered in a word square so that the same words could be read across and down. When this was done, the letters in the square's negative diagonal would spell out the fourth word essential to the case's ultimate solution.

5. The fifth puzzle was a cryptogram that carried instructions for finding the last answer word. But this was a cryptogram with a twist: some of its letters were not enciphered at all, but represented themselves. Solvers thus had to recognize the absence of the usual cryptographic disguise. If they did, they cracked the cipher, and learned that a rearrangement of its unenciphered letters would produce the fifth answer word they needed.

6. Having learned the identity of the five answer words. most solvers proceeded with confidence to the Grand Finale. This was a double-crostic puzzle in which the five answer words served as definitions of words whose letters provided the pool for the message that would reveal Gorby's whereabouts. But the message contained a surprise, and many solvers were nonplussed by it: it said that Gorby was being held south of Berlin and west of Zagreb. A check of the atlases made this seem impossible, given what nearly everyone had assumed to be true. Sooner or later, solvers came to see that trick: Gorby indeed was being held in one of the states originally listed, but in the American Georgia, not the Soviet one. The capital in which he was held hostage was, therefore, Atlanta. The game was now nearly over. To break the original cipher, the keyword ATLANTA was written repeatedly above the ciphertext, thus:

A T L A N T A A T L A N T A A T L A
S N S B A I Z B U Z Z O U L O G B X

Next, the numerical values of pairs of letters, one from the keyboard, the other from the cipher text, were added: A = 1, S = 19, so that A + S = 20 = T, and so on. (This is a cyclic cipher, and so a sum greater than 26, e.g., 28, would indicate 28 − 26 = 2 = B). Gorby's captors, as the decipherment finally revealed, were the Coca-Cola Company.

Joshua M. Kosman invented "Where's Gorby," and is the creator of other games and puzzles of comparable wit and imagination.

Sensory Evaluation

CHAPTER 8

Flavor

Harry T. Lawless

I. FLAVOR MODALITIES

What are the functions of the flavor senses? It seems obvious that they contribute to our enjoyment of foods. However, the deeper question persists of why chemical sensitivity would be so highly developed among humans and other species. Other motivational factors would surely be sufficient to induce appropriate foraging and food consumption, factors such as reinforcing postingestional effects—reduced hunger, satiety, bodily feelings of well-being, correction of ills based on nutritional deficiency. Perhaps time is the key. If these senses provide a gatekeeping function to the alimentary tract, they provide information well before postingestional effects can come into play. Discoveries such as the cephalic-phase insulin response suggest that sensory effects are not only pleasurable, driving acceptance or rejection of appropriate foods, but that chemical sensations also prepare the alimentary tract for what is to come (Nicolaidis, 1977). Although the ecological pressures that gave rise to chemical sensitivity are certainly very old and thus a matter of some speculation, we can still marvel at the complexity of the neural systems that give rise to our overall impression of flavor.

Dusenberry (1992) classified chemical interactions according to their informational functions. He reminds us that chemical interactions among organisms serve several ecological functions. *Semiochemicals* may facilitate communication within a

Cognitive Ecology

species in order to prepare animals for mating, as in the function of pheromones in many species. Among species, he classified food scents as *kairomones,* of primary benefit to the receiver of the information, whereas *allomones,* such as lures for prey or repellents to predators, are of benefit to the sender, and *synomones,* such as floral scents, may be of benefit to both sender and receiver. Although a large part of the chemosensory literature is concerned with pheromonal interactions, primarily among insects, the flavor senses and food-related functions are a primary concern for chemosensory scientists focused on human taste and smell, because pheromonal behaviors in humans are considered to be vestigial at best.

The sensations giving rise to flavor involve several anatomically distinct sensory systems. Taste is mediated by branches of the seventh, ninth, and tenth cranial nerves that innervate papillae, anatomically differentiated structures on the tongue that contain taste buds. Taste buds containing specialized receptor cells exist on the tongue, soft palate, and root of the tongue extending partially down the throat. There are also trigeminal endings, afferents from the fifth cranial nerves, that are sensitive to heat, cold, touch, pressure, pain, and chemical irritation throughout the oral cavity. Olfaction is mediated by the receptors in a small patch of the upper nasal epithelium, where its axons project into the olfactory bulbs. The nasal passages are likewise innervated by an extensive afferent system from the trigeminal nerves, providing additional sensitivity to a class of irritative chemical stimuli. One theme of this chapter is that flavorous stimuli, most of which are mixtures in nature, stimulate many or all of these anatomical subsystems, and it is the integrated impression from this simultaneous stimulation that drives our appetitive responses.

There has been rapid growth in physiological and psychophysical research on taste and smell in the last few decades, to the point where it is no longer appropriate to say that "little is known about the senses of taste and smell," a common utterance in sensory psychology lectures 25 years ago. However, a gap still exists between knowledge from the laboratory and information about chemical sensations in functioning organisms in natural situations. Although the culinary arts are highly developed, there is little interplay between practical flavor engineers such as chefs and academic chemosensory scientists. Also, much applied research on flavors, tastes, and smells goes on within the proprietary confines of the highly competitive flavor and fragrance industry, so there are probably many practical effects that remain undocumented in the scientific literature. For this reason, many of the illustrations used below rely on anecdotal description or firsthand observations within the author's experience.

This chapter is divided into seven sections. Modalities subserving flavor perception are discussed, and attention is drawn to some less than obvious routes of input, including retronasal smell and trigeminal flavors. The various approaches to studying flavor perception including psychophysical and industrial techniques are discussed. Because many readers from psychological backgrounds may be unfamiliar with the battery of test methods used in applied sensory evaluation, they are described in some detail along with alternative approaches from commodity grading and flavor chemistry. In all cases, I discuss some of the limitations of each approach, not so much as criticism, but to suggest room for improvements in the methods and paradigms. Finally, some current issues in taste, smell, and trigeminal chemical sensation are discussed from an ecological perspective. The chapter concludes with some thoughts on cuisine and culture and flavor as art.

A. Cross-Modal Integration

Figure 1 shows the anatomical subsystems mediating chemical sensations in the nose and mouth. One striking characteristic of flavor perception is that information from all these modalities may be synthesized into a unitary experience with a single hedonic response, appetitive or aversive. Consumers taste food and decide immediately whether it is likable or something to reject. As gatekeepers of the alimentary tract, the nose and mouth function together, and do so with speed and seemingly little cognitive effort. This synthesis of sensory inputs is not limited to the chemical senses. Foods are also appreciated for their appearance, visual texture, and color, as well as tactile sensations during active manipulation and destruction in the oral cavity. Cross-modal information is integrated. Influences of color on taste and smell and of texture on taste are common in everyday eating. Food perception is an active process—it cannot be effectively understood when stimuli are imposed upon a passive organism. Humans are by nature foragers and sensation seekers. This provides an important place for studies of flavor perception in intact and behaving organisms.

In contrast to this synthetic sort of perception, it is also possible to focus

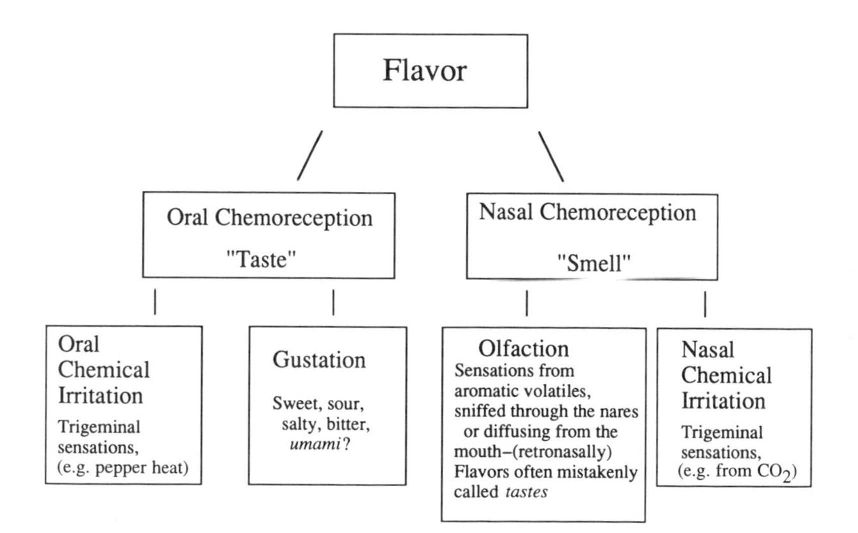

FIGURE 1 Schematic of the anatomical systems mediating perception of flavors.

attention on the individual systems producing chemical sensations. Trained tasters on an industrial descriptive analysis panel are taught to analyze taste sensations from the mouth as distinct from volatile perceptions mediated by the olfactory organs. Sapid substances in the oral cavity give rise to classical taste sensations such as sweet, sour, salty, bitter, and other impressions such as *umami,* an important taste category in Japan and other countries, roughly equivalent to a savory or meaty flavor arising from stimuli such as monosodium glutamate. In contrast to this relatively limited number of major categories for taste experience, there are a much larger number of distinct experiences from volatile chemicals stimulating smell. Sensations as diverse as the smells of citrus, cut grass, mint, various flowers, woods, fruits, herbs, spices, burnt aromas, sulfidic aromas, ethereal, fatty and sweet (like vanilla) smells all seem so distinct as categories as to be unrelated. Thus the qualitative range of olfaction seems quite wide compared to taste. It could be argued that smells provide the majority of the diversity in our flavor experiences.

B. Trigeminal Sensitivity

In addition to these two classical systems for chemical sensitivity, there is a more generalized chemical sensitivity in the nose and mouth, and over the whole body. Mucous membranes such as the anus are also sensitive—an old Hungarian saying has it that the good paprika burns twice, and similar sentiments are known in other countries with hot, spicy cuisines. In the nose and mouth, this more general chemical irritability is primarily mediated by the trigeminal nerves (Fig. 1). These systems have been recently described by the term "chemesthesis" in an analogy to somesthesis (Green & Lawless, 1991). A variety of everyday flavor experiences arise from trigeminal stimulation: the fizzy tingle from CO_2 in soda, the burn from hot peppers, black pepper, and spices such as ginger and cumin, the nasal pungency of mustard, horseradish, the bite from raw onions and garlic, not to mention their lacrymatory effects, to name a few. This important chemical sense is easily overlooked in considerations of taste and smell, because it has received less experimental study than the classical taste and smell modalities. However, a simple demonstration, such as comparing the sensations from warm, flat (decarbonated) soda to warm, fizzy soda should easily convince one of the importance of trigeminal sensations.

Of course, this set of nerves also mediates tactile, thermal, and pain sensations, so the distinction between a chemical sense and a tactile sense becomes blurred somewhat. This blurring is perhaps worst in the sensations of astringency. Tannins in foods are chemical stimuli, and yet the astringent sensations they produce seem largely tactile. They make the mouth feel rough and dry, and cause a drawing, puckery, or tightening sensation in the cheeks and muscles of the face (Bate Smith, 1954). Although scientific analysis would categorize astringency as a group of chemically induced oral tactile sensations, most wine tasters would say that astringency is an important component of wine "flavor." This highlights, once again, the integrative nature of flavor in combining inputs from multiple modalities.

The mechanisms giving rise to these sensations are poorly understood, but one long-standing and popular theory has it that tannins bind to salivary proteins and mucopolysaccharides (the slippery constituents of saliva), causing them to aggregate or precipitate, thus robbing saliva of its ability to coat and lubricate oral tissues. One feels this result as rough and dry sensation on oral tissues, even when there is fluid in the mouth. Note that "roughness" and "dryness" are difficult to perceive unless a person moves the tongue against other oral tissues (which we do all the time when eating). An active perceiver is required for astringent perception. Astringency is well suited to study with active human observers and less well suited to study with immobilized animals in electrophysiological preparations.

The importance of chemesthesis can be defended on two grounds—one anatomical and the other economic. The sheer numbers of trigeminal fibers, relative to the other chemical sense organs are impressive. One study found three times as many trigeminal fibers in the fungiform papillae of the rat as facial nerve fibers innervating taste buds (Farbman & Hellekant, 1978). Students of taste anatomy normally consider a fungiform papilla as an anatomical structure that holds taste buds and thus provides a taste organ, or more accurately, it is thought of as the organ for the perception of chili pepper burn! Even the taste bud itself seems organized to provide trigeminal access to the oral milieu. Trigeminal fibers ascend around the taste bud itself, forming a chalicelike structure (Whitehead, Beeman, & Kinsella, 1985). The trigeminal endings seem to use the specialized structure of the taste bud to find a channel to the external environment. This speculation is consistent with the observation of high responsiveness to pepper chemicals in areas such as the top of the tongue that are rich in fungiform papillae (Lawless & Stevens, 1988).

The economic impact of trigeminal stimulation on the food and flavor industry is easy to underestimate. If CO_2 is considered a trigeminal flavor, the carbonated beverage business—soda, beer, sparkling wines, etc.—amounts to several billion dollars in sales of a trigeminal flavor in this country alone. Putting aside CO_2, we can ask about the economic impact of individual spices or their use in various products. The pepper business amounts to several hundred million dollars annually. Furthermore, so-called ethnic foods are experiencing a period of rapid growth due to a continuing influx of immigrants from cultures with hot spicy cuisines, and a growing trend toward less neophobic and more adventurous dining on the part of many Americans. As evidence of this, the sales of salsa surpassed the sales of ketchup for the first time in 1992.

C. Retronasal Smell

The diversity of experiences from smells is much wider than the qualitative variation in taste. This may seem nonsensical to the unitiated consumer, who defines much of the world of smells as tastes. This opinion arises because volatile chemicals present in the oral cavity can diffuse or be carried into the nose in a backward direction, from the mouth up into the pharynx and then into the nose. In spite of what seems to be a roundabout way of smelling, this is a highly effective route. However, the sensations are poorly localized, and because there are many other sensations ongoing in the mouth, such as taste and tactile experiences while food is being sipped, chewed, or swallowed, the olfactory experience is misreferred or "captured" by the oral sensations and called a taste (Murphy & Cain, 1980). Many so-called flavor enhancers such as ethyl maltol, a reputed sweetness enhancer, probably work this way (Bingham, Birch, de Graaf, Behan, & Perring, 1990).

A simple demonstration makes the retronasal route clear to students—take a sip of a simple fruity beverage while holding the nose pinched shut and the true taste sensations can be sensed without any of the olfactory aromatics. (Beware—breathing is tricky and this takes a little practice!) Next, release the fingers from pinching the nose and breath out. The olfactory sensations will appear as a delayed development over a second or two. This can also be used to separate olfactory and taste sensations in analytical tasting situations when the origin of the sensation is unclear. For example, sometimes the fruity and aromatic volatiles in a wine can be confused with the true sweet tastes from residual sugars—a wine may seem sweeter than it really is just because it has high volatile impact. However, if you pinch your nose shut and stick out the front of your tongue into the wine, you can test for any pure taste stimulation of sweetness. (This trick is best explained to bystanders who may otherwise decide that your wine-tasting techniques have reached new levels of eccentricity.)

Note that retronasal effects highlight the dual function of the sense of smell (P. Rozin, 1982). Olfaction is unique among the senses in being both a proximal and distal sense—one that is capable of responding to sources of stimulation outside the body at a distance (e.g., a fire in the kitchen, and to sources of stimulation at or even inside the body—wine bouquet sniffed from a glass or piney terpenes in a sip of retsina). Within the realm of flavors, one can appreciate smells sniffed and "orthonasally" presented and those that are sipped and retronasally presented. A poorly understood issue is how the aromatic character of a food may change via these two routes, because the physical structure and chemistry of a food will change when it enters the mouth, is mixed with saliva, undergoes temperature changes, and is chewed and swallowed.

D. Limitations of Psychophysical Study

Much systematic work has appeared in the psychological literature in the last 30 years specifying psychophysical relationships in taste and smell. Starting with the simplest relationships, psychologists have studied the limits of detection and recognition thresholds as estimates of the absolute sensitivity of the sensory systems. The second major area has been intensity relationships—how intensity grows with concentration, both in simple solutions and in mixtures. Finally, some hedonic relationships have been assessed, although the human literature on affect and intake is much smaller than the amount of research devoted to understanding rodent feeding.

One can argue that psychophysical studies provide a set of operating characteristics that quantitatively describe the function of a sensory system. Lately, a paradigm of "integration psychophysics" has sprung up that purports to have advantages over simpler descriptive techniques (McBride & Anderson, 1990), and has set some very strict criteria for how information about sensory response should be collected. However, the psychophysical approach can be criticized on a number of grounds, mainly for being limited, incomplete, and of questionable relevance. Intensity-oriented psychophysics provide information on how sensory response first appears and grows as a function of concentration. Is this important information for a foraging organism or one that makes judgments based on chemical information? If milk has gone rancid or bitter as a function of poor shelf life or microbial infestation, does one estimate the degree of bitterness before rejecting it? When one smells a gas leak in the house, does she or he calculate how strong it is before deciding there's a problem? Much more important to the organism is the *qualitative* change in chemical stimulation that has occurred—presence versus absence of certain problem tastes or smells that suggest certain actions. The gas leak itself is important (compared to the scent of boiling cabbage) not the strength of the sulfur aroma.

Early studies of information transmission in olfaction showed that intensity discrimination by the nose was quite poor, compared with the range of qualities that could be perceived (Desor & Beauchamp, 1974; Engen & Pfaffmann, 1959) This is not to say that intensity judgment may be totally absent from applied flavor work. A cook who oversalts a soup is still a poor cook. However, much more culinary energy is devoted to the choice of herbs, spices, and flavors to be used versus not used than in setting appropriate levels. Some recipes are still measured in vague units like "pinches" or "to taste."

A second area of criticism is simply that of using model systems. Although isolation of single variables provides necessary experimental control for scientific study, validity regarding the real world is lost. Psychophysical studies of mixtures are in part justified on the basis of attempting to bridge the gap between single solutions and actual foods. Perhaps the pinnacle of psychophysical irrelevance was achieved in asking for affective reactions to aqueous solutions of single-taste stimuli (e.g., Lawless, 1979a). While I was a visitor to the food industry, my supervisor once remarked, when given aqueous sucrose to rate for its pleasantness, that hedonic ratings were a silly question. The test items were sugar in water. It was not whether she liked or disliked them—they just were not food, so what was the point?

II. HOW FLAVOR IS MEASURED IN FOODS—APPLIED SENSORY EVALUATION

A. Historical Evolution of the Techniques

In both agricultural and industrial situations, the flavor senses are put to another practical use: the sensory assessment of foods and consumer products and the prediction or measurement of consumer acceptability of new products. There is a need to measure perceived product attributes before the costs associated with new product manufacturing, marketing, and sales are incurred. This has led to a battery of applied techniques for sensory evaluation used in the food and consumer products' industries. In food processing and manufacturing, flavor variation occurs as a function of ingredient changes, alterations in processing or manufacturing practices, as a function of shelf-life or time since production, and from interactions with packaging. The sensory quality of a product's flavor is obviously important from a business perspective in maintaining consumer satisfaction. Assessing perceived changes in a quantitative, scientifically valid manner is the goal of the discipline of sensory evaluation.

Historically, the discipline of sensory evaluation can be viewed as having three distinct phases. The first phase was evaluation by a single expert judge, who graded the quality of a product, its safety or integrity, and/or its likely appeal to consumers. Examples of these experts included the brew masters, distillers, and wine makers whose job it was (and often still is) to monitor production phases of their products and recommend appropriate changes in manufacturing in order to maintain expected quality. This tradition continues today, mostly in small industries in which a single individual is called on to be the guru of sensory appeal. It also persists in the traditions of commodity judging, as in the Collegiate Dairy Products Judging Contest, in which panels of college students attempt to duplicate the sensory decisions of experienced dairy judges, and receive awards, publicity, and recognition for doing so (Bodyfelt, Tobias, & Trout, 1988). Presumably these abilities are useful in some future careers in the dairy products industry, especially in situations of quality control or production monitoring. Quality grading is also carried out by government inspectors of commodities such as fish, in which not only sensory quality, but food safety may be an issue. The use of the senses by a single highly experienced individual is common in other scenarios as well. Examples include the sommelier who makes decisions about choices from the wine cellar based on sensory appeal, the chef who creates a new entree, and the industrial flavorist who duplicates a natural flavor with some artificial combination for a specific food use (e.g., a tomato flavor for a powdered Bloody Mary mix).

The second phase of sensory evaluation was the replacement of the individual expert judge with a panel or jury. This approach recognizes that a group of people are sometimes more reliable and more representative than the decisions of a single individual. The individual may have a bad day, be sick with a cold or allergies that prevent adequate tasting, or may be unavailable at the time testing is needed. Furthermore, because individuals differ in the mosaic of their taste and smell acuities to different flavor chemicals, it would be rare to find someone who was highly sensitive to all flavors, off flavors, and taints. Finally, in some expert judging, there is always the liability that personal preference will creep in and cloud objective professional judgment. The use of a panel helps avoid that problem. In this style of sensory evaluation, discussion among judges is the rule, much like a legal jury. The goal is to arrive at a consensus, usually reflecting a quality score for the product. This approach is still common in some wine judgings, typically in the state fair scenario where medals are awarded for quality. It is also seen in cheese-judging "clinics" where a panel of experts decides on quality scores and the defects of a set of samples, and then participants try to match the consensus of the experts.

In the transition to modern sensory evaluation, a panel technique called the Flavor Profile Method was developed in order to specify the sensory character of a product by a trained and objective panel of sensory judges (Caul, 1957). The Flavor Profile is not concerned with overall quality or consumer appeal, but rather providing a sensory specification of the perceived intensity of all flavor attributes using simple category scales. It thus has a psychophysical basis. The panelists make independent evaluations, but then enter a group discussion to arrive at a consensus profile. Averaging is now allowed—if some panelists say a flavor is weak and some say it is strong, this is not scored as moderate. The panel leader holds discussion until agreement is reached. This technique is still in use in some circles today, having provided a useful tool for flavor description. However, the method was criticized even from its earliest stages by psychologists because of the possible influence of social factors upon consensus opinion. Practitioners of the art rejoin that panelist selection avoids overly dominant or submissive individuals, but whether social interaction can ever be removed from this method is a matter of some conjecture.

The third historical phase of sensory evaluation draws much of its rationale from the behavioral sciences and statistics. Techniques are seen as falling into two categories, analytically oriented tests of sensory difference and quantitative description and subjectively oriented tests of consumer appeal. In the first category, laboratory control, reliability, and an analytical orientation of the judges are stressed, while the testing for consumer appeal emphasizes validity in terms of generalizing from the test results to the real world of sensory factors influencing purchase decisions. These techniques are outlined in the next section. The psychological underpinnings of this approach are clear. Regarding the analytic versus hedonic dichotomy, observers may view flavor either in an analytical or an integrative mode, focusing either on individual parts of a flavor complex, or react to it affectively and as an integrated pattern. In either case, the statistical orientation of behavioral data collection demands that observations be independent. Participants may not confer during the test—judgments are rendered via questionnaires or psychophysical scales. To this end, sensory evaluation practitioners developed "booths" or other isolated test chambers in order to avoid distractions and to minimize the interactions of panelists during an evaluation session. This may seem odd to the casual observer in that much food is consumed by humans in social situations. Finally, data are treated with appropriate hypothesis tests and statistical summaries. For example, techniques such as Quantitative Descriptive Analysis (QDA) (Stone, Sidel, Oliver, Woolsey, & Singleton, 1974) commonly use factorial experimental designs (panelists, samples, and replicates would be typical favors), and analysis of variance (ANOVA) is the bread-and-butter of hypothesis testing.

The following section describes common techniques for the current behaviorally oriented version of sensory evaluation. This section is written at the level of an overview or primer. For those who are interested, texts on sensory evaluation include *Sensory Evaluation Practices* (Stone & Sidel, 1985) and *Sensory Evaluation Techniques* (Meilgaard, Civille, & Carr, 1987). These books are written as guides for industrial practitioners. The following discussion may be useful for the curious scientist from other psychological disciplines, for managers or product developers who need to use sensory test data, and for graduate students who need to use sensory tests as part of their data collection, but who have little formal training in sensory evaluation. For those who must actually design and conduct sensory tests, further study and consultation with a sensory professional is recommended. Although the techniques have few arcane mysteries, the design and execution of proper sensory tests is a complex matter, and it cannot be learned effectively from a single book chapter.

B. Sensory Evaluation Basics

The overriding principle of sensory evaluation is to match the sensory technique with the problem at hand. This requires a logical decision-tree approach to test design. Most questions about perception of flavors or products will fall into three categories. First, people ask, Are these two products different? This calls for the overall difference test, also referred to as a discrimination test. These tests usually take the form of a forced-choice procedure, where participants are asked to select one choice from among a set of products in which only one is physically different from some standard sample. The second common question is, How are they different? In other words, the goal is to specify, in perceptual terms, how products differ, in what qualities have they changed and to what extent. This set of procedures is referred to as descriptive analysis. In its most common form, a group of trained individuals examines the products and provides numerical ratings for the perceived intensity of each attribute. This provides quantitative sensory specification of each product that may be compared statistically. The third common question concerns consumer likes and dislikes. These tests are generally conducted with untrained persons who are usually users or purchasers of the product. They are asked to provide quantitative ratings describing the strength of their liking or disliking for the product as a whole, and may also be probed about their opinion regarding specific characteristics. Alternatively, they may simply be asked to pick which product they like best from a set of alternatives.

As noted above, in historical sensory analysis of some standard commodities, the descriptive approach and acceptability testing were combined in the "quality judgments" of expert tasters. The experts were able to identify defects, judge their severity, and produce quality scores that would presumably reflect consumers' rejection of substandard products. This approach has limited applicability to newly developed and processed or engineered foods, where standards for quality are not yet defined. It is also problematic for product in which sensory segments exist, (i.e., consumer groups that have different yet specific profiles of what they like in a certain product, e.g., tastes great? less filling?).

Because these methods involve a controlled stimulus–response measurement scenario using human participants, sensory evaluation borrows some practices from the behavioral sciences. In order to minimize biases that may affect the validity or accuracy of a test, blind coding and control of presentation order are critical. Blind coding is usually achieved by labeling each sample with a meaningless name, such as a randomly chosen three-digit number. Participants are provided with only enough information about the sample to ensure that it is viewed in an appropriate frame of reference or category. Controlling stimulus order may be achieved by fully counterbalancing orders or by using a design such as a Latin square, which would place each product in each position an equal number of times, when viewed across all participants. Alternatively, order might be fully randomized if the number of observations is large enough. Other critical items to control are all the physical variables that would be expected to influence sensory impact: Concentrations, volume, temperature, and so on. These concerns seem second nature or even old hat to behavioral scientists, but they can easily be overlooked in applied situations.

Sensory evaluation has also borrowed from the behavioral sciences in the application of experimental design and statistical analyses. Unfortunately, much industrial product testing still uses simple paired designs of one sample versus a control. However, multilevel and multifactor designs are occasionally applied. Multifactor ANOVA (MANOVA) is common in descriptive studies, in which each attribute is analyzed as a separate dependent measure, as a function of panelists, products, and replicates. Replication is considered mandatory in good sensory practice (Stone & Sidel, 1985). Differences between specific products after ANOVA are then tested by suitable planned comparisons, such as Tukey's or Duncan's tests. There has been some recent interest in application of MANOVA and some multivariate techniques, but they have not caught on to the extent one sees them in the psychological literature. For example, univariate repeated measures is a common analysis. Because foods, like many chemical stimuli, are highly fatiguing, complex sensory tests are good candidates for incomplete designs, such as balanced incomplete blocks. This must be balanced against the power of complete designs, which allow the use of subjects as their own baselines or controls, as in dependent *t* tests and repeated measures ANOVA. The between-person variability can then be partitioned to examine treatment effects (product differences) in purer form (i.e., with between-subject effects removed), or against a smaller residual error term. One of the problems of concern in sensory analysis is whether the data meet the assumptions or requirements for certain statistical tests. Because the data may not be equal-interval in the way numbers are assigned, nonparametric statistics are occasionally employed. A well-trained sensory professional is familiar with the nonparametric alternatives for many parametric tests.

C. Discrimination Testing

1. Goals

A discrimination test is called for when the objective is to determine whether any difference is perceived between two products. The nature of the difference is usually not specified—it is up to the test participants to see if they can find a point of difference. Because a finding of no difference may have important business implications, failure to reject the null is an actionable outcome in these tests. Thus the power and sensitivity of the test is important, and beta risk is an important consideration.

If the difference is studied as a function of different levels (systematically varied) of some ingredient, the experiment resembles a measurement of difference thresholds. For example, the determination of a just noticeable difference (JND) is closely analogous to the discrimination test objective, when several products with different levels of a flavor are compared to some control. This is logically related to historical psychophysical methods such as the constant stimulus method.

2. Variations

The most common forms of the discrimination procedure are the triangle test (Helm & Trolle, 1946) and the duo–trio procedure (Peryam & Swartz, 1950). The triangle test is a three-alternative test in which one sample is different from the other two. The test is counterbalanced for the identity of the odd sample (both ABB and BAA used) and its position in tasting (ABB, BAB, BBA). Chance performance is one-third, and performance in a group above that level provides evidence for a perceivable difference. This differs from traditional forced-choice procedures in which subjects would be directed to choose the strongest or weakest stimulus. Foods are necessarily multivariate. Because it entails comparisons of similarities (or differences), rather than simple intensities, the triangle test is more difficult than choosing the strongest of three samples. In the duo–trio test, a sample designated as a standard or control sample is presented for inspection. Then two samples are presented, and the participant is asked which of the two matches the control. Chance performance is 50%. In neither test are the participants directed to any specific sensory attribute—the test is for the existence of any difference whatsoever.

Participants in these tests should be screened for minimal acuity in discriminating differences in the products or sensory modalities to be tested. Because the tests are often used as a first step in a sequence of tests, there is little attempt to make the selection of participants be representative of consumers as a whole. Rather, discriminative ability is key. The tests are generally conducted in a laboratory environment with control over sample preparation, temperature, lighting, noise, and so on. If no difference is found under such conditions, logic dictates that most consumers would not notice a difference in less controlled situations. However, this logic is not airtight. Consumers have multiple opportunities to interact with a product once it enters the home, and to solicit opinions from other family members. There is always the possibility that what goes unnoticed in a short taste test might be detected once familiarity with the product is gained.

The true forced-choice procedures are also used on some occasions. In these tests, the subject is directed to a specific attribute (e.g., "Choose the sample that is sweeter from this pair."). Such tests are designated as *N*-alternative forced-choice tests (2-AFC for a paired test, 3-AFC for a test with one target and two controls, and so on). There are both theoretical and empirical reasons to believe that they are more sensitive than the overall difference tests (Ennis, 1990; Frijters, 1979). Because the participant's attention is being directed to a sensory attribute that is expected to differ, detection of the difference should be enhanced for attentional reasons. Conversely, overall difference tests such as the triangle may lose sensitivity because they entail the risk that some participants may focus on differences that are artifactual, a function of serving order or even of their momentary state of adaptation, and fail to attend to the attribute or attributes that are expected to differ systematically as a function of the ingredient change.

A variety of other approaches have been applied to discrimination, including signal-detection experiments, yes–no procedures, same–different paired tests, and sorting tests involving multiple targets and multiple blanks (Meilgaard et al., 1987). The overall superiority one test method over another is unclear, and a source of some debate (O'Mahony, Wong, & Odbert, 1986). However, there is a growing recognition that directing attention toward critical attributes will enhance discrimination (MacRae & Geelhoed, 1992). Table 1 shows some original data from sensory evaluation classes asked to perform various discrimination tests on a 10% difference in sucrose concentration. The paired test with directed attention was usually much more sensitive on a chance-corrected basis and more likely to achieve statistical significance.

3. Statistical Analyses

Analyses of triangle and duo–trio tests involve binomial distributions and one-tailed tests. The tests are one-tailed because the alternative hypothesis is that

TABLE 1 Comparison of 9–10% Sucrose in a Beverage

	Proportions correct	
Test type	Group 1	Group 2
Triangle	9/21 (NSD)[a]	12/23 ($p < .05$)
Paired comparison	19/21 ($p < .001$)	24/27 ($p < .001$)

[a]NSD, no significant difference.

population performance would be above the chance level (not above *or* below). Tabled values for probabilities based upon numbers of correct choices from different numbers of participants are available in sensory texts. Most practitioners simply refer to these tables of critical values for a minimum number correct for statistical significance (Stone & Sidel, 1985). The rapid tabulation of results and simplicity of analysis make these tests highly popular and easy to conduct. They are often done as a first step in a larger program of testing, in order to get a rough idea about whether a certain physical change made a difference in sensory impact. A null result in these simple tests can save a great deal of time and expense in further research.

D. Descriptive Analysis

1. Goals

The most generally useful and highly informative class of sensory tests are the descriptive analyses. These techniques attempt to provide a quantitative specification of all the sensory attributes of a food or product. This is typically achieved using a set of scales, each of which provides a numerical response for the perceived intensity of a given attribute. Each sensory attribute represents a (presumably) independent and elemental sensory experience. The results are useful for specifying sensory changes in product development as a function of ingredient, packaging, or processing variables and for shelf-life and quality-control questions. The data are also used for correlation with consumer judgment for purposes of building predictive or explanatory models of factors driving likes and dislikes. Because they are quantitative and analytic in nature, the sensory specifications are also sometimes examined for correlation with instrumental measures of food properties.

2. Variations

The earliest method for descriptive analysis was the Flavor Profile method (Caul, 1957). A group of extensively trained panelists would make judgments about the perceived intensity of all the flavor components of a product, in the order of their appearance. The individual profiles would then be discussed, and a consensus profile was put together under the direction of a panel leader. Although this procedure was a great improvement over the liabilities inherent in using a single expert taster, further advancements were possible. QDA brought aspects of behavioral testing methodology to the descriptive test (Stone et al., 1974). The simple category scale used in Flavor Profile was replaced with an unstructured (presumably less biasing) line scale, anchored with suitable words at the low and high ends. More importantly, this technique was amenable to experimental design and statistical analysis. Replication was a standard feature of the design and allowed for evaluation of test reliability. ANOVA became the routine statistical procedure for these data. Repeated measures analysis could be applied to partition judge effects from product differences. Statistical measures of central tendency rather than consensus values became the framework of the "profile" and statistical significance testing provided the criteria for the existence of product differences.

3. Terminology Issues

A major hurdle in the development of a good descriptive analysis is the selection of useful terms. Although there is some agreement about four basic tastes, other points of view add other taste qualities, such a the *umami* taste of monosodium glutamate (O'Mahony & Ishii, 1986). The realm of olfactory characteristics (Lawless, 1988) and texture words (Szczesniak & Skinner, 1973) are less uniformly agreed upon. Early systems for expert grading of foods centered on physical sources of defects, rather than perceptual description of subjective experiences ("oxidized flavor" as opposed to "blue color"). Such systems are problematic in that many different sensory qualities may be subsumed under a single physical defect (e.g., cardboardlike, painty, fishy, and tallowy notes may all arise from oxidation flavors). Wine tasters sometimes discuss "reduction flavors" such as the mercaptans that arise from microbial degradation. However, "reduction" may lead to a host of different putrid sulfurous odors, so the specificity of this term is lacking. Recent publications have focused on this linguistic problem, and criteria for the practical utility of sensory attributes have emerged. These criteria include simplicity, lack of redundancy with other terms, and the feasibility of finding a physical reference standard to serve as a training example (Civille & Lawless, 1986). Measurement criteria such as reliability (precision) and validity (accuracy) may also be brought to bear (Claassen & Lawless, 1992; Lawless & Claassen, 1993).

Training panels of individuals to think about and to use words in a similar way is obviously a major hurdle for descriptive techniques. Many consumers confuse sourness and bitterness, but this confusion is easily rectified with examples (e.g., citric acid for sour and caffeine and quinine for bitter) (McAuliffe & Meiselman, 1974; O'Mahony, Goldenberg, Stedmon, & Alford, 1979). Training panels to recognize volatile aromatics, (aromas and flavors) is more difficult, but is also aided by use of examples or reference standards. These examples help to categorize and calibrate the qualitative perceptual space for trainees (Rainey, 1986). O'Mahony and co-workers have framed the process as one of concept learning or in their terms, concept alignment (O'Mahony, 1991). A current issue concerns whether a single reference standard is sufficient, or whether the concept learning is enhanced by multiple examples, as apparently happens in other modalities such as visual pattern recognition (Hartley & Homa, 1981; Homa & Cultice, 1984).

Some practitioners have carried the process a step further, and not only calibrated the panels with qualitative references, but then attempted to calibrate the psychophysical intensity curve as well, giving panels examples of low and high levels of each attribute. This kind of intensity anchoring was a critical aspect of the original texture profile method. For example, reference standard for perceived hardness ranged from low end anchors like boiled egg white to high end anchors like peanut brittle (Cardello et al., 1982). Some descriptive analysis practitioners carried this approach over into flavor work, and even went so far as to suggest cross-modal scaling (i.e., a common perceptual intensity scale for tastes and flavors, e.g., Meilgaard et al., 1987). Whether panelists can be so fully calibrated remains a source of some controversy. For example, subjects who eat diets high in red pepper compounds such as capsaicin, become chronically desensitized to those flavors (Lawless, Rozin, & Shenker, 1985). It would seem fruitless to force them into a perceptual intensity scale on the same basis as nonconsumers of hot spicy cuisines, who are more sensitive. Along these lines, panels for evaluation of pepper heat in one Indian laboratory were segregated by pepper use, in order to avoid mixing data from the two types of consumers.

4. Statistical Analysis

As stated previously, most descriptive experiments produce data matrices appropriate for ANOVA. Products, judges, and replicates are the usual factors, with additional factors depending upon the variables manipulated among products, such as ingredient levels or processing treatments. Judges are generally considered random, rather than fixed effects (Lundahl & McDaniel, 1988) and the judge-by-product interaction serves as the appropriate error term for the product *F* ratio in the mixed model ANOVA. Planned comparisons follow the ANOVA, using tests such as Duncan's or Tukey's statistics to compare pairs of mean values. One area in which behavioral scientists seem to have departed from the sensory practitioners is in the use of multivariate *F* tests rather than univariate repeated measures analysis. Although there are violations of the homogeneity of covariance assumption in most sensory data, there has been a reluctance to adopt the MANOVA approach within the sensory evaluation community. At this time, the reasons for this are unclear.

In common practice, the profiles are represented by a line graph in polar coordinates, with the attributes forming equally spaced rays (arrangement otherwise arbitrary) and distances from the origin along each ray representing the mean value for a product on that ray. The points are then connected, forming a spider web or polygon with a sometimes characteristic and recognizable shape for the control product (Fig. 2). This is thought to aid in recognition of product differences due to the human ability to perceive shape, where the differences would be more obscure in a bar graph. Figure 2 shows three distinct profiles for milk samples subjected to two different oxidation processes and one that was subjected to rancidity or lipolysis. Such diagrams can help quality-control personnel recognize defects and their possible sources and thus suggest corrective actions in manufacturing or handling.

E. Affective Tests

The third important type of sensory test involves questioning consumer likes and dislikes. This question is phrased in two ways. One may ask about the liking or disliking for a product, perhaps a single product without reference to another

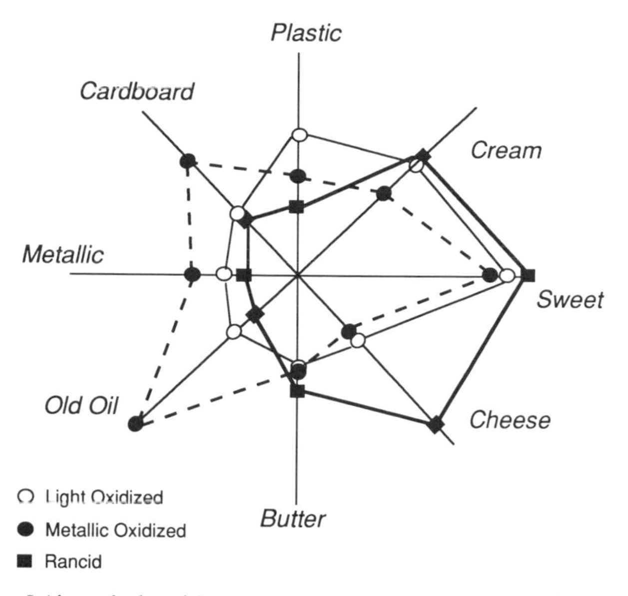

FIGURE 2 Spider web plot of descriptive panel ratings for milk flavors. Distance from the origin represents the mean rating. Open circles represent mean ratings for milk exposed to fluorescent light to induce oxidative reactions of milk lipids. Closed circles represent mean ratings for milk with copper-induced lipid oxidation. Filled squares represent mean ratings for milk samples that had undergone hydrolysis of triglycerides, producing rancidity.

product for comparison. Generally, these data are collected as ratings on a numerical scale, such as the balanced nine-point category scale introduced by the food research section of the U. S. Army Quartermaster Corps in the 1950s (Jones, Peryam, & Thurstone, 1955). This scale runs from "dislike extremely" to "like extremely" with a neutral category at the center of the scale. Data are analyzed using parametric statistics such as ANOVA or *t* tests. This degree of absolute liking and disliking should be referred to as "acceptability."

The second way to phrase the question is in a choice or ranking between two or more products, usually a paired test to see which is liked better. This should be referred to as "preference," although there is widespread misuse of this word in the literature to refer to rated acceptability. Preference or choice data are usually analyzed by means of binomial distribution statistics, because discrete outcomes are counted (numbers of people preferring one item over the other, as a proportion of the total). It is widely believed that preference data are more sensitive than rated acceptability, because two products can get the same acceptability rating on a category scale, but there might be a slight preference for one over the other. Empirical data supporting this belief are not found in the scientific literature. Furthermore, preference tests tell little or nothing about the overall level of acceptance, because one product might be preferred over the other, but both might be unacceptable. On the other hand, acceptability ratings can provide information on the direction of presumed preference.

The emphasis in both procedures is on obtaining a representative sample of consumers for the test. Several principles apply. First, laboratory personnel generally make poor choices for participants. They come with a technical and potentially biased frame of reference for evaluating the products. If consumers are recruited, they should be regular users or purchasers of the product. Finally, because the variability in personal preferences is usually quite high, large numbers of participants are usually needed ($N > 100$, as a rough rule of thumb). With such larger samples, it is possible to look for segments or groups of consumers with different preference patterns, rather than simply looking at overall means for different products or other measures of central tendency.

As always, it is important to blind code samples, use counterbalanced or randomized orders of presentation, and have participants work independently. There are no hard-and-fast rules about how much information to provide panelists about the products, but a simple rule is to provide only enough information to ensure that the product is viewed in the proper frame of reference. An example would be "evaluate the following sample of spaghetti sauce." The sample, labeled with a three-digit code, would be examined from the perspective of how likable that product is when considered a member of the category of spaghetti sauces. Problems arise when additional information is given that might bring in more complex attitudes, such as "low sodium." If possible, such additional information should be withheld so that the sensory attributes are judged on their own merit, without halo effects from other label claims.

F. Limitations of Sensory Evaluation Tests

Although a considerable literature of sanctioned methodologies has been published in the sensory evaluation field, it is not without its practical shortcomings and points of disagreement. Even among sensory professionals, it can be argued that there is a lack of consensus in the sense of a scientific paradigm. Furthermore, two influential groups that are also involved in practical flavor analysis, namely commodity graders and chemists, take a somewhat different view of sensory analysis. These two alternative approaches are discussed below. Some aspects of sensory testing that have received insufficient attention are discussed first.

As noted above, directed attention toward the attributes of interest helps in discrimination of complex stimuli such as foods. From the point of view of trained panelists or expert tasters, the function of training or experience is to provide observers with a mental list of differences to look for when confronted with a food to evaluate. In general, the place of perceptual learning is obvious in the applied world of flavor analysis. This was not lost on early perceptual psychologists such as the Gibsons, who used expert wine tasters as one example of how discrimination and perception improve with learning (Gibson & Gibson, 1955). Focusing on all the important characteristics is one hallmark of a good taster. Perceptual learning also involves learning to direct attention away from the irrelevant (i.e., not to be distracted by what is not important). The case for deleting or ignoring the probably irrelevant sensory attributes is less clear. Nonetheless, halo effects abound in the data of untrained consumers. It is almost fair to say that any attribute may influence the judgment of some other (seemingly independent, unrelated attribute) under some conditions. Figure 3 shows a simple halo effect in which addition of a small amount of vanilla to low-fat milk changed the consumer ratings of perceived sweetness and some texture factors, none of which are actually altered in the product. The confusion between the taste of sweetness and how it is apparently enhanced by volatile (olfactory) flavors such as ethyl maltol has led to a widespread, mistaken belief that ethyl maltol is a sweetness enhancer, and various food ingredient companies have marketed it as such. In fact, the changes in sweetness ratings with addition of ethyl maltol can easily be explained as cases of halo effects with taste–smell confusions due to retronasal stimulation (people call sweet smells sweet tastes) (Bingham et al., 1990; Murphy, Cain, & Bartoshuk, 1977). One person's illusion is another's marketing bonanza.

A second important issue in applied sensory evaluation has to do with the properties attributed to the data. It is all too easy to look at a quantitative profile of a product and interpret that profile as equivalent to the product's flavor. However, the numbers and verbal descriptors are just a model, a convenient shorthand snapshot of reality, not perception itself, which after all is a private matter. The often quoted statement by Box, to the effect that there are no perfectly accurate models, only some that are more useful than others, is appropriate to this field. The

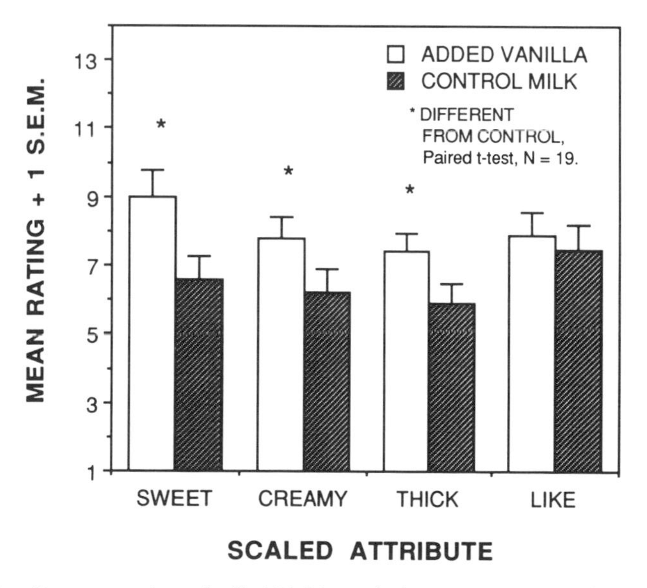

FIGURE 3 Consumer ratings of milk (1% fat) to which a minute amount of vanilla extract had been added and of milk with no added flavor. (Reprinted with permission from Lawless & Clark, 1992.)

limitations of a descriptive profile are well illustrated in the example shown in Table 2. These are the mean ratings for a product by a descriptive panel. The interesting task here is to look at that profile, and try to figure out what the product is. This kind of backwards synthesis rarely yields a correct guess in the classroom about the identity of the product (shown for the curious on page 345). The point is that the descriptive profile is only an approximation of perception. The descriptive data are analytic in nature, whereas the actual apprehension of the product may be more holistic under natural conditions. A simple classroom demonstration is to have students taste various cola beverages of high and low quality. When asked to profile the flavor, they will usually say something integrative like "I like it" or "What does the teacher want? It just tastes like cola" or "Isn't this from the cola nut tree?" After explaining that the cola has three major flavor notes—a brown spicy note, a citrus note, and a vanilla note, they can adopt an analytical perspective and taste the differences in the predominance of the different notes in different brands (easier with cheaper, less well-blended cola brands).

The limitations on analytical profiling are also discussed below in terms of how they may limit perception of tastes that do not neatly fit preestablished categories. As noted previously, there is a major debate over whether the taste of various amino acids such as glutamate salts should comprise a fifth basic taste category, called *umami* in Japanese. Still other flavor psychologists have argued that we should avoid profiling altogether and instead measure only integrated responses, from which we may tease apart the rules of perceptual combination through carefully designed factorial experiments. These advocates borrow heavily from the approach of Anderson's functional measurement (McBride & Anderson, 1990).

The major limitation of affective testing is in individual differences and the need for segmentation. The old statement about *de gustibus non disputandem* is certainly applicable to flavor preferences and to product testing. One approach in practical marketing is simply to engineer products that are the least offensive to a majority (so-called airline food). An opposite approach is to flood the market with enough flavors and line extensions to drive one's competitors market share down, while avoiding cannibalism of one's own products. The idea is to satisfy the demands of differing taste preferences for all population segments. Of course, this is not always

TABLE 2 Descriptive "Profile" of a Common Product

Attribute	Mean rating (15-pt scale)
Cocoa	4.5
Vanillin	2.6
Fruity/cherry	2.8
Nonfat dry milk	6.1
Sweet	10.0
Sour	3.1
Bitter	2.1

practical. A segment may be too small to justify the start-up costs in a new product, bribes for shelf space kicked back to supermarket chains (so-called slotting charges), advertising, and so forth. Nonetheless, the applied sensory practitioner must be aware of the possibility of segments in the data, such as bimodal distributions for product acceptance. Likes and dislikes, of course, are largely a function of experience, familiarity, and habit. A recent study in our lab on an exotic tropical vegetable, hearts of palm, showed a strong association with previous familiarity (Lawless, Torres, & Figueroa, 1993). Sensory reactions may also be determined by personality styles (Stevens, 1991) and one's biological equipment (Amoore, 1971; Bartoshuk, 1979).

These three factors, attention, integration, and segmentation, present important limitations to the use of sensory evaluation data. Users of sensory methods and users of the sensory data would do well to consider these factors when interpreting their results and making business decisions.[1]

III. SENSORY EXPERTS

A. Quality Tests by Expert Judges

A long tradition of sensory folklore has it that some people are much better at sensory perception than others. The legendary nose of Hellen Keller is one example that fits people's intuitions that the blind are somehow better at using their other sensory modalities (Ackerman, 1990). Apparently, Keller could smell the imminence of a coming storm, and would analyze the aroma of a country house by the layers of smell from previous owners, plants, and perfumes. Ackerman also describes in detail the activities of a professional perfumer, a synthetic master of olfactory art. Whether or not such abilities are genetically predisposed, or are a function of perceptual learning, some people are clearly more discriminating than others. This notion of olfactory expertise has been useful not only in the trained panels used for descriptive analysis mentioned above, but in several applied scenarios, including quality grading of food commodities, the work of applied flavorists, and wine tasters, to name a few. Unfortunately, the ways in which olfactory expertise are gained and methods for efficient nasal training are still part of an obscure art. Few data are available, in spite of the fact that experts such as wine tasters were of interest to many researchers in perceptual learning (Gibson & Gibson, 1955).

A fourth category of sensory tests was common in the early history of food quality evaluation. Quality tests were conducted in order to provide consumers with guarantees that certain expectations about product characteristics would be met. These evaluations were generally conducted with highly trained experts or highly experienced persons. Much of the early impetus toward quality testing grew

[1] The profile shown in Table 2 is for the ice-cream novelty called a Fudgcicle.

from the activities of government graders for food commodities, such as the USDA inspection shields of the Food Safety and Quality Service. States could also institute quality standards, as in the 1915 "Iowa Butter" trademark (Bodyfelt et al., 1988).

The rationale behind quality tests is straightforward—products that are free from defects and resemble some desirable standard receive high scores. Products that have evidence of taste, smell, textural, appearance, or packaging problems score lower. A trained judge is thus able to assess the severity of various commonly occurring defects and deduct points from an overall quality score to grade the product. One important consequence is the generation of very useful information about the potential cause of defects, which can suggest actions and remedies to food processors and manufacturers. Because the sensory tests require only the evaluation of a single judge and no data analysis, the method is highly cost efficient. Finally, because there are scientifically consensual terms used for the defects, the vocabulary is to a great extent standardized, at least within national boundaries.

Dairy products judging is one area of sensory expertise in which the route to professional status is clearly laid out. This is because dairy science departments in agricultural or land-grant institutions have often participated in the Collegiate Dairy Products Judging Contest. This contest, in operation since 1916, pits teams of undergraduate students from each school in taste tests of various common products, including fluid milk, cheddar cheese, cottage cheese, vanilla ice cream, yogurt, and butter. A set of products is first graded by a team of experts, usually the faculty "coaches." Defects and quality scores are decided by consensus. Next, students taste the samples in each category (under blind, coded conditions) and make their ratings. Students who come closest to the experts in identifying defects and assigning overall quality scores win awards, including trophies. At certain points in history, the dairy judging performance was even connected to offers of fellowships for graduate study.

The details of this approach are given in *Sensory Evaluation of Dairy Products* by Bodyfelt et al. (1988). For each product type, a chapter discusses the major defects, their origins and chemical nature, and how to recognize them. Recipes for training samples are given in the appendix, so that coaches may train their students with accurate references. With practice, students learn to distinguish the soapy bitter note of hydrolyzed milkfat (a type of rancidity resembling baby vomit) from the cardboardy painty note from lipid oxidation. Students who are very good can distinguish milk oxidized from light exposure from milk exposed to metallic catalyzed oxidation. The book also contains various tasting tips (useful for any taste panelist in any commodity) to help the students focus their attention, take a professional attitude, and maximize their performance. For example, students are warned not to take too small a sample, even though most of the samples will be defective, and some downright unpleasant.

B. Limitations of Traditional Quality Testing

This expert-nose approach to flavor analysis is not without its pitfalls. First, the evaluations combine an overall affective type of judgment—the quality score—

with a more analytic view of product characteristics. This combines activities that would be done separately by a consumer panel and a descriptive analysis panel, respectively, in most current sensory evaluations. Second, the overall quality score may or may not reflect the opinions of consumers, and a growing literature has shown that experts are often unrepresentative of consumer feelings (Lawless & Claassen, 1993). This is especially troublesome with new or engineered products (as opposed to minimally processed agricultural commodities) in which there are no culturally established standards of quality. One person's defect may be another's marketing bonanza. Finally, the terminology used by experts in defect grading may combine complex sets of sensory attributes. For example, the oxidized defect in fluid milk may smell like wet cardboard, paint, or even fish. Current descriptive techniques would provide panelists with separate rating scales for each distinct attribute, rather than combining them into a single score for oxidation.

A common problem in quality grading is how to combine the scores for multiple defects into an overall impression. The principles governing additivity are rarely stated and a matter of some disagreement. A classic example of this problem is in wine evaluation. The longtime standard scale for wine quality was a 20-point scale developed at the University of California at Davis (Amerine & Roessler, 1982). Many variations exist. The goal was to partition the overall score of 20 into various components. Judges would attend to each component and award scores, then total. Components with separate scores included appearance (maximum of 2 points), color (2 points), aroma and bouquet (2), volatile acidity (2), total acidity (2), sweetness (1), body (1), flavor (2), bitterness (2), and general quality (2). This type of scoring has proven difficult to use, except for the most highly practiced observers working in the same laboratory. One problem is that the total quality score may not reflect the overall opinion of the wine's quality—somehow the overall picture got lost in the subdivision according to attributes. This has led to a backwards approach by some wine judges. In the author's experience, judges decide on the overall score, and then apportion it to the various categories if required to do so on the scorecard.

Such bizarre strategies are symptomatic of a scoring method with logical problems. Part of the problem arises from the rule that panelists may approach a sensory test analytically or holistically, but it is difficult to do both. To get around these difficulties, a recent approach has been to have judges simply make overall hedonic ratings as to whether they personally like or dislike the wine (Goldwyn & Lawless, 1991). Presumably the opinions of even a small group of experienced wine evaluators will correlate with sensory quality and provide some guide to consumers when published.

C. Wine Tasters

Nowhere is the cult of the connoisseur stronger and more persistent than in the mythology of wine tasting. The classic Thurber cartoon shows two couples at dinner, with the host informing his guests, "It's a native domestic burgundy without any breeding, but I think you'll be amused at its presumption." That a wine could be described as naive or presumptuous strikes one as silly, but the wine vocabulary is filled with subjective, useless terms that wine writers seem to enjoy using. Amerine and Roessler's *Wines, Their Sensory Evaluation* has an excellent glossary of objective wine terms, including a hit list of words to avoid, including such notables as charming, distinguished, ingratiating, masculine, sensuous, and tame.

One of the more fascinating aspects of wine perception is this specialized vocabulary shared among experts. Some researchers have found this grist for insights into the interplay of language and perception. Adrienne Lehrer (1983) focused on the communication value of wine descriptions. If the sensory vocabulary of wine experts has value, then it must evoke sensory imagery in the reader that bears some relationship to the sensations that would be evoked when actually tasting a wine. To test this she employed a matching task in which descriptions were written for a set of wines. Then a second set of subjects would try to match the descriptions to the set of wines. One advantage of the matching task is that it avoids the need for any external criteria for the accuracy of the description. A description may seem "off-the-wall" to one person, but if another expert recognizes those sensory characteristics, the description has communication value. Although performance in this task is occasionally above chance, people are far worse at it than might be expected. Subsequent attempts at description matching show some advantage to experience (Lawless, 1984), but performance is far from perfect. Even when descriptions are written by English Master's of Wine—an exalted class of experts who must pass various written and tasting tests—performance is quite poor (Williams, Carter, & Langron, 1978).

A systematic look at description matching was undertaken by Solomon (1988) who examined the communication between experts and laypersons, as well as communication within those groups. The results were straightforward. Experts communicated above chance with other wine experts in terms of their matching descriptions to wines, but once again, communication was far from perfect. Furthermore, communication across groups, or among laypersons is virtually nonexistent. This led Solomon (1988) to propose that wine tasters form a *linguistic community,* in which shared perceptions and shared language are learned at various wine-tasting events. One pair of wine experts who achieved perfect communication scores in Lawless's (1984) study was a husband and wife who had drunk wine together for at least 30 years, talking about it and developing their own common set of terms.

In addition to language development, perceptual learning in this domain seems to entail discriminating more features of the stimulus, and possessing a richer categorical structure based on those features (Solomon, 1988). In other words, a good wine taster acts analytically toward the stimulus, pulling apart individual attributes for which she has a vocabulary. She also associates constellations of these attributes (features that tend to covary) with particular regions, styles, varietal grapes, and or vintages. Solomon stated that such categorization serves as an information compression strategy (chunking of sorts) and also that categorization allows quicker inferences ("red Bordeaux with beef") that may be employed with broader ecological utility (saving face when ordering for the boss in a restaurant).

As in dairy judging, wine evaluation can be made objective. Noble and colleagues have developed objectively anchored terminology systems for wine aroma (Noble et al., 1987), cleverly arranged in a wheel format with hierarchical structure (Fig. 4). The outer terms represent fairly distinctive aroma notes. Each outer term has a recipe for an easily concocted flavor standard to act as a prototype or reference for training wine evaluators. However, the system also offers some cognitive econ-

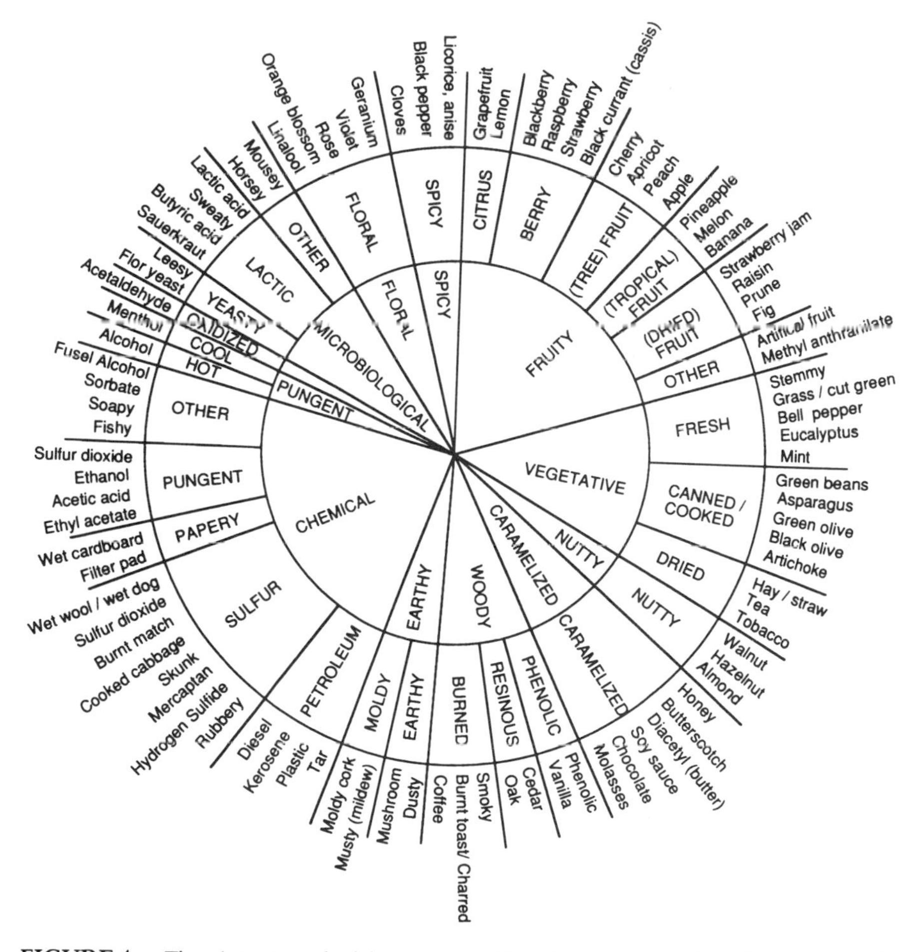

FIGURE 4 The wine aroma wheel showing general (interior tier) terms and specific (exterior tier) terms for wine aroma characteristics. Each third-tier term has a physical reference or recipe for creating that aroma in the original publication, for use in panel training. (Reprinted with permission from Noble, et al., 1987.)

omy. Interior terms act as categories subsuming the more concrete and specific outer terms. The more general terms may have practical value as well in wine evaluation. It is sometimes the case that a wine may have some general fruity character, but this will not be distinct enough to classify the aroma as a particular berry or citrus fruit. In that case there is some utility in having panelists estimate the overall fruity intensity. Different parts of the wheel may apply more or less to different varietal types (a useful exercise for students) and slightly different versions may evolve for different wine types (e.g., for sparkling wines). The approach is simply a clever geometric arrangement of an hierarchical descriptive analysis system.

Part of the perceptual learning that underlies sensory expertise involves mechanical or procedural tricks that are known to expert tasters. Aeration or the slurping that one hears at wine judgings is one example. Drawing air over a sip of wine held in the mouth helps maximize the volatile impact of the aromatic flavor notes and delivers them more efficiently to the olfactory receptors retronasally. Some dairy judges find it helpful to exhale through the nose to check for rancid aromas. The free fatty acids may cause a soapy impression in the nose. Dairy judges may be taught that rubbing a drop of milk onto their palm to evaporate may leave a clearer impression of the aroma of hydrolyzed milkfat. Sometimes wine judges are confused about whether a wine has sweetness from residual sugar or whether the sweetness is a taste illusion from a "sweet" aromatic. In that case, the nose can be pinched and the tip of the tongue inserted into the wine glass. Although this looks a little weird, one can usually sense whether there is any sweet taste on the tongue tip. Most experienced tasters know the value of a warm-up sample to acclimatize the palate. Coffee tasters will sample all three cups from a triangle test before actually starting to judge in seriousness. The preliminary tasting presumably sets a baseline of adaptation from which deviations may be more apparent. Selective adaptation may be a purposeful strategy. Perfumers who recognize one component of a complex mixture may selectively adapt the nose to that substance. Upon resmelling, the other notes of the unknown mixture are now more apparent. This last practice has good precedent in the psychophysical literature because adaptation to one component of a mixture releases the other components from the inhibition or masking that is common in complex flavors. Release from inhibition, of course, raises perceived intensity and thus the discriminability of the other notes (see Section IV.B, Functional Properties).

D. Flavorists and Perfumers

In spite of the common adage in psychology texts that there is no accepted scheme for classifying primary odors, there is quite strong agreement among fragrance professionals about the basic categories for smells (Brud, 1986). Perfumers share a common language, developed in part on the basis of perceptual similarities within categories (Chastrette, Elmouaffek, & Sauvegrain, 1988), and also based upon the sources of their ingredients. These schemes pose several problems to the uninitiated, which is perhaps why the psychological community perpetuates its myth that little is known about odor classification. First, the hierarchy for superordinate versus subordinate categories is relatively flat and broad (Chastrette et al., 1988). This flies in the face of the human disposition to classify by 7 ± 2 categories. Although there have been attempts to simplify odor classes down to less than ten categories, these seem laughable to most applied flavorists. Nonetheless, there is considerable structure and agreement among workers in different fields about smells. For example, Table 3 shows a practical descriptive system for fragrances in consumer products, and a categorization system in tobacco flavors derived from a factor analysis of hundreds of odor terms and aromatic compounds (Civille & Lawless, 1986). The agreement, given the different approaches and product areas, is more parallel than different. A second impediment to the understanding of odor classification outside the flavor and fragrance world is that many of the original categories derive from the source materials of vendors of such ingredients. Thus they have a class for aldehydic (from aldehydes used as perfume fixatives, later an important ingredient in perfumes such as Chanel No. 5) and a class for balsamic fragrances, which seems arcane to the outsider. To me, balsamic fragrances bring to mind a mixture of both pine smells combined with rich and sweet smells like vanilla. Perhaps I have not yet learned to abstract and organize the features of the balsamic category.

Part of creative flavor work is in flavor duplication. In this process, a flavorist (or a perfumer, in his or her job as well), may be asked to create a synthetic mixture to match the sensory impression of a natural product, or the successful product of a

TABLE 3 Odor Category Systems

Functional odor categories[a]	Factor analysis groups[b]
Spicy	Spicy
Sweet (vanilla, maltol)	Brown (vanilla, molasses)
Fruity (noncitrus)	Fruity, noncitrus
Citrus	Citrus
Woody, nutty	Woody, nutty
Green	Green
Floral	Floral
Minty	Cool, minty
Herbal, camphoraceous	Caraway, anise
Other	Animal
	Solvent
	Burnt
	Sulfidic
	Rubber

[a]Derived by principles of nonoverlap and completeness.

[b]Derived statistically from ratings of fragrance compounds.

competitor. To do this, there is first a strong ability to break down a complex smell that may be perceived as a synthetic whole, a pattern to the consumer. Cola flavors are a good example, as noted above. Most consumers react to them as if they were unitary flavors, while a good flavorist can see the spicy, citrus, and vanilla notes quite clearly, and may even dissect the individual spice or citrus fruit contributions. Next, the flavorist has a mental catalogue (as well as various reference works, e.g. Furia & Bellanca, 1971) of available materials and their sensory characteristics. Thus a list might be compiled of possible ingredients to try in compounding the flavor. However, the process is not that simple. There appear to be strong limitations on the channel capacity for perceiving individual components in olfactory mixtures (Laing, Livermore, & Francis, 1991), and it is difficult to work with more than three or four notes at a time. Flavorists get around this by making a number of simple mixtures first. In perfumery, these building blocks are called "accords." then the various accords may be assembled in different forms and levels until the desired overall effect is achieved. This is shown in the diagram in Figure 5. Tomato flavor may be analyzed into 8–10 component sensations. Each may be synthesized from one or more ingredients, and then the accords compounded and checked for overall harmony and as to whether they are on the right track for the overall effect. The final compound is then built from the accords, with various adjustments along

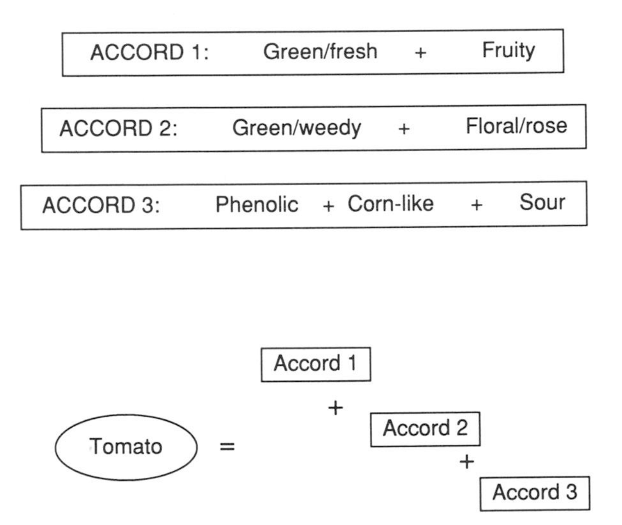

FIGURE 5 Scheme for compounding tomato flavor, based on simpler accords. The flavorist compounds each major note from five or ten aroma chemicals or essential oils, then combines major notes into accords, and then accords into the final flavor.

the way, to emphasize or de-emphasize the contribution of the various subflavors and their colorations.

E. Flavor Analysis and Flavor Perception—The Chemist's View

Perception of flavors, particularly aromatic (olfactory) volatiles, is viewed somewhat differently by practitioners of chemical analysis. For many years, the analytical approach was to identify all the components of a natural product that contribute to flavor. As analytical machines became more sensitive, the exhaustive list of all volatile chemicals detectable in a given product grows to huge proportions. A common statement is that any natural or processed product such as coffee, chocolate, or orange juice has hundreds of identified volatile components. Each of these might contribute to the product's flavor or aroma. As analytical methods continue to improve in their sensitivity and separation power, the number of identified flavor compounds in a product also continues to increase. However, these lists of hundreds of compounds are of little or no utility to the flavorist and of questionable value in telling the important contributing components of the perceived pattern.

A way around this problem is to interface the chemical analysis with some bioassay of the odor potency for each potential contributor. The assay of choice was (and still is, in many circles) the threshold concentration. This approach has great appeal because (1) the threshold is a physical quantity, which hides some of the subjectivity of sensory measurement; (2) the threshold as a concept has intuitive appeal; and (3) it made sense that substances below their threshold would not contribute to a product's odor and conversely, those above their threshold would be good candidates as major players. This last idea was extended to include the notion that the further above threshold a substance was, the more important its potential contribution. The units for such measurement formed a ratio of concentrations—concentration in the product divided by threshold, sometimes called an odor unit or a flavor unit.

Although the notion of odor units has been criticized by psychologists (e.g., Frijters, 1978), its use in flavor analysis persists. One example is CHARM analysis (Acree, Barnard, & Cunningham, 1984), in which effluents from a gas chromatograph are sniffed by subjects, and a dilution series is injected over multiple chromatographic "runs" (a sort of descending method of limits). The longer the odor persists through the dilution series, the higher the supposed impact of the original compound. The referencing of compounds to some component of sensory response is certainly an improvement over the endless cataloging of increasingly minute levels of chemical constituents (Piggott, 1990). However, several problems inherent in threshold measurement seem to have had little impact on the community of analytical flavor chemists. There is little or no appreciation of the response bias or criterion factors that are an important issue in signal-detection theory (i.e., the threshold model that is embraced by flavor analysts is usually a classical one with assumptions about an all-or-none response).

The chemical approach is primarily focused on analysis and separation. However, flavors and aromatics are recognized in food products as mixtures and patterns. Thus there is a major liability in examining even the sensory aspects of single compounds when they may behave differently in mixtures. A number of compounds may be present at less than threshold amounts. However, they may have similar structures and thus stimulate a similar receptor array, the combined effect of which is to produce a suprathreshold effect. Qualitative blending may alter the sensory impact of a mixture as well. If a series of aldehydes are mixed (e.g., C-6 to C-12, C-14, and C-16), the overall impression is one of old wax crayons or laundry soap. This impression is difficult or impossible to see in any one compound, but together they form a new pattern.

Piggott (1990) suggested a possible approach to this synthetic problem to assess the impact of a particular chemical. Rather than examining the individual compounds, one should look at the synthetic mixture both with and without the compound in question. The extent to which the mixture *minus* the compound deviates from the original smell of the whole mixture may indicate the importance of that compound in determining the character of the whole mixture.

IV. FUNCTIONAL PROPERTIES OF THE FLAVOR SENSES

A. Flavors as Mixtures

As noted above, most food flavors occur as complex mixtures of multiple taste stimuli and multiple odorous chemicals. A full appreciation of flavor perception entails understanding how tastes behave in taste mixtures, how odors combine in odor mixtures, and how tastes and smells interact across the two modalities.

In mixtures of simple tastants with one primary quality (sucrose, NaCl, etc.) the tastes remain fairly distinct in a mixture or at very least they seem amenable to psychophysical scaling on the basis of traditional primary taste descriptors (sweet, salty, etc.). The general rule for tastes of different types is that they partially mask one another. Sweetness of sucrose is decreased (holding concentration constant) when bitter quinine is present and vice versa (Bartoshuk, 1975). This effect is commonly called mixture suppression. Most other binary pairs of tastants will show patterns of partial inhibition, especially at higher levels. More complex mixtures of tastants (three or more components) have received scant attention, but once again, suppression is the rule.

A few exceptions to the suppression rule exist (Lawless, 1986). Weak concentrations of NaCl will enhance the sweetness of sucrose, probably due to an intrinsic sweet taste of NaCl (Bartoshuk, Murphy, & Cleveland, 1978). Mixtures of the potent *umami* flavors, monosodium glutamate and the 5′ ribonucleotides, will show greatly enhanced flavors compared to their individually tasted components. The physiological substrates for such synergy appear to reside in enhanced binding at receptors (Cagan, 1981). Finally, mixtures of some sweeteners, particularly the intensive or artificial sweeteners, will show synergy in some combinations (Ayya & Lawless, 1992).

The definition of synergy is a matter of some debate, because tastants with a positively accelerating limb of their psychophysical function would be expected to show "synergy" when added to themselves, although this is nothing more than summation along an expansive psychophysical function for one tastant. To this end, Frank et al. have suggested hyperadditivity beyond self-additivity as a more conservative criterion for taste synergy (Frank, Mize, & Carter, 1989b). This argument is more than academic musing about psychophysical mysteries. Intensive sweetener sales total well over $1 billion annually. The potential for increased sales by sweetener manufacturers on the basis of synergy claims and conversely the potential savings to food and beverage manufacturers from ingredient reductions are significant.

What is the ecological significance of mixture suppression as a general rule? One argument relates to the observation that suppressive interactions tend to parallel (and may be predicated upon) negatively accelerated psychophysical functions (Bartoshuk & Cleveland, 1977). A compressive psychophysical function is exemplified by Fechner's Law in which intensity ratings are proportional to the logarithm of concentration. They are also consistent with power functions such as those generated from magnitude estimation procedures, when the exponent of the power function is less than one. One result of a compressive psychophysical function is the ability of the sensing organism to respond over a broad range of physical energy intensities. In hearing and vision, two highly compressive sensory systems, intensities need to be measured logarithmically (e.g., in decibels) because of the vast range of perceivable energy. In the flavor senses, the range of effective concentrations of single stimuli may not be so broad, but the diversity of mixed chemicals occurring in natural foods and flavors is limited only by the biochemical machinery of their plants or animal sources. Thus it can be argued that mixture suppression provides a kind of across-component compression. This compression across compounds keeps the overall response characteristics in line with the realities of physiological limitations, while providing effective sensitivity and responsiveness for a foraging animal.

In the olfactory sense, the picture is less clear. Certainly, intensity masking is well documented (Lawless, 1986). Odors in mixtures tend to be less intense than they would be perceived as individual stimuli at the same physical concentration. This is the primary mechanism by which so-called air fresheners work. An undesired odor is simply masked by the application of a more intense stimulus. In analyzable mixtures where the components remain more or less perceptually distinct (e.g., pyridine and lavender oil) this can be documented as a decrement in intensity of each component. This phenomenon, similar to mixture suppression in taste, has gone by the name of odor counteraction (Cain, 1975). The total intensity of the mixtures is almost always less intense than the simple sum of the components. If one assumes that intensity ratings have the numerical properties that allow addition (category ratings would not, because they are bounded), then this is another illustration of a compressive and possibly inhibitory mixture effect. Various mathematical descriptions have been applied to this hypoadditive process such as a vector summation model (Berglund, Berglund, Lindvall, & Svensson, 1973; Patte & Laffort, 1979).

However, intensity suppression in mixtures where the components remain discernable is only one class of phenomena in odor mixtures and may be the exception rather than the rule. Odor components may often combine to blend or form perceptually synthetic patterns that are recognized as a unitary whole, rather than a collection of individual features. This type of mixing is probably aided by the tendency of the flavor senses and of olfaction in particular to act as integrators, rather than analyzers. One anecdotal illustration is the strategy employed in some air freshener work when there is a specific target odor to be masked. The common wisdom is to "surround" the target odor (perceptually, not physically) with a complex blend of similar but more pleasant-smelling components, so the target odor loses its identity and is drawn into the pattern-recognition process as the more pleasant-smelling masking agent. For example, a target odor of tobacco smoke might be covered by a blend of spicy smelling components, which bear a family resemblance to some of the characteristics of the empyreumatics. The smoke aromas are then effectively integrated into a more pleasant impression of the spicy fragrances. The odor pattern-recognition mechanisms have been fooled into not recognizing the smoke aroma as distinct.

B. Adaptation, Suppression, and Release

A second important operating characteristic of the flavor senses is their tendency to adapt or to become unresponsive to stimuli that are stable in space and time. This is perhaps most obvious for olfaction in everyday life. When one enters the home of a friend, we often notice the characteristic aroma of the house—the residual smells of their cooking and cleaning, personal care products, of babies or smokers, of pets or perfumes. This mixture tends to characterize and permeate a house in carpets and draperies. After several minutes, these aromas go largely unnoticed by a visitor. The sense of smell has adapted. There is no new information coming in, so both attention and sensory function turn in other directions.

In taste and smell, like the thermal senses, adaptation can be pronounced. It is studied experimentally in two ways. First, thresholds increase following periods of constant stimulation, providing the stimulation is higher in chemical concentration than the preceding baseline. Second, ratings for perceived intensity will decrease over time, and may even approach a judgment of "no sensation" or subjective zero.

Adaptation highlights the flavor senses as detectors of change. The status quo is uninformative—it has already been attended to when changed from the previous baseline, and it makes sense for an organism receiving many simultaneous channels of sensory input to turn attention elsewhere. This has also led to the speculation that sensation intensity may be partly a function of rise time of physical concentration—the total output of a sensory channel may be sensitive to the first or even second derivative of concentration with time. There has been a widespread speculation that if a chemical stimulus could rise slowly in concentration, the change might go undetected, so that the individual might eventually be swimming in a relatively intense atmosphere of smells, as judged by a newcomer to the room entering suddenly from an odor-free area. Such perceptual immunity to slowly increasing smells, coupled with retronasal stimulation, could explain why individuals with halitosis sometimes seem unaware of their problem. A practical concern occurs in situations with slow gas leaks, where a person might be unaware of danger due to slow rise time and ongoing adaptation.

In the sense of taste, adaptation has been used to explain why pure water can sometimes take on various tastes. Under conditions of careful adaptation using stimuli flowed over controlled areas of the tongue, adaptation may be rapid and often complete (McBurney, 1966). Following adaptation to a taste solution and rapid replacement with a flow of water, the water stimulates taste nerves and a small but reliable taste sensation is evoked (Bartoshuk, 1968). For example, after adaptation to NaCl, water takes on a somewhat bitter or sour taste. The lower the concentration of NaCl in the postadapting solution, the stronger the second taste, with pure water evoking the clearest impression. This may explain in part why slightly bitter or sour stimuli (fruit juice, iced tea, beer) seem especially thirst quenching, compared to heartier flavors, the salivary milieu provides adaptation to salt, thus replacement of saliva with a bitter or sour solution may have once signaled the purity (and thus physiologically best hydrating) of fluids.

However, in eating and drinking situations, it is rare to find the degree of stability seen in either the house odor effect or in taste adaptation under controlled flow. Studies of pulsatile stimulation of taste, which are more like real eating and drinking, show less adaptation than the continuous-flow experiments (Meiselman & Halpern, 1973). Human eating and drinking may be characterized at mealtimes as a pattern of rotational or sequential stimulation. In other words, people tend to alternate among food choices and sips of beverages. Most people do not eat all of their mashed potatoes and then all of their steak—they tend to sample back and forth, although children may be an exception. The combination of rare roast beef and dry red wine is more than the sum of their parts (hedonically, at least) when they are sampled in sequence. One function of rotational eating may be to undo adaptation. To the extent that one seeks pleasant flavor stimulation from food, one may wish to maximize this stimulation and minimize adaptation by changing the stimuli from time to time. This is probably not done as a conscious decision, but like mastication itself, is an active strategy for maximizing pleasure during eating that is adopted over a lifetime of practice.

Nonetheless, some adaptation of the palate does probably occur. In the trigeminal senses, the hot and spicy dish that seemed almost intolerable at the beginning of a meal may be quite a bit more manageable by the end. Adaptation can also interact with mixture inhibition to promote exciting contrast effects. Many psychophysical workers in taste have documented the release from suppression effect, wherein adaptation to one component of a taste mixture dulls the responsiveness to that component but also reduces its inhibitory effect on other taste components in the mixture (Lawless, 1979a). The other unadapted components are said to be released from their suppressed (inhibited) state, and become more intense in the mixture.

Release from suppression is easy to demonstrate with a riesling, white zinfandel, or other wine with both good acidity and some residual sugar (or a fruit juice). After biting into a lemon wedge, the wine will seem altogether too sweet. The reverse effect can be seen with a teaspoon of sugar, but be sure to leave it in the mouth to allow adaptation to take effect. This explains why many dry wines that are perfectly acceptable during the main course of a meal, fall into an unacceptable range when paired with sweet desserts. In fact, pairing wines with desserts is inordinately difficult. Acidity contrast is also a problem. The nemesis of many a sommelier is that wine may be brought to the table about the time the salad course arrives. This has the disadvantage of allowing wine to hit the palate right after vinegar may have altered the oral milieu. Often, there has been some adaptation to acid in the salad dressing, so a correct read of the acid balance of the wine is impossible. (Rinsing the mouth is advisable here, if you are the lucky person to get to taste when the waiter pours the sample.)

This sort of contrast also occurs in olfaction (see Fig. 6). In a two-component odor mixture, where the components are distinguishable (i.e., not blended into some inseparable mixture), the release effect appears to operate. Adapting the nose to one component makes the other one stand out (Lawless, 1987). As noted above, this is an analytical strategy used by some perfumers. When trying to analyze a competitor's fragrance, some components may be readily distinguished in the complex mixture and others may be obscured. If the nose is fatigued to the known components, the other components may seem to emerge, allowing them to be more readily identified.

C. Cuisine and the Palate's Balancing Act

Part of the skill in culinary arts may be in achieving a reasonable balance and contrast of flavors. Cuisines of different ethnic groups may have evolved to provide harmony among flavors. For example, the pinelike nature of retsina is difficult to enjoy by itself, but it seems more pleasing when accompanied by Greek food, which has various herbs whose terpene content may harmonize with the resiny notes. Similarly, the Portuguese *vinho verde* is a high-acid wine but can accompany foods of the local style. Contrasts as well as harmonies can be manipulated for effect. Sequential effects are also important in oral texture. Dry red wines are usually thought of as an accompaniment to rare beef, but the oenophile may see beef as the accompaniment to fine red wine. The tannins in the wine render it less than optimal for sipping in isolation. However, the juices and the meat provide a

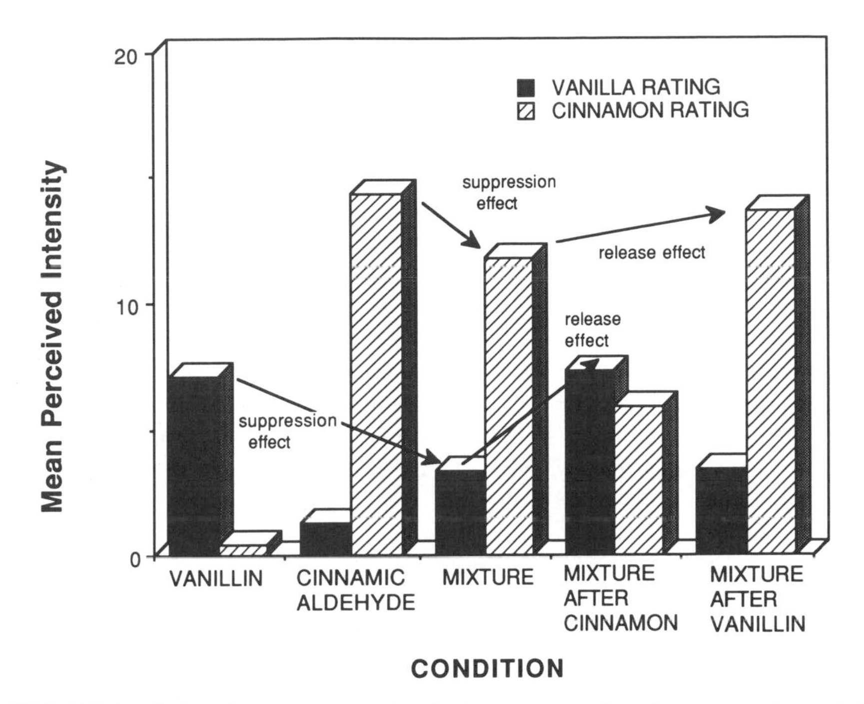

FIGURE 6 Release from odor suppression. In the mixture, vanilla and cinnamon odors partially mask one another, but after adaptation to one component, the other is raised in intensity or partially released from inhibition. (Replotted from data in Lawless, 1987.)

second substrate for the tannins in the wine, providing some relief for the palate from astringency. One wine-tasting competition has gone so far as to provide rare beef slices as an antidote between red wines for their tasters during judging sessions. Contrast may also be important in relief from fatty dishes. Aquavit may be an excellent accompaniment to gravad lox, in part because the strong alcohol seemed to cut through the fish oils, leaving the palate refreshed (among other effects) for the next bite of fish.

V. RESEARCH ISSUES AND ENDURING QUESTIONS

A. *Umami* and the Basic Tastes

Each of the basic taste categories appears to have an ecological function. Sweetness provides information about carbohydrate energy sources, namely sugars. Saltiness is a function of sodium and to lesser extent, other cations, which are necessary to life. Sourness would seem to warn the body that danger from acidity exists. Bitterness is a taste presented by many toxic or pharmacologically active substances. Of course, there are exceptions to each of these rules, both in tasty substances without the implied physiological consequences, and in tasteless substances that may have physiological impact. One important macronutrient with taste properties is missing from this list, and that is a taste category for amino acids, peptides, or proteins.

Some cultures do possess a taste word for this category, most notably the Japanese *umami*. Some semantic overlap occurs with English term *savory*. Although it seems intuitively appealing that meaty or brothy foods have a distinctive flavor, there is much disagreement about whether this is a separate taste category. Some taste psychophysicists seem quite content to profile *umami* stimuli such as monosodium glutamate (MSG) with the classical European taste categories of sweet, sour, salty, and bitter (e.g., Bartoshuk et al., 1974). It has been suggested that the concept of savory flavors is more apparent in a culture like the Australian, where vegemite is a staple food. On the other hand, O'Mahony has provided evidence that even American consumers are able to recognize the *umami* taste concept, and do so in a manner not unlike Japanese when asked about the applicability of *umami* to different foods (O'Mahony, 1991). Thus cultural differences alone cannot dismiss the potential importance of this taste category.

This issue is also clouded by the fact that MSG and several 5′ ribonucleotides act synergistically in mixtures and thus have reported functionality as flavor enhancers. This enhancement takes several forms. One simple demonstration is that very low levels (perhaps even subthreshold) of MSG and a riboside will produce moderate to strong taste intensity in mixtures, more than the sum of their parts (Yamaguchi, 1967). They also appear to act synergistically in sequence. Finally, some food scientists talk about the blending or rounding effects of these flavors in other products (Reaume, 1975). Although this synergy has an established physiological substrate in enhanced binding, both the definitions of synergy (Frank et al., 1989b) as well as the meaning of flavor enhancement for MSG are poorly defined. Although the issue of *umami* as a basic taste is not tightly related to the issue of flavor enhancement, antagonists in this debate often align in support of or opposition to both arguments. Clearly there is an opportunity for further cross-cultural studies and opportunity for debate about the most appropriate definitions for basic tastes, taste categories, and flavor enhancement.

B. Individual Differences in Taste Perception

About the year 1932, a chemist from DuPont named Arthur Fox made a remarkable discovery. While working with the compound, para-ethoxy phenylthiocarbamide, a co-worker remarked about the strong bitter taste from some airborne dust of this compound. However, other lab personnel found the compound to be tasteless (Fox, 1932). A surge of taste testing was undertaken, including genetic studies, and it was determined that insensitive individuals, about one-third of most Caucasian peoples, will possess two recessive genes for this simple Mendelian characteristic (Blakeslee, 1932; Cohen & Ogdon, 1949). The dimorphism extended to phenylthiocarbamide (PTC) and related compounds, although other bitter substances were not strongly related. For example, cross-adaptation studies

(Lawless, 1979b; McBurney, Smith, & Shick, 1972) provided evidence that quinine response was unrelated or activated by a different receptor mechanism. Testing with PTC was later replaced with the less smelly and less toxic compound, 6-n-propylthiouracil or PROP.

Could this effect influence flavor perception for foods? Since PTC-related compounds occur only rarely in nature, this connection depends upon three lines of evidence. First, the PTC dimorphism has been correlated with a variety of other flavor-related compounds. Second, direct tests on food likes and dislikes have been conducted, with marginal results. Third, the hypothesis that PTC sensitivity may offer a defense against goitrogenic vegetables has been examined. Each of these areas will be examined in turn.

Relative sensitivities and responsiveness of PROP tasters versus nontasters to a variety of flavor compounds has been studied. Differences in sensitivity have been noted to the bitterness of caffeine (Hall, Bartoshuk, Cain, & Stevens, 1975) and in suprathreshold responsiveness to saccharin bitterness (Bartoshuk, 1979). Bitterness of potassium salts was also found to be more intense to subjects with high PROP sensitivity (Bartoshuk, Rifkin, Marks, & Hooper, 1988). Others have failed to note any caffeine or potassium connection (reviewed in Schifferstein & Frijters, 1991). An additional puzzle in this class of phenomena are the differences in sensitivity or responsiveness to nonbitter tastes, including the sweetness of sucrose, saccharin, and neohesperidin dihydrochalcone (Gent & Bartoshuk, 1983). One explanation for these diverse effects may be that different numbers of taste papillae are observed among PROP tasters and nontasters. In fact, some suprathreshold responses appear trimodal in nature, leading to the proposal that a highly sensitive "supertaster" group exists, composed of homozygous dominant persons for the PTC gene. In keeping with the previous observations, supertasters appear to have even more taste papillae than tasters (Reedy, Bartoshuk, Miller, Duffy, & Yanagisawa, 1993).

This literature is not without some controversy. Schifferstein and Frijters (1991) noted some failures to reproduce the taster–nontasters differences in other compounds, caffeine being one example. They noted that studies of intergroup differences have rarely controlled for a general sensitivity difference, commonly normalizing individual data to salt responses for comparison. Especially at low concentrations, taster–nontaster differences could arise as a function of some general sensitivity factor. Ambiguities exist in this literature as well. Quinine–PTC correlations are one example. Some studies have noted a small positive correlation between PTC or PROP sensitivity and sensitivity to quinine. Other studies have noted failure to cross-adapt between PTC and quinine, suggesting nonoverlapping or independent receptor mechanisms (Lawless, 1979b; McBurney et al., 1972). If the receptor mechanisms for PTC and quinine are in fact independent and genetically unrelated, a small positive correlation could still arise from general sensitivity factors affecting all of the individual's taste function (age, disease history, medication, smoking, etc.). A serious problem in this literature is the failure to partition general factors from individual paired correlations among compounds. This suggests that a variance component model should be derived to disambiguate these effects. Components would include general taste-sensitivity factors as well as PTC–PROP-specific factors.

The relationship of bitterness sensitivity to food preferences has been examined with mixed results. A weak correlation between numbers of foods disliked and PROP thresholds was reported in two studies (Fischer, Griffin, England, & Garn, 1961; Glanville & Kaplan, 1965). The second study also noted a relationship between preference for strongly flavored foods and PROP, with groups preferring stronger flavors having lower PROP sensitivity. Correlations were higher for PROP than for quinine, although correlations with food scores in both studies were only in the range of .3. To put this in perspective, the correlation between husband–wife pairs for food liking scores was .48. These individuals would have no reason for a close genetic relationship, but rather a common environmental influence for varying lengths of time. In a study of children's food preferences, PROP tasters showed decreased preference for cheese (Anliker, Bartoshuk, Ferris, & Hooks, 1991). This could presumably result either from sensitivity to the bitterness of calcium ions or to the bitterness of various small peptides created in cheese making during proteolysis. If this is a general effect, it should be more pronounced with low-sodium cheeses, because they have less sodium present to mask bitterness as well as often suffering from increased proteolysis during cheese making and aging. Recently, a partial correlation of liking for sucrose and PROP nontasting was observed (Looy & Weingarten, 1992). Although they suggest that this may be due to variation in sweet quality perception, this is difficult to reconcile with their observation that sweet dislikers (often PROP tasters) taste a purer or less complex sucrose sweetness. If sucrose has side tastes, one would predict the opposite effect—that PROP tasters should pick up the side tastes especially if they are partly bitter.

Compounds containing the critical C=S double bond for PTC tasting occur in cruciferous vegetables. Such compounds have also been linked to goitrogenic activity, inhibiting iodine uptake. Thus one hypothesis states that PTC sensitivity may offer a taste-mediated behavioral defense against overconsumption of potentially troublesome dietary threats to the thyroid. Greene (1974) collected data relevant to this hypothesis in the Ecuadorian Andes, where some areas have poor iodine supplementation and run the risk of thyroid disease, reduced thyroid hormone synthesis, and associated cretinism. Greene found a small but positive correlation between test scores for visuomotor maturation and PTC sensitivity in an iodine-deficient community where goiter was endemic. No correlation was observed in a closely matched community that had received iodine supplementation. Although there was only a small proportion of nontasters in the sample, the result is consistent with the idea that sensitive tasters limit their intake of bitter goitrogens and thus are less likely to suffer neurological problems as a function of hypothyroid stress. Along these lines, a higher incidence of nodular goiter was found among nontasters of PTC (Azevedo, Krieger, Mi, & Morton, 1965). However, in the absence of other causal information, such findings are also consistent with a pleiotropic hypothesis that suggests that some underlying basic biochemical mechanism disposes nontasters to thyroid abnormalities (Mattes & Labov, 1989). In other words, a taste–diet correlation is not the only possible explanation.

The ability to observe this relationship may depend strongly on the environment and culture within which it is measured. Two studies of North American college students failed to find strong relationships between PTC status and food habits (Jerzsa-Latta, Krondl, & Coleman, 1990; Mattes & Labov, 1989). In the first case, weak relationships for taster–nontaster differences in bitterness and taste preference for cruciferous vegetables were found. The number of significant relationships were a small percentage of the total number of vegetables and attributes tested. In the second study dietary habits of tasters and nontasters toward foods known to inhibit iodine uptake (including cruciferous vegetables and other foods) showed no differences.

In summary, the dimorphism for phenylthiocarbamide remains a heavily studied phenomenon in the genetics of human chemical sensitivity. However, the ecological reasons for this diversity, its adaptive significance, and potential consequences remain a mystery. Furthermore, why certain racial gene pools (most notably Caucasians) would maintain this dimorphism while most other peoples have primary taster status is a mystery. It seems unlikely that tasting or nontasting would have developed independently in these populations, since at least some apes seem to have nontaster individuals, suggesting that the origin of this effect is evolutionarily very old among primates.

C. Individual Differences in Olfaction

Because a great deal of flavor variation is caused by olfactory input, it is reasonable to suppose that differences in smell sensitivity should affect flavor impressions and therefore food likes and dislikes. The phenomenon of specific anosmia refers to a selective olfactory deficit to a closely related family of odor-active materials, usually with a common characteristic smell. The deficit is selective in that individuals will have thresholds two or more standard deviations above the population mean, but olfactory sensitivity within normal limits to other compounds (Amoore, Venstrom, & Davis, 1968). Specific anosmias have been proposed as a means for identifying primary odor qualities (Amoore et al., 1968), although that view is somewhat controversial. The specific anosmia to androstenone has a strong genetic component (Wysocki & Beauchamp, 1988) as well as developmental and experiential influences (Dorries, Schmidt, Beauchamp, & Wysocki, 1989; Wysocki, Dorries, & Beauchamp, 1989) on the expression of this olfactory "blind spot." Anosmia may prove to be an important tool in understanding olfactory-guided behaviors as well as the interrelationships of olfactory neural channels (O'Connell, Stevens, Akers, Coppola, & Grant, 1989).

In spite of the growing literature on specific anosmia, its functional significance in everyday life remains unclear. Most psychophysical studies have focused on threshold data. Information on responses above threshold concentrations is largely unavailable. Furthermore, most flavors are chemical mixtures, and no information on perception of mixtures by specific anosmics is yet available. It is difficult to find data connecting this phenomenon to the perception of food odors, which are potentially important in food choices and nutritional status. However, a number of specific anosmias would be expected to play a role in food perception: androstenone, the major component of boar taint in pork (Thompson & Pearson, 1977); isobutyraldehyde, a component of malty flavors including various dairy product off flavors (Amoore, Forrester, & Pelosi, 1976; Bodyfelt et al., 1988); isovaleric acid, one of the free fatty acids present in many foods such as cheese and a common by-product of hydrolytic rancidity (Bodyfelt et al., 1988; Brennand, Ha, & Lindsay, 1989); cineole and l-carvone (Pelosi & Viti, 1978), major components in many herbs and spices (Heath, 1981).

A second reason for the potential importance of specific anosmia concerns sensory interactions among flavor materials. Mechanisms of neural inhibition are thought to underlie the suppressive interactions in both taste and odor mixtures. Persons insensitive to the bitter taste of PTC are not only unresponsive to the bitterness of PTC in mixtures, but they show none of the suppressive (inhibitory) effects of the bitter taste on other tastes that may be present (Lawless, 1979a). An analogous demonstration for olfaction has not been performed. Finally, because many odor mixture constituents perceptually "blend" rather than remain distinct percepts, qualitative shifts in odor character of complex mixtures would also be expected if some of the components were not sensed by anosmic individuals.

D. Odor Classification

What are the informational functions of olfaction? What were the environmental pressures or sources of information that guided the development of humans' highly articulated sense of smell? One can only guess at why sensitivity to such a broad array of chemical compounds having a myriad of different sensory effects came about. Certainly some development could be driven by appropriateness in food intake. Although it is generally recognized that tastes form better cues for learned food aversions, smells can also participate as conditioned stimuli. A wide variety of organic compounds are produced as a function of microbial action. Perhaps this was important in prehistory, because microbes can produce enterotoxins. The ability to recognize spoiled meat would depend upon the recognition of the microbial end products of lipid, carbohydrate, and protein degradation, carbonyl oxidation products of lipids and fatty acids, and amines or sulfur compounds from proteins. Somehow, what was once unpleasant may have at some later time become desirable—such as cheeses, that contain both fatty-acid aromas and some sulfur compounds. Of course, it can be argued that as children and adults, we go through a similar process of reorientation. Strong cheeses may be disliked at first, and then the taste preference is acquired as a function of experience. Perhaps the roasted flavors associated with Maillard browning signaled some degree of safety—meats roasted under conditions of high temperature and low water activity were less likely to have microbial problems. However, it is important to note that the flavor senses are imperfect detectors. In some cases microbial toxicity may precede the development of associated sensory cues.

Specifying the aromatic or olfactory qualities of flavors has presented some difficulty to researchers over the years. This is in part due to the lack of agreement on terms for basic or primary odor types. The range of olfactory experiences would seem to be vast, and the recent discovery of a multigene family encoding potential transmembrane receptor proteins for olfaction suggests that the number of different receptor types might be in the hundreds (Buck & Axel, 1991). It is clear that people are generally uncomfortable with a system that requires this level of complexity, and early attempts at defining odor primaries in less than ten categories (e.g., Amoore, 1970) persist in perception textbooks, along with the general admission that such systems are inadequate.

In a more applied vein, techniques such as multidimensional scaling (MDS) have been applied odors to define both perceptual and physicochemical dimensions (Schiffman, 1974). Unfortunately, surveys of large classes of very different odors tend to yield spaces with hedonic and intensity dimensions, which is not a very informative result from the point of view of learning about differences in odor quality. Another approach is to examine delimited categories of odors, in order to study the qualitative dimensions of variation within a subset of the odor space. Examples include the studies of Schiffman on various musks, pyrazines and taste compounds (see Schiffman, 1974, for a review of MDS applied to foods and flavors).

Lawless (1989) examined the transition between two odor classes, because boundary conditions are often revealing regarding the nature of categories. Terpene materials were chosen from woody (pine) and citrus odor classes as well as several ambiguous compounds that seemed to fall somewhat into both categories (Lawless, 1989). This was thought to be analogous to studying formant transitions in categorical perception of speech sounds (e.g., Eimas & Corbit, 1973) except that the underlying physical dimensions are unknown. MDS of these compounds yielded the intuitively predicted groups, with the ambiguous odors plotting between citrus and woody clusters in the MDS output. Furthermore, the experiment was hypothesis generating. Lime odors fell in the ambiguous group, suggesting that some partially woody character might differentiate those odors from other citrus types. Application of this method to judiciously chosen subsets of odor qualities might enable researchers to eventually study large parts of the odor space by putting together a patchwork of adjacent regions.

A major hurdle in this approach is that the plotted spaces for any experiment are two- or three-dimensional, whereas the realm of odors is undoubtedly of higher dimensionality. This suggests that some odors may plot far from each other in some composite model, but actually may be somewhat similar in odor type along dimensions not seen in the plot. Another impediment in this approach is that the transition or ambiguous odors will shift in apparent quality, depending upon context. Contrast effects occur, in which an ambiguous terpene appears more citruslike in the presence of woody odors and more woody in the presence of citrus odors (Lawless, Glatter, & Hohn, 1991). It is tempting to draw a parallel between this effect and the shifts in phoneme category boundaries that occur as a function of preexposure (Eimas & Corbit, 1973).

An important methodological advance in this area has been the application of sorting to generate similarity estimates for MDS analysis. Previously, pairwise similarity ratings were extremely burdensome to subjects, because olfactory adaptation and fatigue is so potent. Sorting tasks are easier for subjects and tap into their subjective notions of how odors should be categorized, which is, after all, one of the goals of this research. Results from sorting of the terpene sets mentioned above have yielded configurations that are very robust and unaffected by levels of training or previous experience (Lawless & Glatter, 1991). Thus the sorting tasks would seem to tap into culturally shared categories and dimensions, which may even have a physiological basis.

Any attempt to define a system for odor-quality classification must recognize that the world of odors, like any other human knowledge domain, may be organized hierarchically (Lawless, 1988). This is implicit in some applied systems for product aroma description, such as the wine aroma wheel (Noble et al., 1987). Hierarchical systems recognize that there are general odor classes that subsume more specific odor types. Such arrangements may serve useful information compression function. The numerous outer tier subclasses may seem overwhelming, but the superordinate categories allow some chunking or organization to the scheme.

As noted previously (Section II.D.3), concept formation or concept alignment may be a critical aspect to the calibration of a descriptive analysis panel. The behavior of experts in odor classification also shows clear examples of concept formation. Trained dairy judges are taught to recognize flavor defects such as rancidity and oxidation (Bodyfelt et al., 1988). The first class of defects derive from hydrolysis of triglycerides (milk-fat globule contents) to their constituent short-chain fatty acids, usually by endogenous lipases that have not been inactivated by pasteurization. The resulting aromatics have been described as cheesy, butyric, like baby vomit, and soapy. Taste sensations may be integrated into this impression as well, particularly bitterness. In the case of lipid oxidation, phopholipids from the fat globule membranes are broken down into unsaturated aldehydes, by the catalytic action of exposure to light or contact with metals such as copper. The resulting flavors cover a range of common associations depending upon the nature and severity of the oxidation. Low levels may be perceived as a cardboardlike flavor, while increasing levels progress to tallowy, painty, and even fishy notes. The important point is that the well-trained dairy judge has learned to combine these diverse associations into a single conceptual category of "oxidized." They then may further differentiate that category into light versus metallic catalyzed oxidation, although that distinction is difficult for many trainees in dairy judging (Claassen & Lawless, 1992).

VI. INTERACTIONS AMONG FLAVOR MODALITIES

A. Odor and Taste

An enduring puzzle in the perception of flavor has been the question of how taste and smell interact. Every cook knows that the perception of flavor is a synthesis of taste and smell impressions, along with effects of texture, temperature, and even appearance. However, under controlled conditions in the psychophysical laboratory, simple mixtures of sucrose (a tastant) and citral (a lemon odorant and flavor), show almost complete addition (90%) and no influence on the intensity ratings of each other, when mixed. This results in an apparent discrepancy between the food literature and chemosensory psychophysics on the issue of how taste and smell interact. It is a common belief among food scientists, as well as consumers, that taste and smell are somehow related. Some of this assumed relationship derives from the use of the word *taste* to mean all aspects of food flavor. However, restricting the word *taste* to mean sensations from nonvolatile substances perceived in the oral cavity, leaves a technical definition that includes such sensations as sweet, sour, salty, and bitter. Regardless of whether these four sensation qualities are taste primaries or merely distinct taste sensations among many other possibilities, one can ask whether taste sensations interact with aromas and volatile flavors that are primarily olfactory.

This literature can be summarized in five major observations.

1. *Sensation intensities are about 90% additive.* Working from a straightforward psychophysical perspective, Murphy et al. (1977) examined perceived odor intensity, perceived taste intensity, and overall perceived intensity of mixtures of sodium saccharin with the volatile flavor compound, ethyl butyrate. A second experiment (Murphy & Cain, 1980) examined the same ratings for sucrose-citral and NaCl-citral mixtures. The pattern of results was consistent in the two studies. Intensity ratings showed about 90% additivity. That is, when framed as a simple question about the summation of gustatory and olfactory stimulation to produce overall impressions of flavor strength, there is little evidence for interactions among the two modalities.

2. *Subjects will misattribute some volatile sensations to "taste."* There was one notable exception to the first rule. The volatile compounds, ethyl butyrate and citral, contributed to judgments of "taste" magnitude, a reliable illusion in both studies. When a flavorous solution is placed into the mouth, untrained subjects have a hard time distinguishing the volatile sensations as odor, and misattribute them to taste. This illusion is eliminated by pinching the nostrils shut during tasting, which prohibits the retronasal passage of volatile materials and effectively cuts off the

volatile flavor impressions. Aside from this mislabeling, the psychophysical evidence points to more independence of taste and smell than interaction, in contrast to popular belief.

3. *Harsh tastes generally suppress and pleasant tastes generally enhance volatile flavor.* A different result is seen in real products, as opposed to the model solutions used in the psychophysical studies above. Von Sydow, Moskowitz, Jacobs, and Meiselman (1974) examined ratings for taste and odor attributes in fruit juices that varied in added sucrose. Ratings for pleasant odor attributes increased and those for unpleasant odor attributes decreased as sucrose concentration increased. No changes in headspace concentrations of volatiles were detected. Von Sydow interpreted this as evidence for a psychological effect as opposed to a physical interaction. For example, attentional mechanisms could influence this shift. Sucrose also suppressed "harsh" tastes such as bitterness, sourness, and astringency. Suppression of other taste components would be expected through simple mechanisms of mixture suppression. However, the overall balance and perhaps the affective reactions to the mixture would be affected in that harshness would be greatly reduced. Because unpleasant tastes may have drawn attention away from volatile characteristics in juices of low sweetness and harsher character, juices that were more "in balance" would allow panelists' attention to focus on other attributes. In that case, they may not have been so captured by the harsh tastes, causing a higher probability for recognition of volatile character and thus higher average ratings. A similar effect was found for blackberry juice at varying levels of sucrose and acidity (Perng & McDaniel, 1989). Sucrose enhanced fruit flavor ratings whereas juices with high acid level showed lower fruit ratings.

This result raises two possibilities. First, an alternative explanation is to invoke the workings of a general halo effect. Stated as a principle, it implies that increases in any pleasant flavor component will increase the ratings of other pleasant components. Conversely, increases in unpleasant flavor notes will cause decreases in intensity ratings for pleasant characteristics (a "horns" effect). The second principle is that panelists may be generally unable to separate the influences of hedonic reaction from their simple intensity judgments, especially when dealing with a real food product. Although it may be possible to adopt a highly detached, analytical and professional attitude in a psychophysical setting, this is much harder when dealing with foods, especially with inexperienced panelists. Even trained descriptive panels may find it hard to put aside their feelings of "yum" versus "yuk." Foods are simply emotional stimuli.

4. *Interactions change with various taste and flavor combinations.* The pattern is potentially complicated by the ways in which interactions may depend upon the particular flavorants and tastants that are combined. Aspartame enhanced fruitiness of orange and strawberry solutions as compared to little or no effect for sucrose, and a somewhat greater enhancement occurred for orange than for strawberry (Wiseman & McDaniel, 1989). In a similar study, sweetness was enhanced by strawberry odor, but not by peanut butter odor (Frank & Byram, 1988). Later studies with

greater numbers of tastants showed general suppression of NaCl saltiness by volatile flavors, but more complex interactions with other tastants and effects of different instructions (Frank, Wessel, & Shaffer, 1990). Further research is needed to see whether these empirical findings can be generalized into coherent rules, or whether tastant–flavorant interactions will remain a matter for case-by-case study with little or no systematic pattern. One potentially profitable avenue for research is the degree of cultural experience panelists have with particular combinations (i.e., a potential influence of learned expectancies).

5. *Apparent interactions change with instructions to subjects.* The instructions that are given to subjects in these studies may have profound effects, as in many other sensory methods. Citral-sucrose mixtures were evaluated using both direct scaling and "indirect" Thurstonian scale values derived from triangle test performance (Lawless & Schlegel, 1984). In the direct scaling, ratings of perceived sweetness and lemon character showed much the pattern of independence noted by previous psychophysical workers. However, in the triangle tests, a pronounced interaction was seen in the case of one pair of mixtures, which was so highly discriminable as to yield a larger than predicted scale value. Furthermore, another pair that was barely discriminable according to the triangle tests received significantly different sweetness ratings. This last result indicates that when subjects' attentions are directed to specific attributes, as in ratings, paired comparisons or forced-choice procedures, they may find products to be much more discriminable than when areas of difference are unspecified, as in traditional discrimination procedures such as the triangle test or duo–trio test. Once again, focused attention produces different results than appreciation of the product as a unitary whole.

The responses that subjects are instructed to make also influence apparent taste–volatile interactions. Strawberry odor could enhance the sweetness of sucrose-strawberry solutions (Frank, Ducheny, & Mize, 1989), an effect reminiscent of the enhancement reported by Wiseman and McDaniel (1989) and also the mislabeling of volatile sensations as taste intensity estimates originally observed by Murphy et al. (1977). Further study of this effect revealed that when subjects were instructed to make total intensity ratings and then partition them into their components, no significant enhancement of sweetness by strawberry odor was seen (Frank et al., 1990). That is, when subjects were able to psychologically "unload" their strawberry impressions on a flavor rating scale, the interaction with sweetness was attenuated. Restriction of ratings to only sweetness in a control experiment restored the enhancement effect. Figure 7 shows a similar result in time-intensity ratings (Lawless & Clark, 1992). Ratings for sweetness in sucrose and strawberry solutions were enhanced when panelists were restricted to a single response (sweetness only) as opposed to when they were allowed to partition their responses between two scales (sweetness and fruitiness).

This finding has broad implications for the ways in which sensory evaluations, particularly descriptive analyses in which multiple attributes of complex foods are

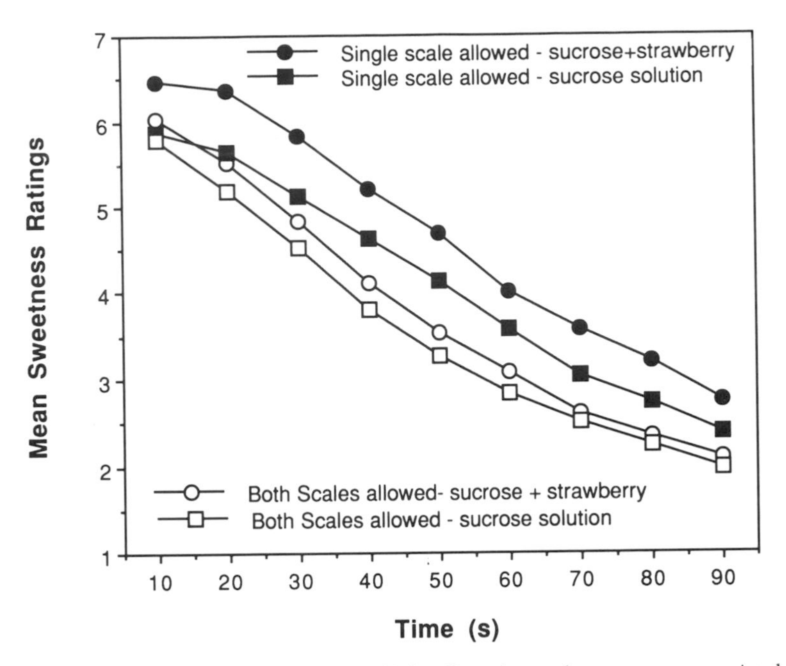

FIGURE 7 "Dumping" or enhancement of halo effect when scale responses are restricted. Aromatic strawberry flavor enhanced sweetness, but to a greater extent when responses were restricted to sweetness ratings only (as opposed to rating both sweetness and fruitiness). (Reprinted with permission from Lawless & Clark, 1992.)

rated, should be conducted. It also suggests some caution in substantiating claims for various synergies or enhancement effects in which ratings are restricted to too few attributes. Respondents may choose to "dump" some of their impressions into the most suitable category or the only allowable response if the attribute they perceive is otherwise unavailable on the ballot. We have termed this a *dumping effect* (Lawless & Clark, 1992). Alleged enhancements such as the effect of ethyl maltol on sweetness should be viewed with caution unless the response biases inherent in dumping responses or in mislabeling smells as tastes can be ruled out. Ethyl maltol is especially problematic in this regard because it has an odor sometimes characterized as "sweet" (Civille & Lawless, 1986). An example of a simple halo effect was shown in Figure 3. As noted above, a minute amount of vanillin (a "sweet" aromatic) was added to lowfat milk. A sample of food science students in a sensory evaluation course (who might have been expected to be more analytical and less influenced) rated the milk as sweeter, creamier, and thicker than a control milk without the added vanillin. Of course, this apparent enhancement may have been exacerbated by the fact that no scale for vanilla flavor was provided, possibly adding an element of dumping to the halo.

B. Trigeminal Interactions

Interactions of taste and smell with the trigeminal flavor senses are poorly understood. However, anyone who has compared flat soda to carbonated soda will recognize that the tingle imparted by CO_2 will alter the flavor balance in a product, usually to its detriment when the carbonation is not present. Flat soda is usually too sweet. Decarbonated champagne is usually very poor wine.

Several psychophysical studies have examined interactions of trigeminal irritation from chemicals with taste and with odor perception. As in most laboratory psychophysics, these studies have focused on simple intensity changes in single chemicals in simple mixtures. The first workers to examine effects of chemical irritation on olfaction found mutual inhibition of smell by CO_2 in the nose (Cain & Murphy, 1980). This seems to occur even though the onset of the sting from CO_2 is delayed somewhat compared to the onset of smell sensations. Because many smells also have an irritative component (Tucker, 1971), it is probable that some of this inhibition is a common event in everyday flavor perception. If a person had decreased sensitivity to nasal irritation the balance of aromatic flavor perception might be shifted in favor of the olfactory components. If irritation impact is reduced, then the inhibitory effects of nasal irritation would also be reduced. This might explain, in part, why smokers who are less responsive to irritation have little or no apparent olfactory deficiency when tested under controlled conditions.

Application of the red pepper compound, capsaicin, to the skin or oral epithelium has profound desensitizing effects (Jansco, 1960; Lawless & Gillette, 1985; Szolscanyi, 1977). This is also known to occur with systemic administration of capsaicin—animals injected with capsaicin become inured to chemical irritants to a remarkable degree. This is believed to reflect a depletion in substance-P, a peptide neurotransmitter for pain (Burks, Buck, & Miller, 1985). Because effects of substance-P have also been linked to functioning of endorphins (Andersen, Lund, & Puil, 1978), there is at least an indirect explanation for the apparent addiction that occurs to spicy foods among some people. High dietary levels of capsaicin also result in a chronic desensitization, as shown in psychophysical tests (Lawless et al., 1985). The extent and influence of this desensitization on other flavor senses should be investigated further.

When confronted with the seemingly shocking level of hot pepper use by the regular spicy food consumer, a common comment by people who do not eat hot spices is, "how can you taste your food?" There are several possible replies, one of which is that the burn or irritation is not unpleasant, and therefore does not command the attention the way other painful stimuli do, and that the other flavors are still present in addition to the pain (you merely have to shift your attention to them—if you want to). In spite of a vast literature on capsaicin in animal models, Nagy once commented that "no detailed quantitation has been conducted on the influence of capsaicin on non-nociceptive sensory stimuli" (Nagy, 1982). This raises the fundamental question of whether chili burn can mask tastes in the mouth, the way that CO_2 sting masks smells in the nose. Partial inhibition of taste responses has been found following pretreatment of oral tissues with capsaicin, particularly of sour and bitter tastes (Karrer & Bartoshuk, 1989; Lawless et al., 1985; Lawless & Stevens, 1984). In contrast, Cowart (1987) observed little or no effect of capsaicin on tastes when capsaicin was mixed with taste stimuli, even though such direct mixing produced equal or higher levels of overall irritation than capsaicin pretreatments. A potential resolution of this paradox is suggested by the finding that capsaicin desensitization takes several minutes to develop (i.e., it depends upon a delay between treatment and test stimuli) (Green, 1989). Such a temporal gap would have occurred to varying degrees in pretreatment experiments with tastants. Conversely, when capsaicin stimuli were given in a more continuous sequence (as in mixture studies) irritation grew over trials.

The potential time dependence of capsaicin inhibition of taste as well as the fact that capsaicin inhibition is most reliably observed for acid and quinine, substances sometimes reported as partially irritative, suggests that the inhibitory effect seen in pretreatment studies may have been due to *desensitization to an irritative component* of the presumed "tastants," rather than a direct effect on gustatory intensity per se. If so, two simple predictions can be made regarding time and concentration variables:

1. Inhibition of taste should parallel the time course of capsaicin self-desensitization, including the need for a hiatus between capsaicin treatment and testing as observed by Green.
2. More intense taste stimuli (strong acid, strong salt) should show proportionally larger decrements after capsaicin treatment, because they have more of an irritative component than weaker, more dilute stimuli.

The reciprocal issue of whether tastes can modulate or ameliorate chili burn is a subject of some speculation. There are folk remedies in various cultures, such as starchy corn (Peru), ghee (India), pineapple (Philippines), sugar (various Latin countries), and beer (Ithaca, New York, among other places). Systematic studies of trying to wash out chili burn with different tasting rinses has shown some effect for sweet (most pronounced), sour, and perhaps salt (Sizer & Harris, 1985; Stevens & Lawless, 1986). Cold stimuli provide a temporary but potent inhibition of pepper burn, as known to many habitués of ethnic restaurants. Since capsaicin is highly lipid soluble, the Indian remedy of ghee (clarified butter) would seem to have some merit. Sour things stimulate salivary flow, which may provide some relief to abused oral tissues. The combination of fatty, sour, cold, and sweet suggests the author's favorite antidote, frozen yogurt. Certainly the Indian culinary practice of alternating cool, sweet chutneys with hot curries would seem to have some merit from this perspective.

C. Halo Effects of Appearance on Flavor

Humans are a visually driven species. In many societies with mature culinary arts, the visual presentation of a food is as important as its flavor and texture characteristics. Japanese cuisine is one good example. A common effect in consumer testing is when foods are more deeply colored, they will obtain higher ratings for flavor intensity. Effects of colored foods on flavor intensity and flavor identification are discussed in Stillman (1993).

A good example of visual influence can be found in the literature on perception of milks of varying fat content. Most people believe that skim milk is easily differentiated from whole milk or even from 2% low-fat milk by appearance, flavor, and texture (mouthfeel). However, most of their perception of fat content is driven by appearance (Pangborn & Dunkley, 1964; Tuorila, 1986). In our laboratory, trained descriptive panelists readily differentiate skim milk from 2% on the basis of appearance (color) ratings, mouthfeel, and flavor. However, when visual cues are removed, the psychophysical functions for flavor and texture as a function of fat content become flattened—discrimination is markedly impaired (Philips, McGiff, Barbano, & Lawless, 1993). When tested in the dark with cold milk (approximately refrigerator temperature), discrimination of skim milk from 2% milk drops almost to chance performance, a result that skim milk drinkers find difficult to swallow. This line of research emphasizes that humans react to the ensemble of sensory stimulation available from a food. Even "objective" trained descriptive panelists are subject to visual bias.

VII. FLAVOR SCIENCE, CUISINE, AND CULTURE

Most of what we understand about flavor perception has arisen from research conducted with physiological, psychophysical, or chemical orientations. There is a dearth of information connecting flavor perception and flavor preferences to the practical culinary arts. Undoubtedly many of our cultural habits arise for social, historical, economic, and religious reasons (Harris, 1985). Any practical discussion of flavor should point out that ethnic cuisine is an important part of each person's cultural identity. Rozin (1983) has explored this notion in a systematic way, noting the particular combinations of recurrent flavors that seem to define certain regional and ethnic styles of cooking. She calls these *flavor principles.* For example, the combination of soy sauce with garlic, brown sugar, sesame seed, and chile will mark a dish as characteristically Korean while soy sauce with garlic, molasses, ground peanuts, and chile will be recognizably Indonesian. She goes on to theorize that these constellations of flavors form attractive blends that tend to resist combinations with other ingredients. Greek cuisine uses a lot of lemon, oregano, cinnamon, and tomato. The lemon–oregano combination recurs, as does the cinnamon–tomato. However, the oregano–tomato does not usually combine, although it can in other cultures.

Flavor harmonies may represent centuries of cultural evolution. In Greek cuisine, the wine retsina is very resiny, reminiscent of pine sap. Pine sap or gum was apparently used as a closure in ancient times on amphora containing wine, so some cultural flavor tradition persists. As mentioned above, this rather unusual wine flavor is difficult for many non-Greeks to accept when tasted in isolation. However, with the use of herbs like oregano in Greek food (many of which have pine-related terpene aromatics), the wine–food match with retsina is eminently palatable. Because wine in most cultures is an accompaniment to food, it is likely that wine styles and flavor principles coevolve.

Consideration of the origin of flavor preferences and culinary combinations is beyond the scope of this chapter. However, one interesting hint is available in the literature both of dairy science and human perceptual studies: Flavors such as garlic are readily transmitted from a mother's diet into milk, and may form an early part of human experience. Human infants, in fact, will spend more time at a nipple that is producing garlic flavored milk than one that is not (Menella & Beauchamp, 1991). It is quite possible that our allegiance to the cuisine of our culture arises from this early but indirect exposure to what the adults in our culture are eating.

References

Ackerman, D. (1990). *A natural history of the senses.* New York: Random House.

Acree, T. E., Barnard, J., & Cunningham, D. G. (1984). A procedure for the sensory analysis for gas chromatographic effluents. *Food Chemistry, 14,* 273–286.

Amerine, M. R., & Roessler, E. B. (1982). *Wines, their sensory evaluation* (2nd ed.). San Francisco: W. H. Freeman.

Amoore, J. E. (1970). *Molecular basis of odor.* Springfield, IL: Charles C. Thomas.

Amoore, J. E. (1971). Olfactory genetics and anosmia. In L. M. Beidler (Eds.), *Handbook of sensory physiology* (pp. 245–256). Berlin: Springer-Verlag.

Amoore, J. E., Forrester, L. J., & Pelosi, P. (1976). Specific anosmia to isobutyraldehyde: the malty primary odor. *Chemical Senses,* 2, 17–25.

Amoore, J. E., Venstrom, D., & Davis, A. R. (1968). Measurement of specific anosmia. *Perceptual and Motor Skills,* 26, 143–164.

Andersen, R. K., Lund, J. P., & Puil, E. (1978). Enkephalin and substance P effects related to trigeminal pain. *Canadian Journal of Physiology and Pharmacology,* 56, 216–222.

Anliker, J., Bartoshuk, L., Ferris, A., & Hooks, L. (1991). Children's food preferences and genetic sensitivity to the bitter taste of PROP. *American Journal of Clinical Nutrition, 54,* 316–320.

Ayya, N., & Lawless, H. T. (1992). Qualitative and quantitative evaluation of high-intensity sweeteners and sweetener mixtures. *Chemical Senses, 17,* 245–259.

Azevedo, E., Krieger, H., Mi, M. P., & Morton, N. E. (1965). PTC taste sensitivity and endemic goiter in Brazil. *American Journal of Human Genetics, 17,* 87–90.

Bartoshuk, L. M. (1968). Water taste in man. *Perception & Psychophysics, 3,* 69–72.

Bartoshuk, L. M. (1975). Taste mixtures: Is mixture suppression related to compression? *Physiology and Behavior, 14,* 643–649.

Bartoshuk, L. M. (1979). Bitter taste of saccharin related to the genetic ability to taste the bitter substance 6-N-Propylthiouracil. *Science,* 205, 934–935.

Bartoshuk, L. M., Cain, W. S., Cleveland, C. T., Grossman, L. S., Marks, L. E., Stevens, J. C., & Stolwijk, J. A. (1974). Saltiness of monosodium glutamate and sodium intake. *Journal of the American Medical Association, 230,* 670.

Bartoshuk, L. M., & Cleveland, C. T. (1977). Mixtures of substances with similar tastes: a test of a new model of taste mixture interactions. *Sensory Processes, 1,* 177–186.

Bartoshuk, L. M., Murphy, C. L. & Cleveland, C. T. (1978). Sweet taste of dilute NaCl. *Physiology and Behavior, 21,* 609–613.

Bartoshuk, L. M., Rifkin, B., Marks, L. E., & Hooper, J. E. (1988). Bitterness of KCI and benzoate: Related to genetic status for sensitivity to PTC/PROP. *Chemical Senses, 13,* 517–528.

Bate Smith, E. C. (1954). Astrigency in foods. *Food Processing and Packaging, 23,* 124–127.

Berglund, B., Berglund, U., Lindvall, T., & Svensson, L. T. (1973). A quantitative principle of perceived odor intensity in odor mixtures. *Journal of Experimental Psychology, 100,* 29–38.

Bingham, A. F., Birch, G. G., de Graaf, C., Behan, J. M., & Perring, K. D. (1990). Sensory studies with sucrose maltol mixtures. *Chemical Senses, 15,* 447–456.

Blakeslee, A. F. (1932). Genetics of sensory thresholds: Taste for phenylthiocarbamide. *Proceedings of the National Academy of Science USA, 18,* 120–130.

Bodyfelt, F. W., Tobias, J., & Trout, G. M. (1988). *Sensory evaluation of dairy products.* New York: Van Nostrand/AVI Publishing.

Brennand, C. P., Ha, J. K., & Lindsay, R. C. (1989). Aroma properties and thresholds of some branched chain and other minor volatile fatty acids occurring in milkfat and meat lipids. *Journal of Sensory Studies, 4,* 105–120.

Brud, W. S. (1986). Words versus odors: How perfumers communicate. *Perfumer and Flavorist, 11,* 27–44.

Buck, L., & Axel, R. (1991). A novel multigene family may encode odorant receptors: A molecular basis for odor recognition. *Cell, 65,* 175–187.

Burks, T. F., Buck, S. H., & Miller, M. S. (1985). Mechanisms of depletion of substance P by capsaicin. *Federation Proceedings, 44,* 2531–2534.

Cagan, R. H. (1981). Recognition of taste stimuli at the initial binding interaction. In R. H. Cagan & M. R. Kare (Eds.), *Biochemistry of taste and olfaction* (pp. 175–204). New York: Academic Press.

Cain, W. S. (1975). Odor intensity: Mixtures and masking. *Chemical Senses and Flavour, 1,* 339–352.

Cain, W. S., & Murphy, C. L. (1980). Interaction between chemoreceptive modalities of odor and irritation. *Nature, 284,* 255–257.

Cardello, A. V., Maller, O., Kapsalis, J. G., Segars, R. A., Sawyer, F. M., Murphy, C., & Moskowitz, H. R. (1982). Perception of texture by trained and consumer panels. *Journal of Food Science, 47,* 1186–1197.

Caul, J. F. (1957). The profile method of flavor analysis. *Advances in Food Research, 7,* 1–40.

Chastrette, M., Elmouaffek, E., & Sauvegrain, P. (1988). A multidimensional statistical study of similarities between 74 notes used in perfumery. *Chemical Senses, 13,* 295–305.

Civille, G. L., & Lawless, H. T. (1986). The importance of language in describing perceptions. *Journal of Sensory Studies, 1,* 203–215.

Classen, M. R. (1991). A comparison of descriptive terminology systems for the sensory analysis of flavor defects in milk. Unpublished master's thesis, Cornell University, Ithaca, NY.

Claassen, M., & Lawless, H. T. (1992). Comparison of descriptive terminology systems for sensory evaluation of fluid milk. *Journal of Food Science, 57,* 596–600, 621.

Cohen, J., & Ogdon, D. P. (1949). Taste blindness to phenyl-thio-carbamide and related compounds. *Psychological Bulletin, 46,* 490–498.

Cowart, B. J. (1987). Oral chemical irritation: Does it reduce perceived taste intensity? *Chemical Senses,* 12, 467–479.

Desor, J. A., & Beauchamp, G. K. (1974). The human capacity to transmit olfactory information. *Perception & Psychophysics, 16,* 551–556.

Dorries, K. M., Schmidt, H. J., Beauchamp, G. K., & Wysocki, C. J. (1989). Changes in the sensitivity to the odor of androstenone during adolescence. *Developmental Psychobiology, 22,* 423–435.

Dusenberry, D. B. (1992). *Sensory ecology.* New York: W. H. Freeman.

Eimas, P. D., & Corbit, T. E. (1973). Selective adaptation of linguistic feature detectors. *Cognitive Psychology, 4,* 99–109.

Engen, T., & Pfaffmann, C. (1959). Absolute judgments of odor intensity. *Journal of Experimental Psychology, 58,* 23–26.

Ennis, D. M. (1990). Relative power of difference testing methods in sensory evaluation. *Food Technology, 44*(4), 114, 116–117.

Farbman, A. I., & Hellekant, G. (1978). Quantitative analyses of fiber population in rat chorda tympani nerves and fungiform papillae. *American Journal of Anatomy, 153,* 509–521.

Fischer, R., Griffin, F., England, S., & Garn, S. M. (1961). Taste thresholds and food dislikes. *Nature,* 191, 1328.

Fox, A. L. (1932). The relationship between chemical constitution and taste. *Proceedings of the National Academy of Sciences USA,* 18, 115–120.

Frank, R. A., & Byram, J. (1988). Taste-smell interactions are tastant and odorant dependent. *Chemical Senses, 13,* 445.

Frank, R. A., Ducheny, K., & Mize, S. J. S. (1989). Strawberry odor, but not red color enhances the sweetness of sucrose solutions. *Chemical Senses, 14,* 371.

Frank, R. A., Mize, S. J., & Carter, R. (1989). An assessment of binary mixture interactions for nine sweeteners. *Chemical Senses, 14,* 621–632.

Frank, R. A., Wessel, N., & Shaffer, G. (1990). The enhancement of sweetness by strawberry odor is instruction dependent. *Chemical Senses, 15,* 576–577.

Frijters, J. E. R. (1978). A critical analysis of the odour unit number and its use. *Chemical Senses & Flavour, 3,* 227–233.

Frijters, J. E. R. (1979). The paradox of the discriminatory nondiscriminators resolved. *Chemical Senses,* 4, 355–358.

Furia, T. E., & Bellanca, N. (1971). *Fenaroli's handbook of flavor ingredients.* Cleveland, OH: Chemical Rubber Co.

Gent, J. F., & Bartoshuk, L. M. (1983). Sweetness of sucrose, neohesperidin dihydrochalcone and saccharin is related to genetic ability to taste the bitter substance 6-n-propylthiouracil. *Chemical Senses,* 7, 265–272.

Gibson, J. J., & Gibson, E. J. (1955). Perceptual learning: Differentiation or enrichment. *Psychological Review, 62,* 32–41.

Glanville, E. V., & Kaplan, A. R. (1965). Food preference and sensitivity of taste for bitter compounds. *Nature, 205,* 851–853.

Goldwyn, C., & Lawless, H. (1991). How to taste wine. *ASTM Standardization News, 19*(3), 32–27.

Green, B. G. (1989). Capsaicin sensitization and desensitization on the tongue produced by brief exposures to a low concentration. *Neuroscience Letters, 107,* 173–178.

Green, B. G., & Lawless, H. T. (1991). The psychophysics of somatosensory chemoreception in the nose and mouth. In L. M. Bartoshuk, R. L. Doty, T. V. Getchell, & J. B. Snow (Eds.), *Smell and taste in health and disease* (pp. 235–253). New York: Raven Press.

Greene, L. S. (1974). Physical growth and development, neurological maturation, and behavioral functioning in two Ecuadorian Andean communities in which goiter is endemic. *American Journal of Physical Anthropology, 41,* 139–151.

Hall, M. L., Bartoshuk, L. M., Cain, W. S., & Stevens, J. C. (1975). PTC taste blindness and the taste of caffeine. *Nature, 253,* 442–443.

Harris, M. (1985). *The sacred cow and the abominable pig: Riddles of food and culture.* New York: Simon & Schuster.

Hartley, J., & Homa, D. (1981). Abstraction of stylistic concepts. *Journal of Experimental Psychology: Human Learning and memory,* 7, 33–66.

Heath, H. B. (1981). *Source book of flavors.* Westport, CT: AVI Publishing.

Helm, E., & Trolle, B. (1946). Selection of a taste panel. *Wallerstein Laboratory Communications, 9,* 181–194.

Homa, D., & Cultice, J. (1984). Role of feedback, category size and stimulus distortion on the acquisition and utilization of ill-defined categories. *Journal of Experimental Psychology: Learning, Memory and Cognition, 10,* 83–94.

Jansco, N. (1960). Role of the nerve terminals in the mechanism of inflammatory reactions. *Bulletin of Millard Fillmore Hospital, Buffalo, 7,* 53–77.

Jerzsa-Latta, M., Krondl, M., & Coleman, P. (1990). Use and perceived attributes of cruciferous vegetables in terms of genetically-mediated taste sensitivity. *Appetite, 15,* 127–134.

Jones, L. V., Peryam, D. R., & Thurstone, L. L. (1955). Development of a scale for measuring soldier's food preferences. *Food Research, 20,* 515–520.

Karrer, T., & Bartoshuk, L. (1989). Oral capsaicin desensitization and effects on taste. In Annual Meeting Abstracts (Abstract #87). Sarasota, FL: *Association for Chemoreception Sciences.*

Laing, D. G., Livermore, B. A., & Francis, G. W. (1991). The human sense of smell has a limited capacity for identifying odors in mixtures. *Chemical Senses, 16,* 392.

Lawless, H. T. (1979a). Evidence for neural inhibition in bittersweet taste mixtures. *Journal of Comparative and Physiological Psychology, 93,* 538–547.

Lawless, H. T. (1979b). The taste of creatine and creatinine. *Chemical Senses, 4,* 249–252.

Lawless, H. T. (1984). Flavor description of white wine by "expert" and non-expert wine consumers. *Journal of Food Science,* 49, 120–123.

Lawless, H. T. (1986). Sensory interactions in mixtures. *Journal of Sensory Studies, 1,* 259–274.

Lawless, H. T. (1987). An olfactory analogy to release from mixture suppression in taste. *Bulletin of the Psychonomic Society,* 25, 266–268.

Lawless, H. T. (1988). Odour description and odour classification revisited. In D. M. H. Thomson (Eds.), *Food acceptability* (pp. 27–40). London: Elsevier Applied Science.

Lawless, H. T. (1989). Exploration of fragrance categories and ambiguous odors using multidimensional scaling and cluster analysis. *Chemical Senses, 14,* 349–360.

Lawless, H. T., & Claassen, M. R. (1993). Validity of descriptive and defect-oriented terminology systems for sensory analysis of fluid milk. *Journal of Food Science, 58,* 108–112, 119.

Lawless, H. T., & Clark, C. C. (1992). Psychological biases in time intensity scaling. *Food Technology, 46*(11), 81, 84–86, 90.

Lawless, H. T., & Gillette, M. (1985). Sensory Responses to Oral Chemical Heat. In D. D. Bills & C. J. Mussinan (Eds.), *Characterization and Measurement of Flavor Compounds* (pp. 27–42). Washington, DC: American Chemical Society.

Lawless, H. T., & Glatter, S. (1991). Consistency of multidimensional scaling models derived from odor sorting. *Journal of Sensory Studies, 5,* 217–230.

Lawless, H. T., Glatter, S., & Hohn, C. (1991). Context dependent changes in the perception of odor quality. *Chemical Senses, 16,* 349–360.

Lawless, H. T., Rozin, P., & Shenker, J. (1985). Effects of oral capsaicin on gustatory, olfactory and irritant sensations and flavor identification in humans who regularly or rarely consume chili pepper. *Chemical Senses, 10,* 579–589.

Lawless, H. T., & Schlegel, M. P. (1984). Direct and indirect scaling of sensory differences in simple taste and odor mixtures. *Journal of Food Science, 49,* 44–47.

Lawless, H. T., & Stevens, D. A. (1984). Effects of oral chemical irritation on taste. *Physiology and Behavior, 32,* 995–998.

Lawless, H. T., & Stevens, D. A. (1988). Responses by humans to oral chemical irritants as a function of locus of stimulation. *Perception & Psychophysics, 43,* 72–78.

Lawless, H. T., Torres, V., & Figueroa, E. (1993). Sensory evaluation of hearts of palm. *Journal of Food Science, 58,* 134–137.

Lawless, H. T., & Skinner, E. Z. (1979). The duration and perceived intensity of sucrose taste. *Perception & Psychophysics, 25,* 249–258.

Lehrer, A. (1983). *Wine and conversation.* Bloomington, IN: Indiana University Press.

Looy, H., & Weingarten, H. P. (1992). Facial expressions and genetic sensitivity to 6-n-propylthiouracil predict hedonic responses to sweet. *Physiology & Behavior, 52,* 75–82.

Lundahl, D. S., & McDaniel, M. R. (1988). The panelist effect—fixed or random? *Journal of Sensory Studies, 3,* 113–121.

MacRae, R. W., & Geelhoed, E. N. (1992). Preference can be more powerful than detection of oddity as a test of discriminability. *Perception & Psychophysics, 51,* 179–181.

Mattes, R., & Labov, J. (1989). Bitter taste responses to phenylthiocarbamide are not related to dietary goitrogen intake in human beings. *Journal of the American Dietetic Association, 89*(5), 692–694.

McAuliffe, W. K., & Meiselman, H. L. (1974). The role of practice and correlation in the categorization of bitter and sour tastes. *Perception & Psychophysics, 16,* 242–244.

McBride, R. L., & Anderson, N. H. (1990). Integration psychophysics. In R. L. McBride & H. J. H. MacFie (Eds.), *Psychological basis of sensory evaluation* (pp. 93–115). London: Elsevier.

McBurney, D. H. (1966). Magnitude estimation of the taste of sodium chloride after adaptation to sodium chloride. *Journal of Experimental Psychology, 72,* 869–873.

McBurney, D. H., Smith, D. V., & Shick, T. R. (1972). Gustatory cross-adaptation: Sourness and bitterness. *Perception & Psychophysics, 11,* 228–232.

Meilgaard, M., Civille, G. V., & Carr, B. T. (1987). *Sensory evaluation techniques.* Boca Raton: CRC Press.

Meiselman, H. L., & Halpern, B. P. (1973). Enhancement of taste intensity through pulsatile stimulation. *Physiology and Behavior, 11,* 713–716.

Menella, J. A., & Beauchamp, G. K. (1991). Maternal diet alters the sensory qualities of human milk and the nursling's behavior. *Pediatrics, 88*(4), 737–744.

Murphy, C., & Cain, W. S. (1980). Taste and olfaction: Independence vs. interaction. *Physiology & Behavior, 24,* 601–605.

Murphy, C., Cain, W. S., & Bartoshuk, L. M. (1977). Mutual action of taste and olfaction. *Sensory Processes, 1,* 204–211.

Nagy, J. I. (1982). Capsaicin: A chemical probe for sensory neuron mechanisms. In L. L. Iversen, S. D. Iversen, & S. H. Snyder (Eds.), *Handbook of psychopharmacology* (pp. 185–235). New York: Plenum Publishing.

Nicolaidis, S. (1977). Sensory-neuroendocrine reflexes. In M. R. Kare & O. Maller (Eds.), *The chemical senses and nutrition* (pp. 124–143). New York: Academic Press.

Noble, A. C., Arnold, R. A., Buechsenstein, J., Leach, E. J., Schmidt, J. O., & Stern, P. M. (1987). Modification of a standardized system of wine aroma terminology. *American Journal of Enology and Viticulture, 38*(2), 143–146.

O'Connell, R. J., Stevens, D. A., Akers, R. P., Coppola, D. M., & Grant, A. J. (1989). Individual differences in the quantitative and qualitative responses of human subjects to various odors. *Chemical Senses, 14,* 293–302.

O'Mahony, M. (1991). Descriptive analysis and concept alignment. In H. T. Lawless & B. P. Klein (Eds.), *Sensory science, theory and applications in foods* (pp. 223–268). New York: Marcel Dekker.

O'Mahony, M., Goldenberg, M., Stedmon, J., & Alford, J. (1979). Confusion in the use of the taste adjectives 'sour' and 'bitter'. *Chemical Senses and Flavour, 4,* 301–318.

O'Mahony, M., & Ishii, R. (1986). *Umami* taste concept: Implications for the dogma of four basic tastes. In Y. Kawamura & M. R. Kare (Eds.), *Umami: A basic taste* (pp. 75–93). New York: Marcel Dekker.

O'Mahony, M., Wong, S. Y., & Odbert, N. (1986). Sensory difference tests: Some rethinking concerning the general rule that more sensitive tests use fewer stimuli. *Lebensmittel Wissenschaft und Technologie, 19,* 93–95.

Pangborn, R. M., & Dunkley, W. L. (1964). Difference-preference evaluation of milk by trained judges. *Journal of Dairy Science, 47,* 1414–1416.

Patte, F., & Laffort, P. (1979). An alternative model of olfactory quantitative interaction in binary mixtures. *Chemical Senses and Flavour, 4*(4), 267–274.

Pelosi, P., & Viti, R. (1978). Specific anosima to I-carvone: The minty primary odour. *Chemical Senses and Flavour, 3,* 331–337.

Perng, C. M., & McDaniel, M. R. (1989). Optimization of a blackberry juice drink using response surface methodology. In Annual Meeting Book of Abstracts (Abstract 513, p. 216). Chicago, IL: Institute of Food Technologists.

Peryam, D. R., & Swartz, V. W. (1950). Measurement of sensory differences. *Food Technology, 4,* 390–395.

Philips, L. G., McGiff, M. L., Barbano, D. M., & Lawless, H. T. (1995). The influence of fat on the sensory properties, viscosity, and color of lowfat milk. *Journal of Dairy Science, 78,* 1258–1266.

Piggott, J. R. (1990). Relating sensory and chemical data to understand flavor. *Journal of Sensory Studies, 4,* 261–272.

Rainey, B. A. (1986). Importance of reference standards in training panelists. *Journal of Sensory Studies, 1,* 149–154.

Reaume, J. (1975). Letter to the Editor. *Journal of the American Medical Association, 233,* 224–225.

Reedy, F. E., Jr., Bartoshuk, L. M., Miller, I. J., Duffy, V. B., & Yanagisawa, K. (1993, April). *Relationships among papillae, taste pores and 6-n-propylthiouracil (PROP) suprathreshold taste sensitivity.* Paper presented at the Fifteenth Annual Meeting of the Association for Chemoreception Sciences, Sarasota, FL.

Rozin, E. (1983). *Ethnic cuisine: The flavor principle cookbook.* Brattleboro, VT: The Stephen Greene Press.

Rozin, P. (1982). "Taste–smell confusions" and the duality of the olfactory sense. *Perception & Psychophysics, 31,* 397–401.

Schifferstein, H. N. J., & Frijters, J. E. R. (1991). The perception of the taste of KCl, NaCl and quinine HCl is not related to PROP sensitivity. *Chemical Senses, 16*(4), 303–317.

Schiffman, S. S. (1974). Physicochemical correlates of olfactory quality. *Science, 185,* 112–117.

Sizer, F., & Harris, N. (1985). The influence of common food additives and temperature on threshold perception of capsaicin. *Chemical Senses, 10,* 279–286.

Solomon, G. E. A. (1988). *Great expectorations: The psychology of expert wine talk.* Unpublished doctoral dissertation, Harvard University, Cambridge, MA.

Stevens, D. A. (1991). Individual differences in taste and smell. In H. T. Lawless & B. P. Klein (Eds.), *Sensory science theory and application in foods* (pp. 295–316). New York: Marcel Dekker.

Stevens, D. A., & Lawless, H. T. (1986). Putting out the fire: Effects of tastants on oral chemical irritation. *Perception & Psychophysics, 39,* 346–350.

Stillman, J. A. (1993). Color influences flavor identification in fruit-flavored beverages. *Journal of Food Science, 58,* 810–812.

Stone, H., & Sidel, J. L. (1985). *Sensory evaluation practices.* New York: Academic Press.

Stone, H., Sidel, J., Oliver, S., Woolsey, A., & Singleton, R. C. (1974). Sensory evaluation by quantitative descriptive analysis. *Food Technology, 28*(1), 24, 26, 28, 29, 32, 34.

Szczesniak, A. S., & Skinner, E. Z. (1973). Meaning of texture words to the consumer. *Journal of Texture Studies, 4,* 378–384.

Szolscanyi, J. (1977). A pharmacological approach to elucidation of the role of different nerve fibers and receptor endings in mediation of pain. *Journal of Physiology (Paris), 73,* 251–259.

Thompson, R. H., & Pearson, A. M. (1977). Quantitative determination of 5 Androst-16-en-3-one by gas chromatography-mass spectrometry and its relationship to sex odor intensity of pork. *Journal of Agricultural and Food Chemistry, 25,* 1241–1245.

Tucker, D. (1971). Nonolfactory responses from the nasal cavity: Jacobson's organ and the trigeminal system. In L. M. Beidler (Eds.), *Handbook of sensory physiology IV(I)* (pp. 151–181). Berlin: Springer-Verlag.

Tuorila, H. (1986). Sensory profiles of milks with varying fat contents. *Lebensmittel Wissenschaft und Technologie,* 19, 344–345.

von Sydow, E., Moskowitz, H., Jacobs, H., & Meiselman, H. (1974). Odor-taste interactions in fruit juices. *Lebensmittel Wissenschaft und Technologie, 7,* 18–20.

Whitehead, M. C., Beeman, C. S., & Kinsella, B. A. (1985). Distribution of taste and general sensory nerve endings in fungiform papillae of the hamster. *American Journal of Anatomy, 173,* 185–201.

Williams, A. A., Carter, C. S., & Langron, S. (1978). The development of a vocabulary for the assessment of wines. In J. M. Broadbent, D. G. Land, H. W. Spencer, M. Harries, A. A. Williams, & M. Sargent (Eds.), *The fourth wine subject day—Sensory evaluation* (pp. 57–67). Bristol, U.K.: Long Ashton Research Station, University of Bristol.

Wiseman, J. J., & McDaniel, M. R. (1989, June). *Modification of fruit flavors by aspartame and sucrose.* Paper presented at Institute of Food Technologists, Annual Meeting. Chicago, IL.

Wysocki, C. J., & Beauchamp, G. K. (1988). Ability to smell androstenone is genetically determined. *Proceedings of the National Academy of Sciences USA, 81,* 4899–4902.

Wysocki, C. J., Dorries, K. M., & Beauchamp, G. K. (1989). Ability to perceive androstenone can be acquired by ostensibly anosmic people. *Proceedings of the National Academy of Sciences USA, 86,* 7976–7978.

Yamaguchi, S. (1967). The synergistic taste effect of monosodium glutamate and disodium 5′inosinate. *Journal of Food Science, 32,* 473–475.

Index

A

B

C

D

E

F

G

H

I

L

M

N

O

P